AF368925

Impacts of Climate Change on Human Health

Impacts of Climate Change on
Human Health

Editor

Clare Heaviside

MDPI • Basel • Beijing • Wuhan • Barcelona • Belgrade • Manchester • Tokyo • Cluj • Tianjin

Editor
Clare Heaviside
University College London
UK

Editorial Office
MDPI
St. Alban-Anlage 66
4052 Basel, Switzerland

This is a reprint of articles from the Special Issue published online in the open access journal *Atmosphere* (ISSN 2073-4433) (available at: https://www.mdpi.com/journal/atmosphere/special_issues/climate_change_human_health).

For citation purposes, cite each article independently as indicated on the article page online and as indicated below:

LastName, A.A.; LastName, B.B.; LastName, C.C. Article Title. *Journal Name* **Year**, *Article Number*, Page Range.

ISBN 978-3-03936-740-5 (Hbk)
ISBN 978-3-03936-741-2 (PDF)

Contents

About the Editor

Clare Heaviside, Associate Professor, is a Natural Environment Research Council (NERC) Independent Research Fellow and Principal Investigator at University College London, UK. Dr Heaviside's research interests lie in the impacts of climate change, particularly on health. She works on urban climate and is interested in the urban heat island, air pollution, heatwaves, and the associated effects on health. She previously worked at Oxford University and before that, was Head of Climate Change at Public Health England, a UK Government Agency, where she led a team and coordinated climate change research, advice, and activities across the agency. Her PhD is in Atmospheric Physics from Imperial College, London. Dr Heaviside embraces a multidisciplinary approach to research, using her knowledge and skills in climatology, environmental science, and public health to produce research outputs that can be applied to improving health and aiding effective climate change adaptation strategies.

Preface to "Impacts of Climate Change on Human Health"

Climate change poses a serious challenge to the health and wellbeing of mankind. Already, human costs are being realised through the increasing frequency of extreme weather events such as floods, droughts, and heatwaves. Changes in temperature have direct and indirect impacts on health, and broader environmental changes will affect infectious disease risk, air pollution, and other forms of exposure. The many different ways in which climate change will affect health are complex, interactive, and different communities are disproportionately affected. International actions such as the Paris Agreement and the Sustainable Development Goals recognise the future risks to society and acknowledge that we are already committed to a certain level of climate change. Future adaptation measures therefore need careful assessment and implementation for us to be able to minimise the potential risks from climate change and, at the same time, maximise the potential health benefits of a cleaner, greener world. This Special Issue comprises original research articles and detailed reviews on the likely impacts of climate change on health in a range of geographical settings, and the potential for adaptation measures to reduce some of these risks. Ultimately, studies like these will motivate policy level action for mitigation and help in determining the most effective methods of adaptation to reduce negative impacts in future through embedding scientific evidence into practice.

Clare Heaviside
Editor

Editorial

Understanding the Impacts of Climate Change on Health to Better Manage Adaptation Action

Clare Heaviside [1,2,3]

1   Environmental Change Institute, University of Oxford, South Parks Road, Oxford OX1 3QY, UK; clare.heaviside@ouce.ox.ac.uk
2   School of Geography Earth and Environmental Sciences, University of Birmingham, Birmingham B15 2TT, UK
3   Department of Social and Environmental Health Research, London School of Hygiene & Tropical Medicine, London WC1E 7HT, UK

Received: 28 February 2019; Accepted: 2 March 2019; Published: 5 March 2019

1. Background

The atmospheric and climate research communities have made significant advances in recent decades in gathering and understanding the scientific evidence supporting the concept of anthropogenic climate change. The IPCC (Intergovernmental Panel on Climate Change) has been instrumental in synthesizing the latest research on the state of our climate and projected future changes [1], as well as reporting on the impacts of climate change [2,3]. The urgency and scale of the challenge of our changing climate is widely accepted, and reflected by several landmark international agreements [4]. Of particular significance was the Paris Agreement of 2016; an international call for action to reduce global temperature rises to 2 °C, and preferably below 1.5 °C [5].

Already the human cost of climate change is being realised through the increasing frequency of floods, heatwaves, storms and forest fires leading to loss of life and injury; there are also long term health impacts following extreme events such as trauma, chronic illness, and mental health effects as a result of displacement [2]. Climate change also affects health through changes in ambient temperatures, air pollution, and changes to local environments which can introduce new disease vectors [3,6–8]. Specific risks will vary for different climatic zones, and impacts for each population group will be felt differently, depending on vulnerability due to social, economic and demographic factors.

The steady growth in the body of research into climate change impacts and adaptation topics has partly been driven by the acknowledgement that until we understand the impacts of various degrees of warming, there is a lack of momentum to aim for a specific target. Quantifying societal impacts can provide incentives to policymakers and others to limit emissions of greenhouse gases and implement adaptation measures, because the devastating effects if we do not take action, become more apparent. It is therefore important that there is continued effort to characterize and quantify the impacts of climate change, particularly at local and regional levels, since this is what will drive improvements in climate change mitigation and adaptation policy [9].

Cross-disciplinary working in the fields of climate change and health allows us to better understand and report the relationships between the environment and health. We need to strengthen the evidence base for impacts, not only of changes to climate, but of interventions which may be put in place to either mitigate against climate change or to adapt to it. Quantification of the human costs of climate change is particularly important when we seek to influence policymakers and the public of the overwhelming need to both reduce the extent of climate change, and to rapidly adapt to changes to which we are already committed due to greenhouse gases emitted since the start of the industrial revolution.

The aim of this Special Issue on *The Health Impacts of Climate Change* is to explore impacts of climate change and adaptation to it, and to promote inter-disciplinary working in the field of atmospheric and health sciences. This type of research enables robust strengthening of the evidence base for climate change impacts on health, which motivates policy level action for mitigation; allows estimation of potential impacts to aid planning and response; and helps assess the most effective methods of adaptation to reduce impacts in future.

This collection of articles draws from a number of international authors and includes original research articles and comprehensive reviews. Summaries of the papers comprising the special issue are given below, categorized into three broad areas: (1) Understanding the complex relationships between the environment and health; (2) Quantification of the health impacts of temperature changes; and (3) Embedding scientific evidence into adaptation practice.

2. Understanding the Complex Relationships between the Environment and Health

The relationship between meteorological variables and health outcomes using empirical methods has been recognized for several decades. More recently, attempts have been made to estimate or project how climate change may affect health in future, and to a greater or lesser extent, these studies have attempted to include modifying factors such as changes in demographics, population health and adaptation measures. Most research has focused on the direct impact of changes in climate, rather than the more indirect pathways through which changes in the environment can have an impact on health, for example, through environmental degradation, and socio-economic changes driven by climate change. There is also significantly more research in high income countries than medium and low-income countries, where data can be sparse [10,11].

Fleming et al. [12] review some of the broad health effects of environmental change, particularly changes to the natural environment, and with reference to infectious and vector-borne diseases, and aeroallergen exposure. Also covered is a growing body of research which aims to quantify the benefits (to health, wellbeing and the economy) as well as the risks of human interactions with the natural environment, focusing mainly on the UK. Benefits to physical and mental health come from interacting with the natural environment, e.g., residing in areas with more greenspace, and visiting parks and recreation areas. The review highlights research which shows that there are differences between the effects of different environments (e.g., coastal, rural, urban) on mental health and exercise capacity [12]. Health-economic assessments of co-benefits to health and the economy from climate change mitigation and adaptation measures are essential to underpin cost-benefit analyses for policy makers.

Water related or water-borne diseases have devastating impacts on health worldwide, especially in developing countries. The relationship between these types of infectious disease and climate variability and change is reviewed by Nichols et al. [13], who highlight that climate change poses a potential risk associated with the incidence of cholera, typhoid, dysentery, leptospirosis, diarrhoeal diseases and harmful algal blooms [13]. Although there are clearly direct links between weather conditions and infectious disease risk, the indirect effects of climate change which may drive population movements and conflict are also likely to affect the incidence of many water related infectious diseases.

Two papers in this special issue address the links between large scale climate dynamics and health: specifically the El Nino Southern Oscillation (ENSO) and the North Atlantic Oscillation (NAO) which are both important modes of global climate variability [14,15]. The first provides an overview of the characteristics and impacts of ENSO, a major driver of inter-annual climate variability, and gives an up to date review of literature seeking to understand the associations between various ENSO indices and health impacts. ENSO may modify parameters such as rainfall, wind, air and sea temperatures in different geographic regions, which has consequences for temperature-related health effects, flooding, and the spread of infectious diseases such as malaria and dengue. One conclusion drawn is that a purely statistical framework is not always sufficient to understand the complex societal impacts of a phenomenon such as ENSO, and that caution should be applied when choosing an ENSO index for impact studies [14].

The second paper looking at atmospheric dynamics specifically investigates the connection between phases of the North Atlantic Oscillation (NAO), the dominant mode of atmospheric variability in the northern hemisphere, and ambulance call outs for elevated blood pressure [15]. The analysis was based on a regression of daily emergency ambulance calls in Lithuania for hypertension, and NAO Indices, adjusting for season, weather and air pollutant confounders. The relationship between high blood pressure and the NAO is likely to be associated with the low temperatures occurring during anticyclonic conditions, and increased emergency ambulance calls for elevated arterial blood pressure were found for both the positive and negative values of the NAO index, varying depending on season [15].

3. Quantification of the Health Impacts of Temperature Changes

The environment, weather and climate all have profound impacts on health, either through direct associations or through less obvious pathways, which can make quantification difficult. However, one of the most widely studied, and therefore the most well understood impacts of climate change on health is through direct changes in temperature, which can increase the risk of illness and hospitalization, or lead to premature mortality.

Kinney [16] presents a comprehensive, up to date review of the relationship between temperature and health impacts (from both heat and cold), from morbidity to mortality, in different global regions, as well as how these relationships have changed over time. Heat-mortality relationships have been shown to vary by geographic region and population group, but less well studied is how relationships may have changed over time. Kinney reviews the emerging literature to try to answer this question, and gives guidance on taking account of trends over time and space when projecting future impacts of climate change and the associated change in temperature [16].

There are further research papers addressing the effects of temperatures on health in this issue, including analyses on heatwaves and cold spells, and reduced productivity in relation to heat. Two of the papers in this special issue focus on the incidence of heatwaves [17] and cold spells [18] over a 50 year period (1966–2015) in the city of Poznan, Poland. The authors highlight how the diversity of land type is reflected by diverse thermal conditions across the city compared with the reference monitoring stations, and the increased frequency of heatwave and reduced frequency of cold spells over the period studied.

Muthers et al. [19] pinpoint specific heatwave events which occurred during the years of 2003 and 2015, and estimate the resulting heat-related mortality in South West Germany. Results showed daily mortality anomalies reached +70% and +56% during the summers of 2003 and 2015, respectively, and climate projections indicate the increasing probability of comparable heatwaves occurring every second summer in the near future (2021–2050s) [19].

As well as leading to premature mortality, high temperatures and heat stress can also be detrimental to productivity. This can be a difficult effect to quantify, and is a subject tackled by Dear [20]. Here, a modelling approach is adopted which aims to build on existing evidence and which can express productivity as a function of environmental heat, and considers variations between individual workers [20].

4. Embedding Scientific Evidence into Adaptation Practice

The final set of papers address possible adaptation and planning responses to the effects of climate change on health. For example, Orru et al. [21] argue that resilience in the face of climate change is necessarily driven by the functioning of health systems and the drivers which shape system effectiveness. The authors use an example of the Estonian health system, and find that the health impacts of climate change have not been mainstreamed into policy. There is currently an opportunity to combine information from various sources, on health systems, the environment and vulnerability, although this prospect has not been embraced fully due to various reasons, including lack of knowledge and understanding of extreme events and how to incorporate projected risks into policy, and unclear division of responsibility [21]. Careful resource planning is an important consideration for the health sector, especially in the face of an increased risk of extreme weather events. Papadakis et al. [22]

analyse emergency ambulance calls during episodes of extreme temperatures in London, UK, and propose a statistical physics based method to more accurately estimate the true burden on the health system during these events, which has potential for the ambulance service to more efficiently manage resources [22].

In the Netherlands, the issue of the urgent need for additional housing within existing urban boundaries is addressed [23]. Koopmans et al. test a range of urban planning strategies and their potential impacts on heat stress up to 2050, in the Hague. In particular, the impact of urban fabric on the urban heat island intensity is considered, and it was found that increasing vegetation fraction is a critical parameter in reducing excess heat stress in denser neighbourhoods, even if high rise buildings are preserved. However, the effects of climate change are thought to be likely to overwhelm the effects of increased urban densification in future [23].

Taylor et al. [24] use building models to quantify indoor temperatures in different dwelling types, and investigate the potential impacts on heat-related mortality of a range of adaptation measures under future climate change scenarios. Adaptation measures included mitigation of heat, improved energy efficiency, and changes in occupant behaviour, finding that external shutters provided the greatest benefits in terms of reducing heat-related mortality (by around 40%) and closed windows led to increased risk (of 29–64%) [24]. The authors highlight the importance of appropriate building design to both save energy and reduce heat-related mortality, especially for dwelling types associated with heat-vulnerable populations.

5. Discussion

The research published in this special issue has touched on three broad areas: (1) a discussion of the complex interactions between our environment and health, including the influence of large-scale dynamical systems; (2) quantitative estimates of the impacts of climate change in health, particularly through changes in temperature; and (3) an examination of the potential for adaptation and mitigation methods to modify the risks to health from climate change. Clearly, multi-disciplinary research on all these areas can help move us forward in understanding the risks and minimising the impacts of climate change on health.

There are more challenges for the future, however. Any mitigation and adaptation action on climate change needs to follow a 'co-benefits' approach to maximise public health benefits at the same time as reducing health risks [25]. Economic costs of impacts based on a lack of action should be calculated, but to motivate policy decisions, economic benefits from taking mitigation or adaptation action are essential. A multi-disciplinary approach is essential to minimise the risk of unintended consequences that are a danger when considering impacts in isolation.

As this inter-disciplinary field of research grows, and climate impacts on health outcomes become more widely understood, the scientific community has an opportunity to instigate positive societal changes, but further thought should be given as to how and where to take action in order to maximise health benefits. To reduce health inequalities, resources need to be applied to the populations mostly at risk, and where benefits are likely to be felt most widely. Historically, the balance of research has disproportionately focused on high income countries, where risks are generally lower [26]. The rollout of global research programmes on environment and health which embrace innovative approaches has the potential to shift this balance towards lower and middle income countries, and to focus attention where it is most needed [27].

Acknowledgments: We are grateful to all authors and reviewers who contributed to this special issue. C.H. is funded by NERC (grant number NE/R01440X/1).

Conflicts of Interest: The authors declare no conflict of interest.

References

1. Stocker, T.F.D.; Qin, G.-K.; Plattner, M.; Tignor, S.K.; Allen, J. (Eds.) Cambridge University Press: New York, NY, USA, 2013.
2. Field, C.B.; Barros, V.R.; Stocker, T.F.; Qin, D.; Dokken, D.J.; Ebi, K.L.; Mastrandrea, M.D.; Mach, K.; Plattner, G.; Allen, S.; et al. Managing the Risks of Extreme Events and Disasters to Advance Climate Change Adaptation. In *A Special Report of Working Groups I and II of the Intergovernmental Panel on Climate Change*; Intergovernmental Panel on Climate Change: Geneva, Switzerland, 2012; pp. 1–594.
3. Smith, K.; Woodward, D.; Campbell-Lendrum, D.; Chadee, Y.; Honda, Y.; Liu, Q.; Olwoch, B.; Revich, B.; Sauerborn, R. Human health: Impacts, adaptation and co-benefits. In *Climate Change 2014: Impacts, Adaptation and Vulnerability. Part A: Global and Sectoral Aspects. Contribution of Working Group II to the Fifth Assessment Report of the Intergovernmental Panel on Clima*; Intergovernmental Panel on Climate Change: Geneva, Switzerland, 2014.
4. Murray, V.; Waite, T. Climate Change and Human Health—The Links to the UN Landmark Agreement on Disaster Risk Reduction. *Atmosphere* **2018**, *9*, 231. [CrossRef]
5. IPCC Summary for Policymakers. *Global Warming of 1.5 °C*; Intergovernmental Panel on Climate Change: Geneva, Switzerland, 2018.
6. Hajat, S.; Vardoulakis, S.; Heaviside, C.; Eggen, B. Climate change effects on human health: Projections of temperature-related mortality for the UK during the 2020s, 2050s and 2080s. *J. Epidemiol. Community Health* **2014**, *68*, 641–648. [CrossRef] [PubMed]
7. Heal, M.R.; Heaviside, C.; Doherty, R.M.; Vieno, M.; Stevenson, D.S.; Vardoulakis, S. Health burdens of surface ozone in the UK for a range of future scenarios. *Environ. Int.* **2013**, *61*, 36–44. [CrossRef] [PubMed]
8. Medlock, J.M.; Leach, S.A. Effect of climate change on vector-borne disease risk in the UK. *Lancet. Infect. Dis.* **2015**, *15*, 721–730. [CrossRef]
9. Patz, J.A.; Campbell-Lendrum, D.; Holloway, T.; Foley, J.A. Impact of regional climate change on human health. *Nature* **2005**, *438*, 310–317. [CrossRef] [PubMed]
10. Gasparrini, A.; Guo, Y.; Hashizume, M.; Lavigne, E.; Zanobetti, A.; Schwartz, J.; Tobias, A.; Tong, S.; Rocklöv, J.; Forsberg, B.; et al. Mortality risk attributable to high and low ambient temperature: A multicountry observational study. *Lancet* **2015**, *386*, 369–375. [CrossRef]
11. Health Protection Agency. *Health Effects of Climate Change in the UK 2012*; Health Protection Agency: London, UK, 2012.
12. Fleming, L.E.; Leonardi, G.S.; White, M.P.; Medlock, J.; Alcock, I.; Macintyre, H.L.; Maguire, K.; Nichols, G.; Wheeler, B.W.; Morris, G.; et al. Beyond climate change and health: Integrating broader environmental change and natural environments for public health protection and promotion in the UK. *Atmosphere* **2018**, *9*, 245. [CrossRef]
13. Nichols, G.; Lake, I.; Heaviside, C. Climate Change and Water-Related Infectious Diseases. *Atmosphere* **2018**, *9*, 385. [CrossRef]
14. McGregor, G.; Ebi, K. El Niño Southern Oscillation (ENSO) and Health: An Overview for Climate and Health Researchers. *Atmosphere* **2018**, *9*, 282. [CrossRef]
15. Vencloviene, J.; Braziene, A.; Zaltauskaite, J.; Dobozinskas, P. The Influence of the North Atlantic Oscillation Index on Emergency Ambulance Calls for Elevated Arterial Blood Pressure. *Atmosphere* **2018**, *9*, 294. [CrossRef]
16. Kinney, P. Temporal Trends in Heat-Related Mortality: Implications for Future Projections. *Atmosphere* **2018**, *9*, 409. [CrossRef]
17. Półrolniczak, M.; Tomczyk, M.A.; Kolendowicz, L. Thermal Conditions in the City of Poznań (Poland) during Selected Heat Waves. *Atmosphere* **2018**, *9*, 11. [CrossRef]
18. Tomczyk, A.; Półrolniczak, M.; Kolendowicz, L. Cold Waves in Poznań (Poland) and Thermal Conditions in the City during Selected Cold Waves. *Atmosphere* **2018**, *9*, 208. [CrossRef]
19. Muthers, S.; Laschewski, G.; Matzarakis, A.; Półrolniczak, M.; Tomczyk, M.A.; Kolendowicz, L.; Papadakis, G.; Chalabi, Z.; Thornes, J.; Taylor, J.; et al. The Summers 2003 and 2015 in South-West Germany: Heat Waves and Heat-Related Mortality in the Context of Climate Change. *Atmosphere* **2017**, *8*, 224. [CrossRef]
20. Dear, K. Modelling Productivity Loss from Heat Stress. *Atmosphere* **2018**, *9*, 286. [CrossRef]

21. Orru, K.; Tillmann, M.; Ebi, K.; Orru, H. Making Administrative Systems Adaptive to Emerging Climate Change-Related Health Effects: Case of Estonia. *Atmosphere* **2018**, *9*, 221. [CrossRef]
22. Papadakis, G.; Chalabi, Z.; Thornes, J. Ambulance Service Resource Planning for Extreme Temperatures: Analysis of Ambulance 999 Calls during Episodes of Extreme Temperature in London, UK. *Atmosphere* **2018**, *9*, 182. [CrossRef]
23. Koopmans, S.; Ronda, R.; Steeneveld, G.-J.; Holtslag, A.; Klein Tank, A. Quantifying the Effect of Different Urban Planning Strategies on Heat Stress for Current and Future Climates in the Agglomeration of The Hague (The Netherlands). *Atmosphere* **2018**, *9*, 353. [CrossRef]
24. Taylor, J.; Symonds, P.; Wilkinson, P.; Heaviside, C.; Macintyre, H.; Davies, M.; Mavrogianni, A.; Hutchinson, E. Estimating the influence of housing energy efficiency and overheating adaptations on heat-related mortality in the West Midlands, UK. *Atmosphere* **2018**, *9*, 190. [CrossRef]
25. Haines, A.; McMichael, A.J.; Smith, K.R.; Roberts, I.; Woodcock, J.; Markandya, A.; Armstrong, B.G.; Campbell-Lendrum, D.; Dangour, A.D.; Davies, M.; et al. Public health benefits of strategies to reduce greenhouse-gas emissions: Overview and implications for policy makers. *Lancet* **2009**, *374*, 2104–2114. [CrossRef]
26. Gasparrini, A.; Guo, Y.; Sera, F.; Vicedo-Cabrera, A.M.; Huber, V.; Tong, S.; de Sousa Zanotti Stagliorio Coelho, M.; Nascimento Saldiva, P.H.; Lavigne, E.; Matus Correa, P.; et al. Projections of temperature-related excess mortality under climate change scenarios. *Lancet Planet. Heal.* **2017**, *1*, e360–e367. [CrossRef]
27. Costello, A.; Abbas, M.; Allen, A.; Ball, S.; Bell, S.; Bellamy, R.; Friel, S.; Groce, N.; Johnson, A.; Kett, M.; et al. Managing the health effects of climate change: Lancet and University College London Institute for Global Health Commission. *Lancet* **2009**, *373*, 1693–1733. [CrossRef]

Editorial

Climate Change and Human Health—The Links to the UN Landmark Agreement on Disaster Risk Reduction

Virginia Murray [1,2] and Thomas David Waite [3,*]

[1]   Public Health Consultant in Global Disaster Risk Reduction, Public Health England, London SE1 8UG, UK;
      virginia.murray@phe.gov.uk
[2]   Member of Integrated Research on Disaster Risk (IRDR) Scientific Committee, Beijing 100094, China
[3]   Interim Head of Extreme Events and Health Protection, Public Health England, London SE1 8UG, UK
*   Correspondence: thomas.waite@phe.gov.uk

Received: 22 May 2018; Accepted: 23 May 2018; Published: 15 June 2018

Keywords: climate change; disaster risk reduction; Sendai Framework

This special issue addresses the impact of climate change on human health—an enormous topic that, at many levels, is proving challenging to address. The recent (2015–2016) synchronous adoption of the five UN landmark agreements—comprising the Sendai Framework for Disaster Risk Reduction 2015–2030 [1]; the Agenda 2030 for Sustainable Development Goals (SDGs) [2]; COP21's Paris Climate Conference [3]; World Humanitarian Summit [4]; and Habitat III New Urban Agenda [5]—provides a new opportunity to build coherence across different but overlapping policy areas that all link the impact of climate change on health. Disasters, many of which are exacerbated by climate change and are increasing in frequency and intensity, significantly impede progress towards sustainable development.

Specifically, the Sendai Framework for Disaster Risk Reduction states that 'more dedicated action needs to be focused on tackling underlying disaster risk drivers, such as the consequences of poverty and inequality, climate change and variability'. The Sendai Framework goes on to state that 'addressing climate change as one of the drivers of disaster risk while respecting the mandate of the United Nations Framework Convention on Climate Change represents an opportunity to reduce disaster risk in a meaningful and coherent manner throughout the interrelated intergovernmental processes'. Furthermore, the Framework calls for action on many wider factors that can determine health outcomes, including unplanned and rapid urbanization, poor land management and compounding factors such as demographic change, weak institutional arrangements, non-risk-informed policies, lack of regulation and incentives for private disaster risk reduction investment, complex supply chains, limited availability of technology, unsustainable use of natural resources, declining ecosystems, pandemics and epidemics.

Taken together, these UN General Assembly-adopted frameworks make for a complete resilience agenda requiring action that spans development, humanitarian, climate and disaster risk reduction spheres to reduce the impacts on human health. To reduce the impact of climate change on health, the synergies between policies, programmes and institutions need to be highlighted and supported by the alignment of actions. These frameworks serve to strengthen the existing risk and resilience frameworks for multi-hazard assessments and aim to develop a dynamic, local, preventive, and adaptive urban governance system at the global, national, and local levels that can help to address climate change issues.

The five Agreements represent a major turning point in global efforts to tackle existing and future challenges in all countries. Specific emphasis to support resilience-building measures and a shift away from managing crises to proactively reducing their risks is apparent. In order to respond efficiently to all the Agreements, the effective use of best available knowledge, innovative thinking, leadership, coordination mechanisms and partnerships are vital. A systems approach, complemented by a new style of research thinking and leadership, can help align the needs of this emerging field with existing policy agendas [6]. This will require the engagement of stakeholders at all levels, as well as the breaking down of traditional silos to be replaced by more integrated partnerships that reflect a more holistic approach to risk management. Scientific methods, networks and communication offer critical assistance to the development of well-informed policies and decisions across all countries [7].

Yet, stronger links between evidence and decision making in policy and planning are also needed to ensure delivery of the 2030 agenda for sustainable development. The UN Office for Disaster Risk Reduction of the UNISDR has acknowledged this and has issued a roadmap to support the implementation of the Sendai Framework [8]. Understanding disaster risk is a priority action in the framework, with four expected outcomes:

1. To assess and update the current state of data, scientific and local and indigenous knowledge and technical expertise availability on disaster risk reduction and fill the gaps with new knowledge.
2. To synthesize, produce and disseminate scientific evidence in a timely and accessible manner that responds to the knowledge needs of policy-makers and practitioners.
3. To ensure that scientific data and information support are used in monitoring and reviewing progress towards disaster risk reduction and resilience building.
4. To build capacity to ensure that all sectors and countries have access to scientific information for better-informed decision-making.

Many roles exist for practitioners, scientists and researchers to achieve these outcomes, including the promotion of crucial coherence with disaster risk reduction and the UN landmark agreements (in particular, the SDGs and climate change) in data collection and indicators to assist in monitoring and evaluation. In addition, the development and promotion of multi-hazard early warning systems, which incorporate improved climate information, aerial and spatial data, emergency response services and communication to end users, will support the needs of policy- and decision-makers at all levels for scientific data and information to strengthen preparedness, response and to "Build Back Better" in the 'three Rs': recovery, rehabilitation and reconstruction.

By reducing losses and the impacts on the most vulnerable communities and locations around the world, the scientific community can help achieve the Sendai Framework goal, which is to prevent new and to reduce existing disaster risk by implementing integrated and inclusive measures that minimise hazard exposure and vulnerability to disaster, increase preparedness for response and recovery and, thus, strengthen resilience.

Author Contributions: T.W. and V.M. contributed equally to the concept, creation and editing of this editorial.

Funding: This research received no external funding.

Conflicts of Interest: The authors declare no conflict of interest.

References

1. United Nations International Strategy for Disaster Reduction. Sendai Framework for Disaster Risk Reduction 2015–2030. 2015. Available online: http://www.preventionweb.net/files/43291_sendaiframeworkfordrren.pdf (accessed on 6 April 2018).
2. UN General Assembly Transforming Our World: The 2030 Agenda for Sustainable Development Resolution Adopted by the General Assembly on 25 September 2015 A/RES/70/1. Available online: http://www.un.org/en/development/desa/population/migration/generalassembly/docs/globalcompact/A_RES_70_1_E.pdf (accessed on 6 April 2018).
3. United Nations Framework Convention on Climate Change. Conference of the Parties Twenty-First Session Paris, 30 November to 11 December 2015. Adoption of the Paris Agreement FCCC/CP/2015/L.9/Rev.1. Available online: https://unfccc.int/resource/docs/2015/cop21/eng/l09r01.pdf (accessed on 6 April 2018).

4. UN General Assembly Outcome of the World Humanitarian Summit 23 August 2016 A/71/353. Available online: https://www.agendaforhumanity.org/sites/default/files/A-71-353%20-%20SG%20Report%20on%20the%20Outcome%20of%20the%20WHS.pdf (accessed on 6 April 2018).
5. UN General Assembly. New Urban Agenda. United Nations Conference on Housing and Sustainable Urban Development (Habitat III), Resolution Adopted by the General Assembly on 23 December 2016. A/RES/71/256. Available online: http://www.un.org/en/development/desa/population/migration/generalassembly/docs/globalcompact/A_RES_71_256.pdf (accessed on 6 April 2018).
6. Berry, H.L.; Waite, T.D.; Dear, K.B.G.; Capon, A.G.; Murray, V. The case for systems thinking about climate change and mental health. *Nat. Clim. Chang.* **2018**, *8*, 282–290. [CrossRef]
7. Aitsi-Selmi, A.; Blanchard, K.; Al-Khudhairy, D.; Ammann, W.; Basabe, P.; Johnston, D.; Ogallo, L.; Onishi, T.; Renn, O.; Revi, A.; et al. *Report: Science Is Used for Disaster Risk Reduction*; United Nations Office for Disaster Risk Reduction, Scientific and Technical Advisory Group: Geneva, Switzerland, 2015. Available online: http://preventionweb.net/go/42848 (accessed on 6 April 2018).
8. United Nations Office for Disaster Risk Reduction. *UNISDR the Science and Technology Roadmap to Support the Implementation of the Sendai Framework for Disaster Risk Reduction 2015–2030*; United Nations Office for Disaster Risk Reduction: Geneva, Switzerland, 2015. Available online: https://www.preventionweb.net/files/45270_unisdrscienceandtechnologyroadmap.pdf (accessed on 6 April 2018).

 *atmosphere*

Article

Quantifying the Effect of Different Urban Planning Strategies on Heat Stress for Current and Future Climates in the Agglomeration of The Hague (The Netherlands)

Sytse Koopmans [1],*, Reinder Ronda [2], Gert-Jan Steeneveld [1], Albert A.M. Holtslag [1] and Albert M.G. Klein Tank [1,2]

[1] Meteorology and Air Quality Section, Wageningen University, P.O. Box 47, 6700AA Wageningen, The Netherlands; gert-jan.steeneveld@wur.nl (G.-J.S.); bert.holtslag@wur.nl (A.A.M.H.); albert.kleintank@wur.nl (A.M.G.K.T.)

[2] Royal Netherlands Meteorological Institute, P.O. Box 201, 37371 GA De Bilt, The Netherlands; reinder.ronda@knmi.nl

* Correspondence: sytse.koopmans@wur.nl

Received: 29 May 2018; Accepted: 29 August 2018; Published: 13 September 2018

Abstract: In the Netherlands, there will be an urgent need for additional housing by the year 2040, which mainly has to be realized within the existing built environment rather than in the spatial extension of cities. In this data-driven study, we investigated the effects of different urban planning strategies on heat stress for the current climate and future climate scenarios (year 2050) for the urban agglomeration of The Hague. Heat stress is here expressed as the number of days exceeding minimum temperatures of 20 °C in a year. Thereto, we applied a diagnostic equation to determine the daily maximum urban heat island based on routine meteorological observations and straightforward urban morphological properties including the sky-view factor and the vegetation fraction. Moreover, we utilized the Royal Netherlands Meteorological Institute's (KNMI) climate scenarios to transform present-day meteorological hourly time series into the future time series. The urban planning strategies differ in replacing low- and mid-rise buildings with high-rise buildings (which reduces the sky-view factor), and constructing buildings on green areas (which reduces the vegetation fraction). We found that, in most cases, the vegetation fraction is a more critical parameter than the sky-view factor to minimize the extra heat stress incurred when densifying the neighbourhood. This means that an urban planning strategy consisting of high-rise buildings and preserved green areas is often the best solution. Still, climate change will have a larger impact on heat stress for the year 2050 than the imposed urban densification.

Keywords: urban heat island; urban planning; heat resilience; climate scenarios

1. Introduction

Climate change is expected to affect society substantially in terms of increased temperatures and precipitation, as well as more frequent weather extremes [1,2]. Sophisticated climate-change projections were carried out for many countries including the Netherlands. However, these projections mostly focus on rural areas, while policy makers are also interested in how they can make future cities more heat-resilient. The Netherlands has the challenge of realizing one million new houses by the year 2040, which mainly need to be created within the existing built environment. This study aims to clarify how climate change evolves in Dutch cities, and which urban morphological factors need to be modified in order to reduce heat stress.

Temperatures in cities are typically significantly higher for most of the day compared to their rural surroundings [3]. The heat stress due to these higher temperatures has a significant impact on health [4,5]. These elevated urban temperatures are known as the urban heat island (UHI), which is defined as the difference in 2 m of air temperature difference between a city and the surrounding countryside. Many factors contribute to a warmer urban environment. In general, the radiation balance of a city is altered due to its complex canyon structure, which results in a relatively small albedo, and thus, a large energy gain for the urban surface [6]. Furthermore, the energy balance of urban areas is different from rural areas due to the lower sky-view factor [3,7], enhanced impervious surfaces (initiating a higher sensible heat flux at the expense of the latent heat flux), and anthropogenic heat release [8]. Because of the relatively large heat conductivity and thermal capacity of the urban fabric, cities can store more heat during the daytime than rural environments. This energy is released after sunset creating the largest UHI in the evening or onset of the night. The elevated temperatures in urban areas affect human health and are a significant contributor to the excess mortality observed during heat waves [4,5]. High night-time temperatures have adverse effects on human health by preventing an adequate recovery from daytime heat and by inducing sleep deprivation [9]. The best known examples for western Europe were the high number of excess deaths in France during the heat waves of 2003 and 2006 [10,11]. In the Netherlands, the number of excess deaths was estimated to be between 1400 and 2200 in the summer of 2003 [12].

The UHI is difficult to quantify directly with observations, since the availability of long-term climatologic observational data in cities is limited. This is because official weather stations are preferably located in open rural or natural areas with the aim of obtaining a substantial spatial representativeness [13]. Consequently, long-term climatological observations to monitor urban heat are rarely available for cities. Other observational sources from urban areas such as amateur weather station data or field campaign data generally do not cover long periods, and the data often represent only small areas due to the high variability in temperature within the urban area setting [14,15]. To overcome this lack of observations, the UHI is often modeled using sophisticated atmospheric mesoscale models coupled to an urban canopy [16–19]. However, those models require large amounts of initial urban morphological data, which are often inaccurate and sometimes not available [20]. Another approach is to measure the UHI with remote sensing techniques [21,22]. However, correlations between remote-sensing-derived surface temperatures and air temperatures are weak, due to the absence of advection and flux divergence in the air volume [21]. In addition, the intervals of the revisiting times restrict a dynamical analysis of UHIs [22].

As an alternative approach, in this study, we diagnosed the daily maximum UHI by applying an empirical formula designed by Theeuwes et al. (2017) [20]. With this approach, we calculated urban minimum temperatures and compared them with rural minimum temperatures. The data needed in this equation consist of routine meteorological observations and morphological city characteristics. A novel concept is the application of this equation to determine the UHI for current and future climates in an urban agglomeration. As a test bed, we computed the UHI for the agglomeration of The Hague in the Netherlands (Figure 1). Previous research shows that cities of varying size in a maritime climate such as the Netherlands may experience severe heat stress [14]. This also holds for near-coastal cities like Rotterdam where nocturnal UHIs are up to 7 °C during clear heat-wave days [23]. Molenaar et al. [24] estimated that the number of hours of experienced heat stress, here defined as the hours with physiologically equivalent temperatures above 23 °C, will double from the current average of about 250 hours per year to slightly more than 500 hours per year under the warmest Royal Netherlands Meteorological Institute (KNMI) climate scenario.

Also a novel concept, we investigated the impact of construction plans on the magnitude of the UHI. In the future, the Dutch government will face the challenge of building an additional one million residences before the year 2040 [25]. A substantial part of this new construction has to be developed within existing urban areas. This means that the population density will increase, which will apply pressure to the heat resiliency of cities. For instance, Steeneveld et al. [14] found a good correlation between the higher percentiles of UHI and population densities for Dutch cities of various sizes. Within the current research, we investigated the effect of different urban planning strategies on heat stress for current and future climate scenarios by creating heat maps with a resolution of 100 m. The urban planning strategies refer to the redevelopment and the construction of additional residences within the urban neighborhoods. The construction differs between high-rise buildings and low- and mid-rise buildings. The choice has consequences for the urban morphological characteristics of an urban area, as vegetation fractions and sky-view factors are important parameters in determining the UHI. The aim of this study was to estimate urban night-time temperatures for current and future climate in order to obtain measures which reduce the UHI.

This paper is organized as follows: Section 2 summarizes the modeling approach and introduces the climate scenarios and urban planning strategies we explored. Section 3 presents the results we obtained. Finally, in Sections 4 and 5, we reflect on our modeling approach and outcomes, and we draw conclusions.

2. Methods

This section presents the followed research methodology, the utilized KNMI climate scenarios, and the study area.

2.1. Study Area

The Ministry of Infrastructure and Environment of the Netherlands was interested in the consequences of climate change and additional housing on heat stress in cities. The Hague was chosen, because it is located in a large urbanized area where there is a great demand for housing [25]. The study area (referring to rectangle in Figure 1A) includes the cities of The Hague, Delft, and Zoetermeer, as well as the green agricultural areas that surround these cities. These cities have 532,561, 102,253, and 124,695 inhabitants, respectively [26], spread over 98.12, 24.06, and 37.06 km^2, respectively. Other cities were included in our study to allow for a reliable verification and to enable a robust statistical relationship between residence density and urban morphological characteristics (see Section 2.2). Taking into account the population of the smaller villages and hamlets, the agglomeration of The Hague accommodates about one million inhabitants. It is one of the most densely populated regions in the Netherlands. The study area has a population density of 1270 inhabitants per km^2 [27].

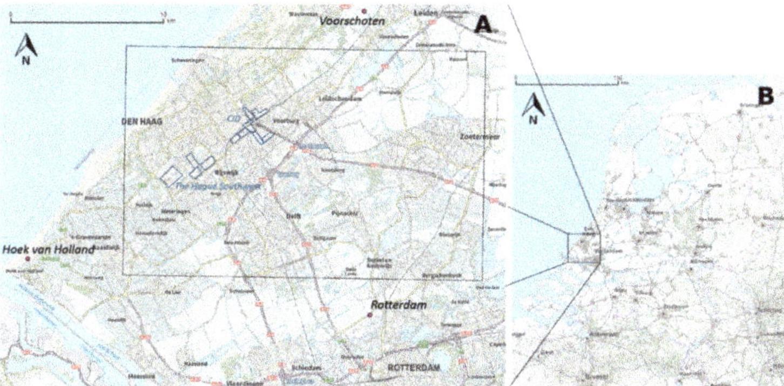

Figure 1. The agglomeration of The Hague (panel **A**) and an overview map of the Netherlands (panel **B**) [28]. The black rectangle in panel A indicates the study area for which the heat map was calculated. The three WMO stations are indicated in italics (Voorschoten, Hoek van Holland, and Rotterdam). Rotterdam acts as a rural reference station. The blue shapes in The Hague (Den Haag) represent the urban districts, Central Innovation District (CID), and The Hague Southwest, for which different urban planning strategies were evaluated for heat stress.

The climatological conditions of these regions are characterized by the vicinity of the North Sea. The prevailing wind direction is from the southwest. The temperature distribution is rather mild with maximum and minimum temperatures of 21 °C and 16 °C in August. In January, the daily maximum and minimum temperatures amount to 6 °C and 3 °C. The yearly precipitation measures 666 mm, with the largest monthly precipitation in November.

2.2. Diagnostic Equation to Compute Minimum Temperatures in Cities

Urban heat islands and urban temperatures can be estimated looking at the combined effects of weather conditions and urban morphological characteristics. From these factors, a daily maximum urban heat island (UHI_{max}) can be estimated using a diagnostic equation created by Theeuwes et al. (2017), henceforth referred as T17 [20] (Equation (1)).

$$\overbrace{\phantom{(2 - SVF - f_{veg})}}^{\text{urban morphology}} \overbrace{\phantom{\sqrt[4]{\frac{S^{\downarrow}DTR^3}{U}}}}^{\text{meteorology}}$$
$$UHI_{max} = \left(2 - SVF - f_{veg}\right)\sqrt[4]{\frac{S^{\downarrow}DTR^3}{U}}. \tag{1}$$

This equation was validated using observational data from 14 cities across northwestern Europe that vary in size. The equation appears to be robust. The UHI_{max} expresses the maximum temperature difference between urban and rural environments on a given day in Equation (1). SVF denotes the sky-view factor and f_{veg} denotes the vegetation fraction of the urban area. $S^{\downarrow}$ denotes the mean downward shortwave radiation ($K \cdot ms^{-1}$), DTR denotes the diurnal temperature range (K), and U denotes the mean wind speed ($m \cdot s^{-1}$) measured at a rural station nearby the city. The average measurement period for each of the meteorological parameters can be found in T17 [20]. The heat maps are computed on grid cells with a resolution of 100 m. The SVF and vegetation fraction were determined using a source area of 500 by 500 m around the grid cell, which was designated to a 100 m resolution. The SVF originates from a 5-m-resolution dataset [29], derived from a digital elevation model based on airborne lidar measurements from aircraft measurements made in 2008 [30]. Upscaling from a 5-m to a 500-m resolution was performed by taking the median of street level SVF data points. The vegetation fraction dataset originates from a normalized difference vegetation index map (NDVI) [31].

The spatial contrast of the UHI_{max} across the city is a good measure to estimate which parts of the city suffer more from heat stress. Nevertheless, when climate change is incorporated, the UHI_{max} is not an adequate measure of heat stress anymore, since climate change is heating up the world regardless of whether an area is classified as rural or urban. Therefore, it is useful to make an evaluation based on the absolute values of urban temperatures, such as minimum temperatures (Tn_{city}), which can be computed as follows:

$$Tn_{city} = Tn_{rural} + UHI_{max} \times 0.46. \tag{2}$$

When minimum temperatures are observed at the rural reference station, the UHI is typically substantially smaller than the UHI_{max}. On a typical cloudless day, the minimum rural temperature increases by 46% of the UHI_{max} [3]. This fraction of the UHI_{max} is also known as the UHI_{TMIN}. The WMO station in Rotterdam acts as a rural reference station (provides Tn_{rural} in Equation (2)) (Figure 1A). Measurements from other nearby WMO stations were discarded, because they were located too close to the sea (Hoek van Holland) or were recently relocated (Voorschoten). In order to compute climatologies, we used the data from 15 summer half-years (1 April–30 September) covering the years 2002 to 2016 from the WMO station Rotterdam. Finally, we present heat maps showing the average number of nights per year that exceed the minimum temperature of 20 °C. This metric is consistent with ongoing climate adaption policies in the Netherlands [32]. These nights are often referred to as tropical nights. The heat maps were generated with the QGIS software [33].

2.3. Future Climate Scenarios

For future climate scenarios, the meteorological time series from the rural station Rotterdam during the period 2002–2016 were transformed into the projected climate of 2050. The Netherlands weather services (KNMI) developed four different scenarios to depict how the climate may evolve, namely the G_L, G_H, W_L, and W_H scenarios. The scenarios differ in the estimated global temperature rise (G or W) and the degree of change in air circulation patterns (L and H suffix), all of which are schematically represented in Figure 2.

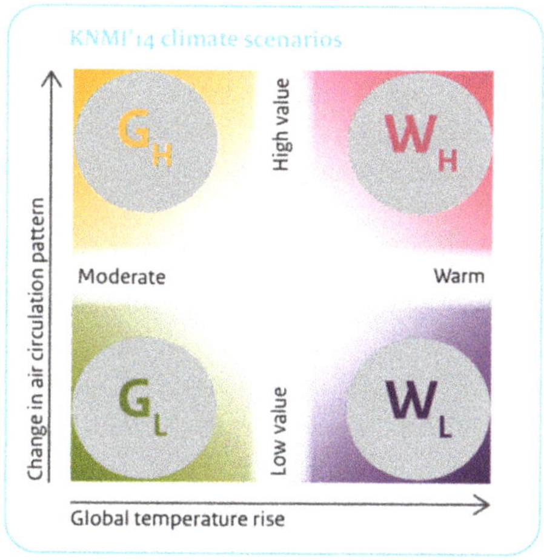

Figure 2. Four scenarios were used to calculate future heat stress. They differ in global temperature rise and change in air circulation pattern. Reproduced with permission from KNMI, Brochure KNMI'14 climate scenarios, 2015 [34].

These climate scenarios are based on the Intergovernmental Panel on Climate Change (IPCC) global climate model calculations, on global climate model EC-Earth results, and on a downscaling step within the regional climate model, Racmo [34]. All four scenarios have an equal chance of occurrence. As indicated in Figure 2, the scenarios differ in the degree of change in air circulation patterns and the global temperature rise. The differences in global temperature rise are due to different projections of greenhouse-gas emissions adopted from the emission scenarios from the IPCC [34]. The scenarios G_L and G_H match the lower end of the emission reduction scenarios (RCP4.5 and RCP6), and the scenarios W_L and W_H match the high-emission scenario (RCP8.5) which does not include specific climate mitigation [35,36]. Within all of the model simulations, a distinction can be made between models showing a large and small precipitation response [37]. Simulations with a large precipitation response foresee wet humid winters and dry summers, and are assigned to H-scenarios. The change in precipitation response is linked to circulation change, and H-scenarios show more frequent westerly circulation in the winter and domination of high pressure in the summer. In the summer, this means an increase in solar radiation and more easterly winds, which implies higher temperatures. This weather type also favors urban heat islands.

The scenarios provide monthly temperature increments for daily percentiles [38]. These increments were relative to the climate period of 1981–2010. There is already a climate signal in the time series (2002–2016) compared to this climate period, and this bias was subtracted from the results. Using the procedure of Molenaar et al. [24], these increments were interpolated for all of the days in a month. In this study, the transformation was performed on hourly values, because the time frame in which minimum temperatures were determined differs from the standard, i.e., from 8:00 a.m. universal coordinated time (UTC) to 8:00 a.m. UTC the next day. After the creation of a future hourly time series, a proper minimum temperature can be derived. As such, we use

$$T^f(h) = T^c(h) + \Delta T(d) + \Delta HTD(h) - \overline{\Delta HTD(h)}, \tag{3}$$

where

$$\Delta HTD(h) = \frac{\left(T^c(h) - T^c_{min}\right)}{\left(T^c_{max} - T^c_{min}\right)} \times \Delta DTR. \tag{4}$$

The symbols are defined by the following:

$T^c(h)$ = hourly temperature current climate
T^c_{min} = minimum daily temperature, 8:00 a.m. to 8:00 a.m. the next day
T^c_{max} = maximum daily temperature, 8:00 a.m. to 8:00 a.m. the next day
$T^f(h)$ = hourly temperature for future climate
$\Delta T(d)$ = daily temperature increment to future climate
ΔDTR = change in average diurnal temperature range to future climate
$\Delta HTD(h)$ = hourly temperature deviation
$\overline{\Delta HTD(h)}$ = 24-hour average bias of hourly temperature deviation

For a transformation to hourly values, the daily temperature increments, $\Delta T(d)$, were added to individual hours (see Equation (3)). This increment, $\Delta T(d)$, was the outcome of the procedure of Molenaar et al. [24]. A novel aspect is that the change in diurnal temperature range (DTR), ΔDTR, between the future and current climate was also taken into account (Equations (3) and (4)). With Equation (3), a new dataset of minimum temperatures for 2050 can be directly calculated using the hourly temperature dataset of $T^f(h)$.

The UHI_{max} also changes in a future climate (see Equation (1)). The scenarios generally show an increase in global radiation and generally show a decrease in DTR, which influences the UHI_{max}. On the contrary, however, the H-scenarios (G_H and W_H) show a small increase in DTR for the warmest months of July and August. In all climate scenarios, these months show an increase in global radiation from 2% in the G_L scenario to 7% in the W_H scenario, as well as a change in DTR between −3% in the G_L scenario and 2% in the W_H scenario. For the transformation in global radiation, we followed the method of Bakker [39]. The climate scenarios provided only monthly changes in global radiation. In our study, the change in radiation was distributed over all days without exceeding the maximum realistic daily radiation sum, which was set at 75% of the radiation at the top of the atmosphere [39]. Changes in wind speed were not considered, because they fell within the natural variation range [34]. Transformation numbers can

2.4. Urban Planning Strategies

The impact of the redevelopment and the construction of additional residences was evaluated for two urban districts. One district was a residential area consisting of low- and mid-rise buildings located in the southwestern part of The Hague, referred to as "The Hague Southwest" (Figure 3A). The other district was located near the center of The Hague and mainly consisted of a relatively old business area and newer high-rise commercial buildings. This area is scheduled for redevelopment, and is referred to as the Central Innovation District (CID) (Figure 3B). The densification of this district is much larger than that of The Hague Southwest, because of the larger building assignment.

The building assignments of the districts could either be achieved by building on currently green spaces or by replacing the existing low- and mid-rise buildings with high-rise buildings. This will either reduce the vegetation fraction or raise the SVF, which will consequently increase the UHI_{max}. This choice and the alternative of building residences in surrounding green corridors were examined using five distinct urban planning strategies shown in Table 1.

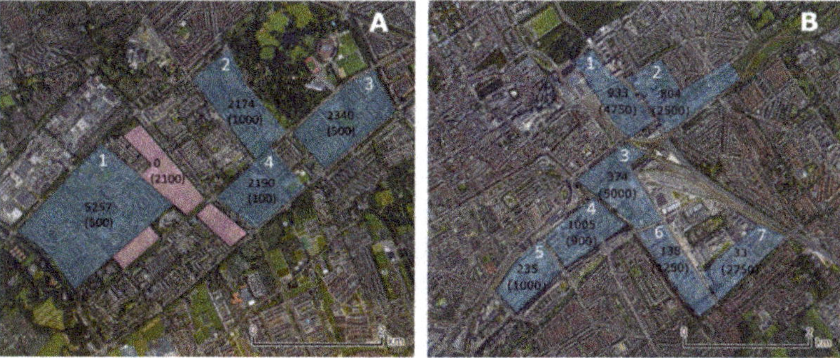

Figure 3. Theoretical building assignment for districts of The Hague Southwest (panel **A**) and the CID (panel **B**). The building assignment is indicated for different neighborhoods within brackets, and the number above the brackets indicates the current number of residences in the neighborhood. The pink areas in panel A indicate green corridors which were built using strategy C. The white numbers label the different neighborhoods.

Table 1. Five urban planning strategies that complete the building assignments for The Hague Southwest and the Central Innovation District (CID). The enumeration starts with B, in accordance with the panels in Figures 8 and 10. Panel A in these figures represents the current housing.

	The Hague Southwest		CID (Central Innovation District)
B.	Build on green corridors around the urban neighborhoods with low- and mid-rise buildings.	B.	Construct high-rise buildings whereby green spaces are preserved as much as possible.
C.	Construct low- and mid-rise buildings on green spaces within the urban neighborhoods.	C.	Construct high-rise buildings whereby the vegetation fraction is increased.
D.	Construct high-rise buildings within the urban neighborhoods to preserve existing green spaces.		

In this study, the building assignment was expressed as a change in the SVF and/or vegetation fraction in order to estimate the changes in the UHI (see Equation (1)). Hence, as a first step, we had to determine the numerical relationship among the density of the residences [27], the SVF, and the vegetation fraction (f_{veg}) for the current building volume of the agglomeration of The Hague. This is expressed as

$$\text{Residences per km}^2 = -213(\pm19)\text{SVF} - 200(\pm30)f_{veg} + 1.85(\pm0.04)\left(\text{SVF} \times f_{veg}\right) + 22687. \quad (5)$$

Figure 4 shows the empirical relationship among the density of the residences, the SVF, and the f_{veg} in a three-dimensional (3D) plot. Herein, the SVF and the f_{veg} are denoted in percentages. The ranges used for this relationship vary between 52–99% and 11–93% for the SVF and the f_{veg}, respectively. For the CID, it was necessary to extrapolate to a minimum SVF of 22% and a vegetation fraction of 5% to meet the requirements of the high density of residences. Note that the population density does not currently occur in the agglomeration of The Hague. The UHI_{max} equation (Equation (2)), however, was tested for highly urbanized areas [20]; thus, it can deal with the planned densification for the CID. The values used for the SVF, the f_{veg}, and the density of residences for the five urban planning strategies can be found in Tables A1 and A2 in Appendix A.

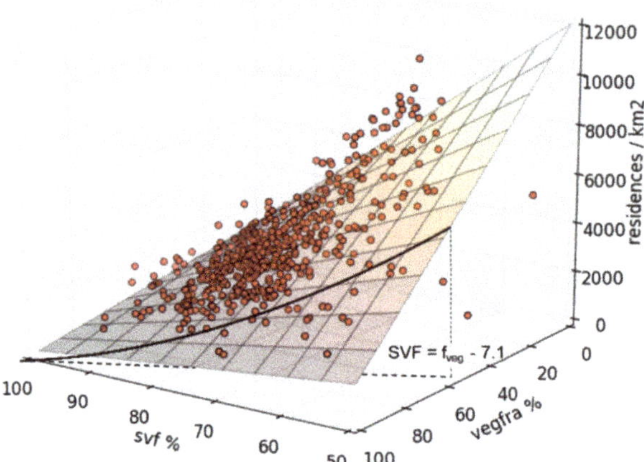

Figure 4. Empirical relationship among the density of the residences [27], the sky-view factor (SVF), and the vegetation fraction for the agglomeration of The Hague presented on a three-dimensional (3D) surface (R^2 = 0.64). Areas with more than one-third non-residential buildings were removed from the analysis. Areas with a population density below 750 per km^2 were also removed. The black line and the black dotted line indicate the equilibrium between the slopes of the vegetation fraction and the SVF. The grid resolution used is 500 m.

3. Results

3.1. Model Validation for the Agglomeration of The Hague

The diagnostic equation for the UHI applied here (Equation (2)) was designed and tested using observational data from cities of variable sizes in northwestern Europe [20]. It was crucial to verify the results presented here with citizen weather stations in the agglomeration of The Hague. Firstly, minimum temperatures are a different metric from the maximum UHI used for verification by T17 [20]. Secondly, the retrieval and integration of morphological data and meteorological data were slightly different compared with the procedure of T17 [20]. Finally, spatial differences may occur because the area is bordered by the coast, where seasonality in sea-surface temperatures and the presence of a sea breeze may play a role.

A quality assessment of crowdsourced weather data is indispensable, because weather stations may have issues and are not always properly installed [40,41]. For the verification, we used data from citizen weather stations in the agglomeration of The Hague, and data were also obtained from the Weather Underground platform. We selected only Davis Vantage Pro and Oregon scientific stations, since they show small biases in the night [40]. This added up to nine stations in total. The time series from these stations comprised two years of data (2015–2016) restricted by the summer period (April–October). Minimum temperatures were discarded if more than two hourly values were missing (as in the analysis by Hopkinson et al. [42]). After evaluating this constraint, the minimum availability was 48% and the average availability was 78% for a single citizen station.

The citizen weather stations and the diagnostic equation were compared in a quantile–quantile plot in Figure 5A, and the bias and standard deviation of the stations grouped by percentile are shown in Figure 5B. Only the lower percentiles with low minimum temperatures had a substantial cold bias. It is expected that the lack of anthropogenic heat led to the underestimation in the diagnostic equation (Equation (2)). During cold weather in the spring, the anthropogenic heat source is larger than in the summer due to heating of buildings [43]. Furthermore, in the lowest and highest percentile, there seemed to be more variance in minimum temperatures of the citizen weather stations than in the modeled minimum temperatures of the diagnostic equation. The other percentiles showed good agreement between the model and the observations, with only slight cold biases, and supported the reliability of the equation.

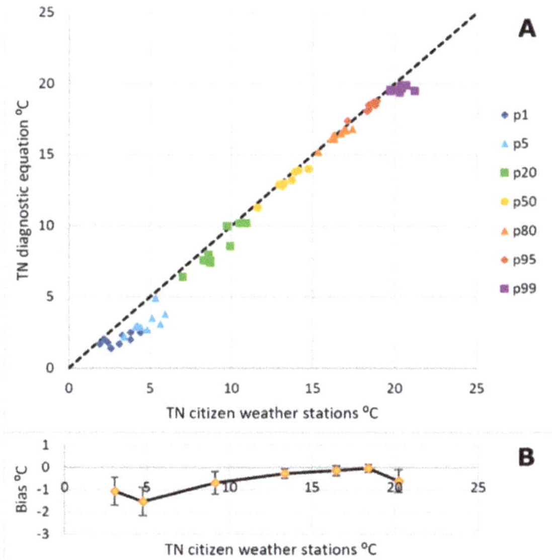

Figure 5. (**A**) Quantile–quantile plot for modeled and observed (Davis and Oregon citizen weather stations) night-time minimum temperatures for the years 2015–2016. (**B**) The bias between the data from these citizen weather stations and the diagnostic equation are presented as a function of temperature. Error bars indicate the standard deviation.

3.2. Heat Stress for Current and Future Climate in 2050

As a starting point, a heat map of the current climate was constructed for The Hague (Figure 6). Note that this heat map covered a smaller area than the study area displayed in Figure 2. In the current climate, tropical nights are not common. On average, rural areas experience 1.25 tropical nights per year, while the urbanized parts of The Hague face 3.5 to a maximum 4.5 tropical nights per year.

Note that, in rural areas, the number of tropical nights was distributed quite uniformly over the 15 years. During a few (20%) of the years, there were no tropical nights and the warmest year in the series, 2006, had three tropical nights. The tropical nights typically occurred when weather conditions were governed by a warm humid cyclonic southerly flow with an upper level trough west of the European continent and approaching fronts, sometimes referred to as a Spanish plume [44]. These situations are unstable with high chances of thunderstorm and squall lines bringing in colder air masses; therefore, such situations do usually not last long. These situations prevent rural areas from developing a cool stable boundary layer at night. Urban areas may experience tropical nights particularly during relatively long warm episodes or heat waves, which is more often accompanied by an anticyclonic weather type. For the year 2006, a maximum of 14 tropical nights occurred in the center of The Hague. Six tropical nights occurred within a period of eight days. Note that a warm summer like 2006 is not unlikely in the current climate. The return time of such a summer is estimated at eight years [45].

Figure 6. Modeled average number of nights per year above 20 °C for the current climate (2002–2016) for The Hague.

Concerning the climate in 2050, we foresee that the number of tropical nights will increase substantially. Depending on the scenario, rural areas may experience 2.1–5.8 tropical nights per year, and the center of The Hague may experience 6.5–16 tropical nights per year (Figure 7). For the city, this means a three- to fourfold increase in tropical nights for the warmest W_H scenario compared to the current climate. The observed warm year 2006 would fall in between the W_L and the W_H scenarios for an average year around 2050. The transformed year 2006 would lead to 15 tropical nights in rural areas and up to 32 tropical nights in the center of The Hague for the W_H scenario.

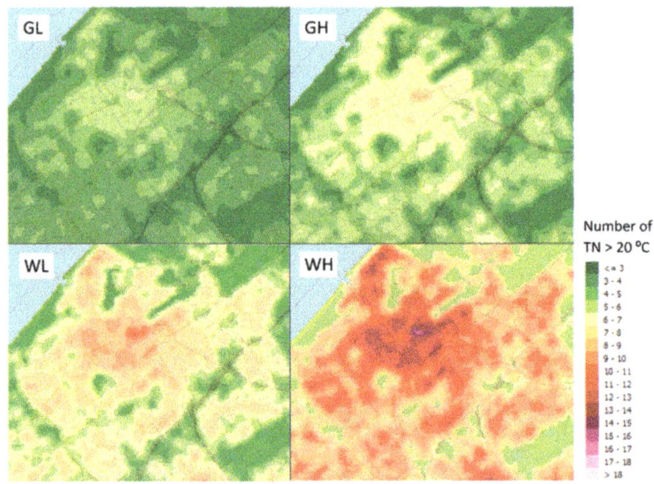

Figure 7. Modeled average number of nights per year above 20 °C for the four Royal Netherlands Meteorological Institute (KNMI'14) scenarios representing the year 2050.

The absolute difference in tropical nights between rural and urban areas increases in the warmer scenarios, because more prevalent weather conditions will enable more tropical nights in cities in a warmer climate. Minimum temperatures simulated in cities of around 18 °C in the current climate will exceed 20 °C in the W_H scenario, and those nights will occur much more frequently than the tropical nights of today. Interestingly, the average UHI is likely to decrease a little for urban tropical nights in the future, because the more prevalent weather conditions are less related to strong UHIs.

In any case, there will be very little change in the number of tropical nights for all future climate scenarios (less than 3%). This is elaborated in the discussion.

3.3. Urban Planning Strategies for The Hague Southwest

For the urban district, The Hague Southwest, three urban planning strategies (abbreviated to strategies) were examined, and the impact on the number of tropical nights and UHI_{max} was evaluated. The current climate does not indicate clear absolute differences in tropical nights between urban areas, due to the scarcity of these nights in low- and mid-rise urban areas. Therefore, the different strategies were shown for the future W_H scenario only. The additional heat stress (measured in number of tropical nights per year) will be mostly scattered over the urban district in the strategy where new residences will be built on the currently green corridors around the neighborhoods (Figure 8A,B). The other strategies concentrate the heat stress particularly within the neighborhoods. The largest increase in the number of tropical nights takes place in neighborhood 2, which has the highest density of residences with the largest building assignment (Figure 3). For the hottest place in this district, the average number of tropical nights per year will increase from 3.1 to 3.5 in the current climate (not shown), and from 11 to 13 nights in the W_H scenario shown in Figure 8C,D (the mildest G_L scenario shows an increase from 4.7 to 5.4–5.5 tropical nights per year). Thus, the maximum expected numbers of tropical nights for this neighborhood in strategy C and strategy D are very similar. However, strategy C (construct low- and mid-rise buildings) shows larger patterns of these maxima than strategy D (construct high-rise buildings to preserve green spaces). Strategy C is moderately warmer than strategy D for the entire urban district, especially in neighborhood 2. This is best seen in the continuous scale of the 95th percentile of UHI_{max} in Figure 9 for the current climate. The maximum increase in the 95th percentile of UHI_{max} is 1.2 °C for strategy C and 0.8 °C for strategy D, and remains nearly constant for the future scenarios. Note that these relative differences in the scenarios are also reflected in the expected minimum temperatures. In summary, the realization of high-rise buildings to preserve existing green spaces appears to be a better strategy than constructing low- and mid-rise buildings on existing green spaces for The Hague Southwest.

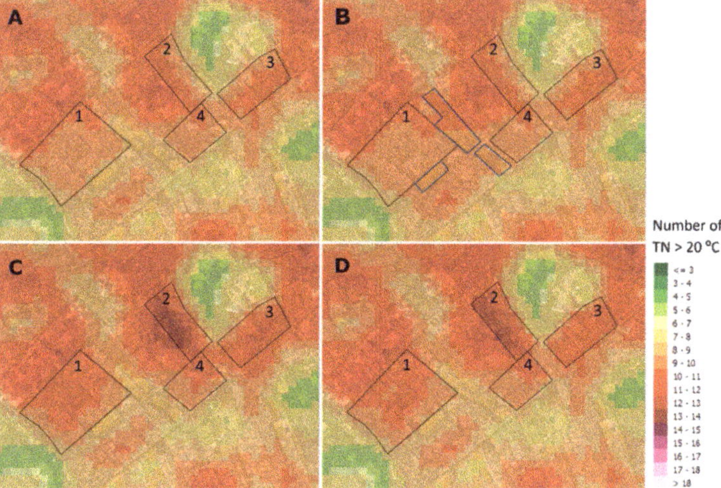

Figure 8. Modeled number of tropical nights per year for the W_H climate scenario in 2050 for The Hague Southwest for: (**A**) current housing, (**B**) building on green corridors around the neighborhoods with low- and mid-rise buildings, (**C**) constructing low- and mid-rise buildings on green spaces within the urban neighborhoods, and (**D**) constructing high-rise buildings within the urban neighborhoods to preserve existing green spaces.

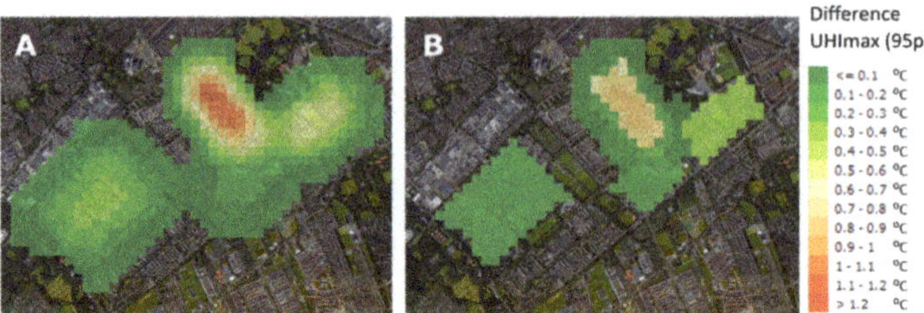

Figure 9. Difference in the modeled 95th percentile of UHI_{max} between (**A**) strategy C (constructing low- and mid-rise buildings) and (**B**) strategy D (constructing high-rise buildings). Both are compared with the current housing for the current climate.

3.4. Urban Planning Strategies for the CID

For the CID (Central Innovation District), we looked at the frequency of tropical nights and UHI_{max} to examine the effects of the urban planning strategies (Figures 10 and 11). In contrast to The Hague Southwest, the effect of the urban planning strategies is larger for the current climate, and therefore, is presented in Figure 10. The most urbanized neighborhoods 1 and 3 show the largest increase in tropical nights in both strategies, from 3.5 tropical nights per year up to more than five tropical nights per year in the current climate. The future climate scenarios foresee 8–15 tropical nights per year in the G_L and W_H scenarios, respectively. When compared to the district, The Hague Southwest, it is not clear which strategy is best. However, differences between the two strategies for the 95th percentile in UHI_{max} reveal that the strategy where green spaces are preserved most (strategy C) is the best choice (Figure 11). The neighborhoods 1, 2, 3, and 7 show mixed results, which is due to the small sizes of the areas. If the neighborhoods were larger, then strategy C would result more clearly in less heat stress. In the Supplementary Materials the other climate scenarios are displayed for the urban planning strategies for CID and The Hague Southwest, the number of tropical nights for the warmest year 2006, and monthly transformation tables for temperature, DTR and global radiation from the current climate to the future climate in 2050.

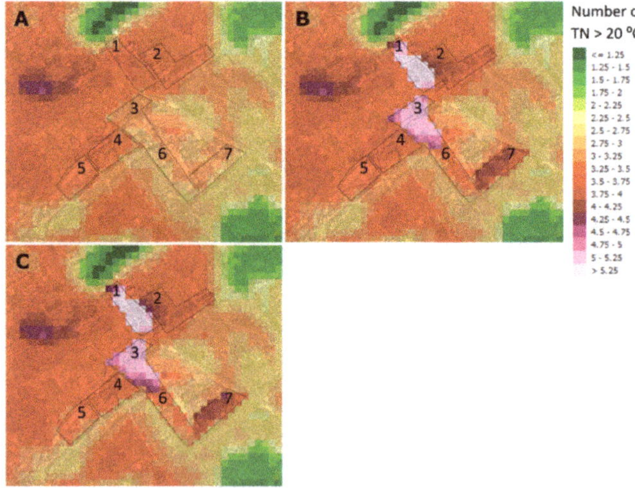

Figure 10. Modeled number of nights above 20 °C per year for the CID in the current climate for (**A**) current housing, (**B**) constructing high-rise buildings whereby green spaces are preserved as much as possible, and (**C**) constructing high-rise buildings whereby the vegetation fraction is increased.

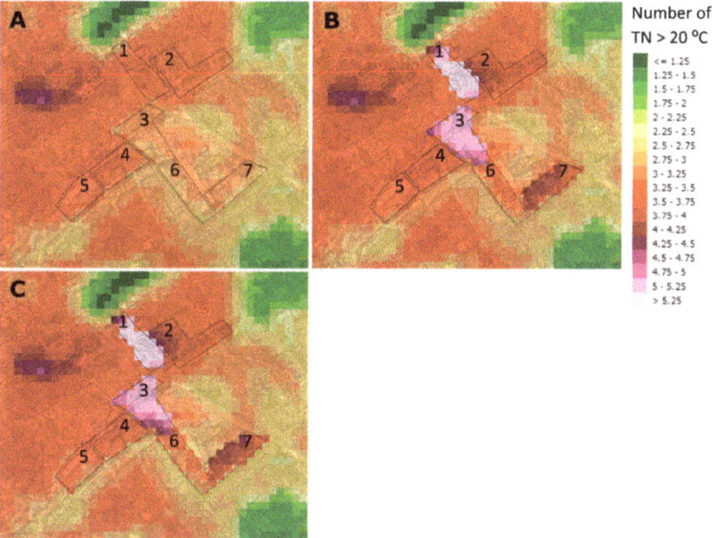

Figure 11. Difference in the modeled 95th percentile of the UHI_{max} between strategy B and strategy C for the W_H climate scenario.

4. Discussion

4.1. Thresholds in Urban Planning Strategies

In this paper, we studied the effects of different urban planning strategies on heat stress for current and future climates in the agglomeration of The Hague. It appears that preserving green spaces was clearly the best urban planning strategy in order to avoid heat stress for The Hague Southwest. The difference between the strategies for the CID was much smaller. An explanation for this difference appears when examining the statistical relationship among the density of residences, the SVF, and the vegetation fraction (Equation (5)). The larger the slope or partial derivative is in Equation (5), the more residences could be built per a certain decrease in SVF or decrease in vegetation fraction. Thus, the effectiveness of the strategy depends on the specific urban conditions. The vast majority of the data points in Figure 4 are positioned above the black line, which means that more residences could be built per percentage point decrease in the SVF than per percentage point decrease in the vegetation fraction. Note that a percentage point decrease in the SVF has the same effect on UHI_{max} as a percentage point decrease in the vegetation fraction (Equation (1)). This means that, for the majority of neighborhoods, a decrease in the SVF due to high-rise buildings causes less heat stress than building low- and mid-rise buildings on green spaces.

$$SVF = f_{veg} - 7.1. \tag{6}$$

The threshold which determines the favorable strategy can be derived from Equation (5) and is denoted in Equation (6). It is also represented as the dotted black line at the bottom of Figure 4. Note that this relationship is specific for The Hague agglomeration and might alter slightly for other urban areas, even in the similar climates of other countries in northwestern Europe.

This threshold indicates that reducing the SVF by constructing high-rise buildings is a better measure, unless the SVF is considerably low and the vegetation fraction is high. This is relatively rare for the agglomeration of The Hague. The differences in effectiveness between the strategies can be attributed to the cross product in Equation (5), which appeared to be significant. The density of residences can be imagined roughly as a product of urban fraction (inverse of vegetation fraction) and height of the buildings (more or less correlated with the inverse of SVF). Two intermediate values in this product result in a substantial larger density of residences rather than a low and a high value for equal levels of heat stress.

4.2. Applicability and Limitations of the UHI Diagnostic Equation

Spatial variation in UHI_{max} and night-time minimum temperatures within cities were investigated several times using multi-year observation data in cities. Montavez et al. [46] found that the urbanization of Granada in Spain resulted in increased night-time minimum temperatures predominantly in the city center. Eliasson [47] concluded that urban–park temperature differences were on the same order of magnitude as urban–rural temperature differences, which corresponds to large parks in The Hague. Heusinkveld et al. [23] found that spatial differences in vegetation cover are paramount in determining the spatial variation in UHI, and the correlation coefficients were best for an upwind fetch area of 700 m. In the current study, the wind direction was not taken into account, and therefore, a smaller area of 500 by 500 m was an appropriate choice. When considering SVF, an aerial mean is used to relate the SVF to UHI to incorporate advection effects on air temperature, which was performed in studies such as those by Goh et al. [48] and Unger [49]. Similar to our study, Unger [49] extracted the SVF on a scale of 500 by 500 m.

The verification of two years of crowdsourced city temperature data reveals that the diagnostic method used in this study does not indicate substantial biases in minimum temperatures in urban areas. This supports the use of the current diagnostic equation. Although the diagnostic equation is a physically meaningful equation with conservation of fundamental dimensions, not all potentially important contributing factors were taken into account. One can think of contributing factors such as anthropogenic heat, albedo, thermal admittance, which is related to the heat storage of buildings, distance to open water from a city's edge, and a vertical temperature gradient, as a measure for stability at the rural reference station. The latter two appeared to have less significance [20]. Caution is advised when applying the method to places that have large anthropogenic heat production like industrial areas or highly densely populated areas. Industrial areas were not sampled in the validation of T17 [20]. Anthropogenic heat can be included into the equation by adding it to the incoming solar radiation, or it can be added as a separate variable. When added to the incoming solar radiation, it has no substantial influence, apart from the cautioned areas with large anthropogenic heat production. Since the proximity to open water is not considered in the equation, the urban areas close to the sea have larger uncertainties for our study area. Open water is difficult to implement in a time-invariant UHI_{max} equation, because the effect is strongly dependent on the season [50]. For albedo and thermal admittance, it is a challenge to retrieve reliable area-wide data on building properties. Albedo data can be straightforwardly retrieved from satellite observations, although a top view would over-represent roof albedo, which is less decisive on the street level than wall and street albedo [51].

A conversion factor of 0.46 was used to estimate the UHI_{TMIN} from the UHI_{max} (see Equation (2)). The ratio between UHI_{max} and UHI_{TMIN} (conversion factor) is probably not fixed for all the percentiles, as demonstrated in a crowdsourced data study conducted in the Netherlands [14], and for city weather stations in Rotterdam [23]. The last study reported UHI_{max} to UHI_{TMIN} ratios in the range of 0.63–0.87 for the 95th percentile. However, it might be a safe choice to have a conservative conversion factor, since the rural reference station Rotterdam is affected by UHI mainly due to advection from Rotterdam and the airport itself [23]. Spatially, the conversion factor may change between areas with dominant narrow or wide street canyons [52].

4.3. Climate-Change Projections on Heat Stress

As demonstrated in Figure 7, the projected climate change could have a reasonably large-to-huge effect on heat stress levels. Regional climate models for Europe show similar outcomes. The high consistency in worsening health indices seem alarming for Europe [9]. Gasparrini et al. [53] and Huynen and Martens [54] found an increase in heat-related excess mortality for future climate scenarios. For the Netherlands, the population-attributable fraction of mortality is estimated to increase by 44–119%, depending on the KNMI'14 climate scenarios [54]. Furthermore, mortality increases when heat waves are combined with air pollution [55]. Air pollution also increases with higher temperatures, due to the associated higher chemical reaction rates and elevated emissions of biogenic ozone precursors [56]. Nevertheless, the direct effect of elevated temperatures during heat waves seems to primarily be the effect of excess mortality [5,55].

Specific to the number of tropical nights, Fischer and Schär [9] showed a three- to fourfold increase in the number of tropical nights for France and Central Europe between 1961–1991 and 2021–2050. For downtown Paris, an increase from six to 35 tropical nights per year was estimated between current and future climate (A1B scenario 2071–2099) [57]. Apart from the different methodology used, these results are in line with our findings.

4.4. Comparison of Urban Heat Island and Climate Scenario Contributions on Heat Stress

Urban heat islands, as well as the projections of the climate scenarios for 2050, show significantly raised temperatures compared to rural areas in the current climate (see Figure 7). More specifically, rural areas in the W_L and W_H scenarios showed approximately the same number of tropical nights for the center of The Hague in the current climate. Nevertheless, climate change appears to create more significant heat stress than the imposed urbanization by the applied building assignments. This applies especially to The Hague Southwest. The neighborhoods 1 and 3 in the CID have large building assignments, and the increase in the number of tropical nights falls within the single climate-change effect of the G_L and G_H scenarios. In particular, the higher minimum temperatures showed a large increase in the climate scenarios. For the 5% warmest nights, an increase in night minima of 0.7 °C is foreseen for the G_L scenario, and up to 2.2 °C for the W_H scenario (1.2 °C and 2.7 °C compared to the reference climate period of 1981–2010). This climate-change effect is about the same for rural and urban areas, which is discussed in the next paragraph. In contrast, the impact of urban planning strategies is smaller, but still relevant. The largest relative temperature differences among the urban planning strategies were found in The Hague Southwest. The preserving green strategy led to a 42% lower increase in UHI_{max} in the 95th percentile than the building on green spaces strategy, i.e., 0.33 °C and 0.57 °C, respectively. For the CID, the building assignment was much larger, and the increased vegetation fraction strategy noted an increase of 1.09 °C compared to 1.36 °C for the preserving green strategy (difference of 20%). Tropical nights in urban areas were generally related to a lower UHI_{max} percentile, which corresponds to a lower UHI_{max}, and consequently, the numbers increased by 25%.

The climate-change effect is nearly the same for rural and urban areas since the UHI will be very similar in the future climate. In the diagnostic equation for UHI_{max} (Equation (1)), this is represented by an offset between a projected decrease in DTR and a projected increase in incoming short-wave radiation, except for the W_L scenario. This scenario shows the largest decrease in DTR with an unchanged amount of sunshine. However, this results in a minor reduction in UHI by 2–3%. In July and August, the UHI slightly increased (2–3%) for the H-scenarios, according to the increase in DTR. Most other studies confirmed the small or non-significant average UHI changes for mid-latitude cities in western Europe for the future climate [58–60]. There are climate models even hinting at a decrease in UHI_{max} and UHI_{TMIN} due to larger projected precipitation deficits and associated drier soils in the summer time [57,59]. In such situations, rural temperatures increase, because less energy is used for evapotranspiration. Hamdi et al. [59] found that, for a coupled urban and regional climate model, the higher percentiles of UHI_{TMIN} could decrease by 1 °C. For our study area, the effect of soil dryness does not play a substantial role in the UHI, because the land is below sea level and ground water levels are controlled.

5. Conclusions

In this study, we investigated the effect of additional housing in constructed zones on heat stress, worked out using different urban planning strategies for the current and future climates for the city of The Hague (the Netherlands). The heat stress is expressed based on the frequency of tropical nights, where minimum temperatures are above 20 °C, and on the 95th percentile of the maximum daily urban heat island, UHI_{max}. The proposed additional housing was added near the city center and in a residential area in The Hague. The urban planning strategies were applied in delineated urban neighborhoods and differed in replacing low- and mid-rise buildings with high-rise buildings, or constructing buildings on vegetated areas. The temperature projection was computed using a validated diagnostic equation which combined weather data and urban morphological characteristics. The vegetation fraction appeared to be a more critical parameter than the sky-view factor, which was reduced by the tall buildings for the vast majority of urban configurations. This means that the combination of mid-rise and high-rise buildings with a preservation of vegetated areas was the best strategy. There is, however, an empirically determined optimum between vegetation fraction and sky-view factor. The most favorable green strategy mitigated the heat stress by 42% and 20% for the two urban districts tested.

In general, climate change will cause a larger increase in heat stress than the extra heat stress caused by the imposed urbanization. Only the largest imposed building assignments could compete with the colder climate scenarios. The most urbanized area of the city has on average 4.5 tropical nights per year. For this area, we found a range of 6.5–16 tropical nights per year for the coldest and warmest climate scenarios. For the warmest summer in the data series (year 2006), the number of tropical nights would increase from 14 in the current climate to 32 in the warmest climate scenario. The results were verified with a selection of high-quality citizen weather stations. The model results were in good agreement with observations and showed only a slight cold bias. The prescribed method based on a diagnostic equation is a fast and efficient way of determining climatologies in minimum temperatures, and it is directly applicable for other cities across northwestern Europe.

Supplementary Materials: The following are available online at http://www.mdpi.com/2073-4433/9/9/353/s1, Figure S1: Modelled number of tropical nights per year for the year 2006, Figure S2: Modelled number of tropical nights per year for year 2006 transformed to the four KNMI'14 scenarios, Figure S3: Modelled average number of nights per year above 20 °C for The Hague Southwest for the current climate, Figure S4: Modelled average number of nights per year above 20 °C for The Hague Southwest for the G_L climate scenario, Figure S5: Modelled average number of nights per year above 20 °C for The Hague Southwest for the G_H climate scenario, Figure S6: Modelled average number of nights per year above 20 °C for The Hague Southwest for the W_L climate scenario, Figure S7: Modelled number of nights above 20 °C per year for the CID for the G_L climate scenario, Figure S8: Modelled number of nights above 20 °C per year for the CID for the G_H climate scenario, Figure S9: Modelled number of nights above 20 °C per year for the CID for the W_H climate scenario, Table S1: Transformation table G_L climate scenario from current climate to future climate in 2050, Table S2: Transformation table G_H climate scenario from current climate to future climate in 2050, Table S3: Transformation table W_L climate scenario from current climate to future climate in 2050, Table S4: Transformation table WH-scenario from current climate to future climate in 2050.

Author Contributions: Data curation, S.K.; Formal analysis, S.K.; Funding acquisition, R.R., G.-J.S. and A.M.G.K.T.; Investigation, S.K. and R.R.; Methodology, S.K., R.R. and A.M.G.K.T.; Project administration, G.-J.S.; Software, S.K.; Supervision, G.-J.S.; Validation, S.K.; Visualization, S.K.; Writing—original draft, S.K.; Writing—review & editing, S.K., R.R., G.-J.S., A.A.M.H. and A.M.G.K.T.

Funding: This study was funded by the Ministry of Infrastructure and Environment.

Acknowledgments: The authors acknowledge feedback from Jan Willem Notenboom, Raymond Sluiter (KNMI Datalab), Rien Bout and Nieske Bisschop (Ministry of Infrastructure and Environment), Erik de Haan (Province of South Holland), and Arno Lammers (Municipality of The Hague). In addition, the authors acknowledge Andrea Pagani (KNMI Datalab) for the provision and feedback of the sky-view factor data. The authors acknowledge Weather Underground® for making the weather data available, which were used for verification, and acknowledge the weather enthusiasts for sharing their data to the platform of Weather Underground®. We acknowledge Robin Palmer for language editing.

Conflicts of Interest: The authors declare no conflict of interest.

Appendix A

Here, the values for the SVF and the vegetation fraction are presented, which were used for the urban planning strategies shown in Tables A1 and A2. According to strategies C and D, only one characteristic, the SVF or f_{veg}, was changed. The green corridor in strategy B has the urban characteristics of neighborhood 1, which allows it to meet the criteria of the building assignment.

For the CID strategy C, both the SVF and the f_{veg} were changed for neighborhoods which became heavily urbanized in order to maintain realistic values. For instance, neighborhood 1 has urban characteristics similar to downtown Paris in strategy C [20].

Table A1. Values used for the sky-view factor (SVF) and the vegetation fraction for urban planning strategies in The Hague Southwest. In addition, the density of residences is indicated. Neighborhoods are displayed in Figure 3.

Southwest	A. Current Situation			B. Building Green Corridors			C. Building Low–Mid-Rise			D. Preserve Existing Green Spaces		
Neighborhood	Res/km²	f_{veg}	SVF	Res/km²	f_{veg}	SVF	Res/km²	f_{veg}	SVF	Res/km²	f_{veg}	SVF
1	5344	0.51	0.73	5344	0.51	0.73	5852	0.43	0.73	5852	0.51	0.69
2	5606	0.49	0.72	5606	0.49	0.72	8185	0.11	0.72	8185	0.39	0.58
3	5305	0.48	0.70	5305	0.48	0.70	6439	0.32	0.70	6439	0.48	0.61
4	6703	0.51	0.72	6703	0.51	0.72	7009	0.47	0.72	7009	0.51	0.69
5	0	n/a	n/a	5344	0.73	0.51	0	n/a	n/a	0	n/a	n/a

Table A2. Values used for the SVF and the vegetation fraction for the urban planning strategies in urban district CID. In addition, the residence density is indicated. Neighborhoods are displayed in Figure 3.

CID	A. Current Situation		B. Green Spaces Preserved			C. Increased Vegetation Fraction		
Neighborhood	f_{veg}	SVF	Res/km	f_{veg}	SVF	Res/km²	f_{veg}	SVF
1	0.20	0.71	17010	0.05	0.23	17010	0.06	0.22
2	0.30	0.66	9811	0.13	0.55	9811	0.26	0.47
3	0.21	0.87	15643	0.06	0.29	15643	0.10	0.26
4	0.21	0.67	7706	0.17	0.64	7706	0.35	0.54
5	0.08	0.80	5805	0.08	0.77	5805	0.47	0.60
6	0.15	0.86	5907	0.15	0.74	5907	0.46	0.60
7	0.19	0.82	11919	0.09	0.46	11919	0.19	0.39

References

1. Intergovernmental Panel on Climate Change (IPCC). Managing the Risks of Extreme Events and Disasters to Advance Climate Change Adaptation. In *A Special Report of Working Groups I and II of the Intergovernmental Panel on Climate Change*; Field, C.B., Baros, V., Stocker, T.F., Qin, D., Dokken, D.J., Ebi, K.L., Mastrandrea, M.D., Mach, K.J., Plattner, G.-K., Allen, S.K., et al., Eds.; Cambridge University Press: Cambridge, UK; New York, NY, USA, 2012; p. 582. ISBN 978-1-107-02506-6.
2. Intergovernmental Panel on Climate Change (IPCC). Annex I: Atlas of Global and Regional Climate Projections. In *Climate Change 2013: The Physical Science Basis. Contribution of Working Group I to the Fifth Assessment Report of the Intergovernmental Panel on Climate Change*; Stocker, T.F., Qin, D., Plattner, G.-K., Tignor, M., Allen, S.K., Boschung, J., Nauels, A., Xia, Y., Bex, V., Midgley, P.M., Eds.; Cambridge University Press: Cambridge, UK; New York, NY, USA, 2013.
3. Oke, T.R. The energetic basis of the urban heat island. *Q. J. R. Meteorol. Soc.* **1982**, *108*, 1–24. [CrossRef]
4. Clarke, J.F. Some effects of the urban structure on heat mortality. *Environ. Res.* **1971**, *5*, 93–104. [CrossRef]
5. Basu, R. High ambient temperature and mortality: A review of epidemiologic studies from 2001 to 2008. *Environ. Health* **2009**, *8*, 1–40. [CrossRef] [PubMed]
6. Aida, M. Urban albedo as a function of the urban structure—A model experiment. *Boundary-Layer Meteorol.* **1982**, *23*, 405–413. [CrossRef]

7. Holmer, B. A simple operative method for determination of sky view factors in complex urban canyons from fisheye photographs. *Meteorol. Z.* **1992**, *1*, 236–239. [CrossRef]

8. Allen, L.; Lindberg, F.; Grimmond, C.S.B. Global to city scale urban anthropogenic heat flux: Model and variability. *Int. J. Climatol.* **2011**, *31*, 1990–2005. [CrossRef]

9. Fischer, E.M.; Schär, C. Consistent geographical patterns of changes in high-impact European heatwaves. *Nat. Geosci.* **2010**, *3*, 398–403. [CrossRef]

10. Fouillet, A.; Rey, G.; Laurent, F.; Pavillon, G.; Bellec, S.; Ghihenneuc-Jouyaux, C.; Hémon, D. Excess mortality related to the August 2003 heat wave in France. *Int. Arch. Occup. Environ.* **2006**, *80*, 16–24. [CrossRef] [PubMed]

11. Joe, L.; Hoshiko, S.; Dobraca, D.; Jackson, R.; Smorodinsky, S.; Smith, D.; Harnly, M. Mortality during a Large-Scale Heat Wave by Place, Demographic Group, Internal and External Causes of Death, and Building Climate Zone. *Int. J. Environ. Res. Public Health* **2016**, *13*, 299. [CrossRef] [PubMed]

12. Garssen, J.; Harmsen, C.; de Beer, J. The effect of the summer 2003 heat wave on mortality in the Netherlands. *Eurosurveillance* **2005**, *10*, 165–167. [CrossRef] [PubMed]

13. WMO. Guide to Meteorological Instruments and Methods of Observation. 2014 edition WMO-No. 8. Available online: https://library.wmo.int/opac/doc_num.php?explnum_id=4147 (accessed on 9 May 2018).

14. Steeneveld, G.J.; Koopmans, S.; Heusinkveld, B.G.; van Hove, L.W.A.; Holtslag, A.A.M. Quantifying urban heat island effects and human comfort for cities of variable size and urban morphology in the Netherlands. *J. Geophys. Res.* **2011**, *116*. [CrossRef]

15. Brandsma, T.; Wolters, D. Measurement and Statistical Modelling of the Urban Heat Island of the City of Utrecht (The Netherlands). *J. Appl. Meteorol. Climatol.* **2012**, *51*, 1046–1060. [CrossRef]

16. Chen, F.; Kusaka, H.; Bornstein, R.; Ching, J.; Grimmond, C.S.B.; Grossman-Clarke, S.; Loridan, T.; Manning, K.W.; Martilli, A.; Miao, S.; et al. The integrated WRF/urban modelling system: Development, evaluation, and applications to urban environmental problems. *Int. J. Climatol.* **2011**, *31*, 273–288. [CrossRef]

17. Salamanca, F.; Martilli, A.; Tewari, M.; Chen, F. A study of the urban boundary layer using different urban parameterizations and high-resolution urban canopy parameters with WRF. *J. Appl. Meteorol. Climatol.* **2011**, *50*, 1107–1128. [CrossRef]

18. Kusaka, H.; Kondo, H.; Kikega, Y.; Kimura, F. A simple single-layer urban canopy model for atmospheric models: Comparison with multi-layer and slab models. *Bound. Layer Meteorol.* **2001**, *101*, 329–358. [CrossRef]

19. Ronda, R.J.; Steeneveld, G.J.; Heusinkveld, B.G.; Attema, J.J.; Holtslag, A.A.M. Urban finescale forecasting reveals weather conditions with unprecedented detail. *Bull. Am. Meteorol. Soc.* **2017**, *98*, 2675–2688. [CrossRef]

20. Theeuwes, N.; Steeneveld, G.J.; Ronda, R.J.; Holtslag, A.A.M. A diagnostic equation for the daily maximum urban heat island effect for cities in Northwestern Europe. *Int. J. Climatol.* **2017**, *37*, 443–457. [CrossRef]

21. Roth, M.; Oke, T.R.; Emery, W.J. Satellite-derived urban heat islands from three coastal cities and the utilization of such data in urban climatology. *Int. J. Remote Sens.* **1989**, *10*, 1699–1720. [CrossRef]

22. Yuan, F.; Bauer, M.E. Comparison of impervious surface area and normalized difference vegetation index as indicators of surface urban heat island effects in Landsat imagery. *Remote Sens. Environ.* **2007**, *107*, 375–386. [CrossRef]

23. Heusinkveld, B.G.; Steeneveld, G.J.; van Hove, L.W.A.; Jacobs, C.M.J.; Holtslag, A.A.M. Spatial variability of the Rotterdam urban heat island as influenced by urban land use. *J. Geophys. Res. Atmos.* **2014**, *119*, 677–692. [CrossRef]

24. Molenaar, R.E.; Heusinkveld, B.G.; Steeneveld, G.J. Projection of rural and urban human thermal comfort in The Netherlands for 2050. *Int. J. Climatol.* **2016**, *36*, 1708–1723. [CrossRef]

25. EIB, Investeren in Nederland. Available online: https://www.eib.nl/pdf/investeren_in_nederland.pdf (accessed on 12 June 2015). (In Dutch)

26. CBS (Centraal Bureau voor de Statistiek) StatLine. Bevolking; Geslacht, Leeftijd, Regio. Available online: https://opendata.cbs.nl/statline/portal.html?_la=nl&_catalog=CBS&tableId=03759ned&_theme=299 (accessed on 7 August 2018). (In Dutch)

27. CBS (Centraal Bureau voor de Statistiek) Kaart van 100 Meter bij 100 Meter Met Statistieken. Available online: https://www.cbs.nl/nl-nl/dossier/nederland-regionaal/geografische%20data/kaart-van-100-meter-bij-100-meter-met-statistieken (accessed on 8 May 2018). (In Dutch)

28. OpenTopo Achtergrondkaart, Retrieved via PDOK-Services Plug-in in QGIS. (PDOK are the Dutch Public Geo Services). Available online: http://pdokviewer.pdok.nl (accessed on 8 May 2018).

29. KNMI. Sky View Factor of the Netherlands. Available online: https://data.knmi.nl/datasets/sky_view_factor_netherlands/1.0?q=sky+view (accessed on 26 February 2018).

30. AHN. Actueel hoogtebestand Nederland. Version 2. Available online: www.ahn.nl (accessed on 1 April 2012). (In Dutch)

31. Attema, J.J.; Heusinkveld, B.G.; Ronda, R.J.; Steeneveld, G.J.; Holtslag, A.A.M. Summer in the city: Forecasting and mapping human thermal comfort in urban areas. In Proceedings of the IEEE 11th International Conference on e-Science, Munich, Germany, 31 August–4 September 2015.

32. Climate Adaptation Services. Climate Impact Atlas. Available online: www.klimaateffectatlas.nl/en (accessed on 4 August 2018).

33. QGIS Development Team. QGIS Geographic Information System. Open Source Geospatial Foundation Project. Available online: https://qgis.org/en/site/ (accessed on 31 August 2018).

34. Klein Tank, A.; Beersma, J.; Bessembinder, B.; van den Hurk, B.; Lenderink, G. *KNMI'14 Climate Scenarios for the Netherlands, KNMI Publication: Brochure KNMI'14 Climate Scenarios*; KNMI: De Bilt, The Netherlands, 2015; p. 34. Available online: http://www.klimaatscenarios.nl/brochures/images/Brochure_KNMI14_EN_2015.pdf (accessed on 1 July 2015).

35. Riahi, K.; Rao, S.; Krey, V.; Cho, C.; Chirkov, V.; Fischer, G.; Kindermann, G.; Nakicenovic, N.; Rafaj, P. RCP 8.5—A scenario of comparatively high greenhouse gas emissions. *Clim. Chang.* **2011**, *109*, 33–57. [CrossRef]

36. Cubasch, U.; Wuebbles, D.; Chen, D.; Facchini, M.C.; Frame, D.; Mahowald, N.; Winther, J.G. Introduction. In *Climate Change 2013: The Physical Science Basis. Contribution of Working Group I to the Fifth Assessment Report of the Intergovernmental Panel on Climate Change*; Cambridge University Press: Cambridge, UK; New York, NY, USA, 2013.

37. Van den Hurk, B.; Siegmund, P.; Klein Tank, A. *KNMI'14: Climate Change Scenarios for the 21st Century—A Netherlands Perspective*; Royal Netherlands Meteorological Institute (KNMI): De Bilt, The Netherlands, 2014. Available online: http://bibliotheek.knmi.nl/knmipubWR/WR2014-01.pdf (accessed on 26 May 2014).

38. KNMI, Toelichting Transformatie Tijdreeksen. Available online: http://www.klimaatscenarios.nl/toekomstig_weer/transformatie/Toelichting_TP.pdf (accessed on 6 October 2015). (In Dutch)

39. Bakker, A. *Time Series Transformation Tool Version 3.1: Description of the Program to Generate Time Series Consistent with the KNMI'14 Climate Scenarios*; Technical Report; KNMI: De Bilt, The Netherlands; p. 40. Available online: http://bibliotheek.knmi.nl/knmipubTR/TR349.pdf (accessed on 17 November 2015).

40. Bell, S.; Cornford, D.; Bastin, L. How good are citizen weather stations? Addressing a biased opinion. *Weather* **2015**, *70*, 75–84. [CrossRef]

41. Meier, F.; Fenner, D.; Grassmann, T.; Otto, M.; Scherer, D. Crowdsourcing air temperature from citizen weather stations for urban climate research. *Urban Clim.* **2017**, *19*, 130–191. [CrossRef]

42. Hopkinson, R.F.; McKenney, D.W.; Milewska, E.J.; Hutchinson, M.F.; Papadopol, P.; Vincent, L.A. Impact of Aligning Climatological Day on Gridding Daily Maximum–Minimum Temperature and Precipitation over Canada. *J. Appl. Meteorol. Climatol.* **2011**, *50*, 1654–1665. [CrossRef]

43. CBS (Centraal Bureau voor de Statistiek) Statline. Aardgas; Aanbod en Verbruik. Available online: https://opendata.cbs.nl/statline/portal.html?_la=nl&_catalog=CBS&tableId=00372&_theme=200 (accessed on 7 August 2018). (In Dutch)

44. Hamid, K. Investigation of the passage of a derecho in Belgium. *Atmos. Res.* **2012**, *107*, 86–105. [CrossRef]

45. KNMI, Warme en Zonnige Zomer 2006. Available online: https://www.knmi.nl/over-het-knmi/nieuws/warme-en-zonnige-zomer-2006 (accessed on 31 August 2006). (In Dutch)

46. Montavez, J.P.; Rodriguez, A.; Jiménez, J.I. A study of the urban heat island of Granada. *Int. J. Climatol.* **2000**, *20*, 899–911. [CrossRef]

47. Eliasson, I. Urban nocturnal temperatures, street geometry and land use. *Atmos. Environ.* **1996**, *30*, 379–392. [CrossRef]

48. Goh, K.C.; Chang, H.C. The relationship between height to width ratios and the heat island intensity at 22:00 h for Singapore. *Int. J. Climatol.* **1999**, *19*, 1011–1023. [CrossRef]

49. Unger, J. Intra-urban relationship between surface geometry and urban heat island: Review and new approach. *Clim. Res.* **2004**, *27*, 253–264. [CrossRef]

50. Theeuwes, N.; Solcerova, A.; Steeneveld, G.J. Modelling the influence of open water surfaces on the summertime temperature and thermal comfort in the city. *J. Geophys. Res. Atmos.* **2013**, *118*, 8881–8896. [CrossRef]

51. Hamdi, R.; Schayes, B. Sensitivity study of the urban heat island intensity to urban characteristics. *Int. J. Climatol.* **2008**, *28*, 973–982. [CrossRef]

52. Theeuwes, N.; Steeneveld, G.J.; Ronda, R.J.; Heusinkveld, B.G.; van Hove, L.W.A.; Holtslag, A.A.M. Seasonal dependence of the urban heat island on the street canyon aspect ratio. *Q. J. R. Meteorol.* **2014**, *118*, 8881–8896.

53. Gasparrini, A.; Guo, Y.; Sera, F.; Vicedo-Cabrera, A.M.; Huber, V.; Tong, S.; Coelho, M.S.Z.; Saldiva, P.H.N.; Lavigne, E.; Correa, P.M.; et al. Projections of temperature-related excess mortality under climate change scenarios. *Lancet. Planet. Health* **2017**, *1*, e360–e367. [CrossRef]

54. Huynen, M.T.E.; Martens, P. Climate change effects on heat- and cold-related mortality in the Netherlands: A scenario-based integrated environmental health impact assessment. *Int. J. Environ. Res. Public Health* **2015**, *12*, 13295–13320. [CrossRef] [PubMed]

55. Analitis, A.; Michelozzi, P.; D'Ippoliti, D.; de'Donato, F.; Menne, B.; Matthies, F.; Atkinson, R.W.; Iñiguez, C.; Basagaña, X.; Schneider, A.; et al. Effects of heat waves on mortality: Effect modification and confounding by air pollutants. *Epidemiology* **2014**, *25*, 15–22. [CrossRef] [PubMed]

56. Athanassiadou, M.; Baker, J.; Carruthers, D.; Collins, W.; Girnary, S.; Hassel, D.; Hort, M.; Johnson, C.; Johnson, K.; Jones, R.; et al. An assessment of the impact of climate change on air quality at two UK sites. *Atmos. Environ.* **2010**, *44*, 1877–1886. [CrossRef]

57. Lemonsu, A.; Kounkou-Arnoud, R.; Desplat, J.; Salagnac, J.-L.; Masson, V. Evolution of the Parisian urban climate under a global changing climate. *Clim. Chang.* **2013**, *116*, 679–692. [CrossRef]

58. McCarthy, M.P.; Best, M.J.; Betts, R.A. Climate change in cities due to global warming and urban effects. *J. Geophys. Res.* **2010**, *37*. [CrossRef]

59. Hamdi, R.; van de Vyver, H.; de Troch, R.; Termonia, P. Assessment of three dynamical urban climate downscaling methods: Brussels's future urban heat island under an A1B emission scenario. *Int. J. Climatol.* **2014**, *34*, 978–999. [CrossRef]

60. Grossman-Clarke, A.; Schubert, S.; Fenner, D. Urban effects on summertime air temperature in Germany under climate change. *Int. J. Climatol.* **2017**, *37*, 905–917. [CrossRef]

 *atmosphere*

Article

The Influence of the North Atlantic Oscillation Index on Emergency Ambulance Calls for Elevated Arterial Blood Pressure

Jone Vencloviene [1,*], Agne Braziene [1], Jurate Zaltauskaite [1] and Paulius Dobozinskas [2]

[1] Department of Environmental Sciences, Vytautas Magnus University, Donelaicio St. 58, Kaunas LT-44248, Lithuania; braziene.agne@gmail.com (A.B.); jurate.zaltauskaite@vdu.lt (J.Z.)

[2] Department of Disaster Medicine, Lithuanian University of Health Sciences, Eiveniu St. 4, Kaunas LT-50028, Lithuania; paulius@smp.lt

* Correspondence: j.vencloviene@gmf.vdu.lt; Tel./Fax: +370-650-27090

Received: 31 May 2018; Accepted: 26 July 2018; Published: 28 July 2018

Abstract: The North Atlantic Oscillation (NAO) is the most prominent pattern of atmospheric variability over the middle and high latitudes of the Northern Hemisphere, especially during the cold season. It is found that "weather types" are associated with human health. It is possible that variations in NAO indices (NAOI) had additional impact on human health. We investigated the association between daily emergency ambulance calls (EACs) for exacerbation of essential hypertension and the NAOI by using Poisson regression, adjusting for season, weather variables and exposure to CO, particulate matter and ozone. An increased risk of EACs was associated with NAOI < −0.5 (Rate Ratio (RR) = 1.07, p = 0.013) and NAOI > 0.5 (RR = 1.06, p = 0.004) with a lag of 2 days as compared to −0.5 ≤ NAOI ≤ 0.5. The impact of NAOI > 0.5 was stronger during November-March (RR = 1.10, lag = 0, p = 0.026). No significant associations were found between the NAOI and EACs during 8:00–13:59. An elevated risk was associated during 14:00–21:59 with NAOI < −0.5 (RR = 1.09, p = 0.003) and NAOI > 0.5 (RR = 1.09, p = 0.019) and during 22:00–7:59 with NAOI < −0.5 (RR = 1.12, lag = 1, p = 0.001). The non-linear associations were found between the NAO and EACs. The different impact of the NAO was found during the periods November–March and April–October. The impact of the NAOI was not identical for different times of the day.

Keywords: North Atlantic Oscillation; weather; emergency ambulance calls; exacerbation of essential hypertension

1. Introduction

The North Atlantic Oscillation (NAO) is the most prominent and recurrent pattern of atmospheric variability over the middle and high latitudes of the Northern Hemisphere, especially during the cold season months (November–March) [1]. In Northern Europe, during winter, the positive NAO phase was associated with a stronger westerly wind flow, a higher temperature and increased storminess and precipitation, whereas the negative NAO phase leads to weakened westerly wind and a lower temperature and decreased storminess and precipitation [1]. A positive NAO phase is accompanied by a statistically significant increase in the frequency of cyclones and cyclone depths in the North Atlantic between 55° N and 75° N and in the south-eastern Mediterranean region [2]. As the largest amplitude anomalies in sea level pressure occur during the winter months [3], the highest impact on surface temperature, precipitation and atmospheric pollution indices are also observed during winter months rather than during summer months [1]. The observations during the period 1950–2000 show that during winters with a positive NAO precipitation amount is increased in north-eastern Europe, whereas the precipitation is supressed under the anticyclonic conditions during the summer [4].

The same results were observed during the period 1961–2010 in Lithuania [5]. The NAO exerts a strong control on Europe's climate interannual variability and pollution. Oscillations between high and low NAO phases produce large changes in the wind speed and direction over the Atlantic, the heat and moisture transport between the Atlantic and other continents [1]. Thus, NAO influences hemispheric-scale and continent-scale pollution transport [6–8]. The backward trajectories analysis in Lithuania indicates that the predominant origin of air masses is South-western Europe and North Atlantic region [9,10].

It is found that "weather types" (synoptic events categorized by pressure patterns and wind fields) are associated with human health. According to the studies in Italy, significant increases in hospital admissions for myocardial infarction were evident 24 h after a day characterized by an anticyclonic continental air mass and 6 days after a day characterized by a cyclonic air mass [11]; an increase in ambulatory blood pressure followed a sudden day-to-day change of weather pattern from anticyclonic to cyclonic days [12]. It is possible that variations in NAO indices (NAOI) had additional impact on human health, after adjusting for weather variables and air pollutants. A statistically significant association between mortality from ischemic heart disease and the NAO was found in England in winter time [13] and an association between the incidence of and mortality from acute myocardial infarction and the Arctic Oscillation (AO, which is close to the NAO) indices was found in northern Sweden [14].

In our previous studies, we found the associations between emergency ambulance calls (EACs) for elevated blood pressure (EABP) and weather and space weather variables [15] and exposure to air pollution [16]. The aim of the study was to determine the complex association between daily EACs for EABP occurring in the morning until the early afternoon (8:00–13:59), in the afternoon until the evening (14:00–21:59) and at night until the early morning (22:00–7:59) and NAO indices, adjusting for weather and space weather variables and air pollution.

2. Methods

The study was conducted from 1 January 2009 to 30 June 2011. The description of data of EACs is presented in our previous work [15]. The values of daily NAO indices were obtained from the National Oceanic and Atmospheric Administration (NOAA) database ftp://ftp.cpc.ncep.noaa.gov/cwlinks/. We used the NAOI on the day of the call and on two previous days. As the quintiles of the daily NAOI were -1.01, -0.49, -0.06 and 0.43 during the study period and the non-linear impact of the NAO is probable, we used the NAOI as categorical variable with categories: ≤-1, $(-1, -0.5]$, $(-0.5, 0]$, $(0, 0.5]$ and >0.5. As confounders, we used the non-linear variables of air temperature (T, °C), wind speed (WS, kt), barometric pressure (BP, hPA) and relative humidity (RH, %), low and active-stormy geomagnetic activity levels, high-speed solar wind (HSSW), daily concentrations of ozone and PM_{10} and the highest 8-h moving average of CO concentration. These variables are described in more detail in previous publications [15,16].

Statistical Analysis

Continuous variables are presented as the mean value (standard error). The univariate associations between NAOI categories and the daily number of EACs for EABP were evaluated by applying ANOVA and Kruskal-Wallis test.

As the numbers of EACs Y_t are count variable, we suppose that Y_t followed a Poisson distribution with mean λ_t, depending on predictor variables. We applied multivariate Poisson regression to evaluate the association between daily NAOI variables and daily EACs for EABP, which was specified as:

$$\ln(\lambda_t) = \beta_0 + \beta X_t + \gamma Z_t \tag{1}$$

where X_t is a NAOI variable, β is vector of coefficients for X_t, Z_t—vectors of confounding factors—years, seasonality, week days, day length, weather, space weather [15] and air pollution variables, as these may also affect the daily EACs rate; γ—is vector of coefficients for Z_i. The daily incidence of EACs is defined as $E(Y_t) = \lambda_t$. Poisson regression coefficients were interpreted as the difference between the log of expected counts.

Researchers in the topics of epidemiology and public health have been used the term "the relative risk." It is the ratio of two risks (or, informally, of rates or odds) comparing the risk of disease or death among the exposed to the risk among the unexposed. We could also interpret the Poisson regression coefficients as the log of the rate ratio. This explains the "rate" in incidence rate ratio. In our investigation, it is important for health professionals to evaluate how many times the high or low NAOI value increases the λ_i (the daily mean value of EACs) as compared to reference category ($-0.5 \leq$ NAOI ≤ 0.5). Let $X_i = 1$, when NAOI with a lag of 2 days > 0.5, $X_i = 2$, when NAOI with a lag of 2 days < -0.5 and $X_i = 0$ otherwise. Then $\ln(\lambda_i) = \beta_0 + \beta_1(X_i = 1) + \beta_2(X_i = 2) + \gamma Z_i$, $\ln(\lambda_i \mid X_i = 1) = \beta_0 + \beta_1(X_i = 1) + \gamma Z_i$, $\ln(\lambda_i \mid X_i = 0) = \beta_0 + \gamma Z_i$ and the ratio of daily incidences when $X_i = 1$ and $X_i = 0$ ($\lambda_i \mid X_i = 1)/(\lambda_i \mid X_i = 0) = \exp(\beta_1)$ is defined as Rate Ratio (RR). We presented adjusted rate ratios (RRs) in the multivariate Poisson regression model. The RRs are presented with 95% confidence interval (CI) and p-value. The analysis was performed separately for the number of calls during the whole day, in the morning until the early afternoon, in the afternoon until the evening and at night until the early morning during the colder (November–March) and the warmer (April–October) periods. For a sensitivity analysis, we evaluated the association between EACs for EABP and the NAOI separately for older (>65 years) and younger patients. Statistical analysis was performed using SPSS 19 software.

3. Results

There were 17,114 emergency calls for EABP during the 911 days of the study: 26% of the calls were received in the morning until the early afternoon (8:00–13:59), 44.5%—in the afternoon (14:00–21:59) and 29.5%—at night (22:00–7:59); 60.2% of the patients were older than 65 years. The mean number of EACs during the whole day was 18.8, during the time period of 8:00–13:59 it was 4.9, during the period of 14:00–21:59—8.4 and during the period of 22:00–7:59—5.5 [15]. The mean daily number of EACs for elderly was 11.3 and for younger patients it was 7.5. No significant difference was observed in the distribution of NAOI categories during the colder and the warmer periods. During the colder period, more days of NAOI < −0.5 were observed in December and more days of NAOI > 0.5 in March (48.7%). During the warmer period, about 30% of days of NAOI < −1 was in June and 58.4% of days of NAOI > 0.5 during the period of April–May (Figure 1).

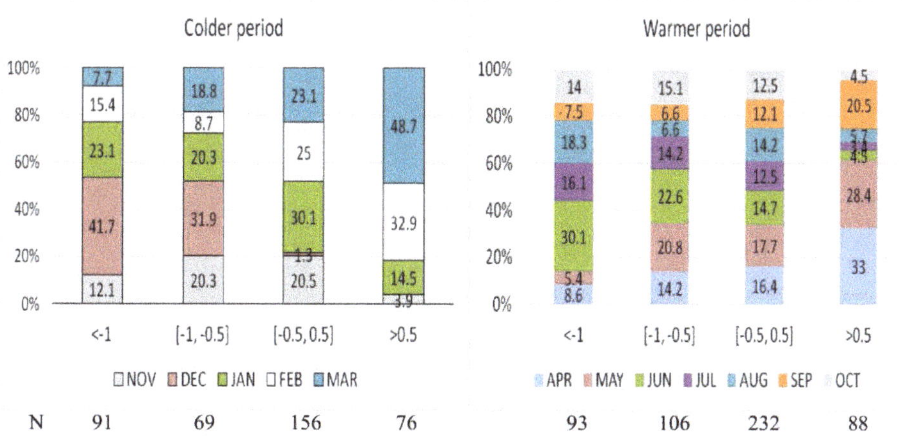

Figure 1. The monthly distribution in the categories of the NAO index during the colder period and the warmer period.

A non-linear impact of the NAOI on the risk of EACs was observed (Figure 2). During the whole day, an increase in the risk of EACs was associated both with NAOI < −0.5 on the day of the call and on two previous days and NAOI > 0.5 with a lag of 2 days (Table 1). A protective impact of the NAOI between −0.5 and 0 was seen especially on the same and on the previous day (Figure 2). A negative impact of NAOI < −1 or NAOI > 0.5 on the same day was stronger during the period of November–March (Figure 2). No any significant associations between the NAOI and daily EACs were found during the period of April–October).

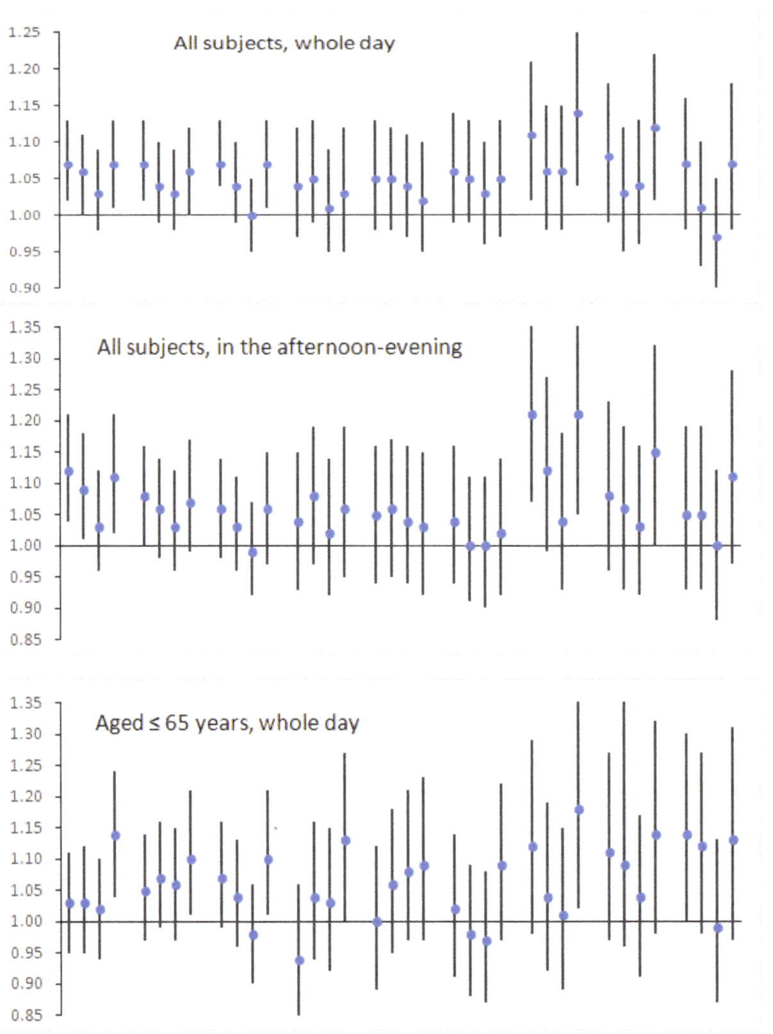

Figure 2. *Cont.*

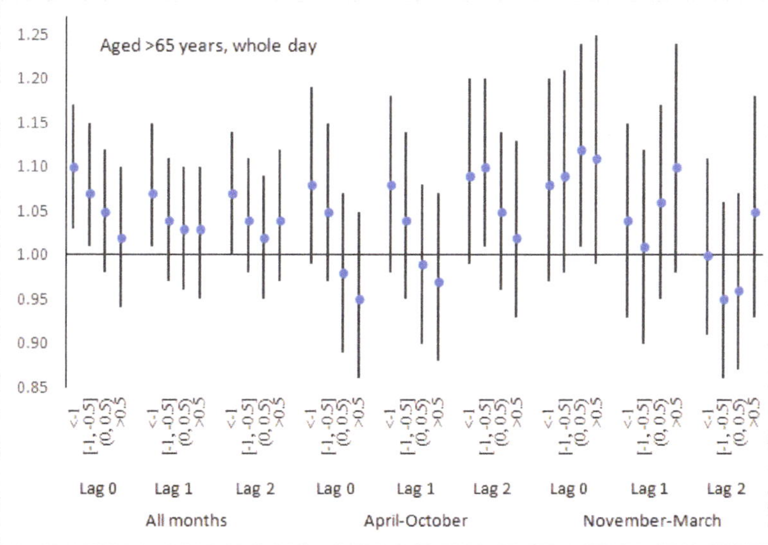

Figure 2. Rate ratios of emergency ambulance calls for elevated arterial blood pressure in the categories of the NAO index (reference category the NAO index between −0.5 and 0), adjusting for years, month, the day of the week, day length, weather and space weather variables.

Table 1. Significant associations between the NAO index and daily emergency calls for elevated arterial blood pressure in rate ratios (RR), adjusting for day length, month, years, the day of the week, space weather variable, weather variables and exposure to CO, PM_{10} and ozone.

Month	Subjects	NAOI	RR (Lag 0)	p	RR (Lag 1)	p	RR (Lag 2)	p
1−12	All whole day	<−1 vs. [−1, 0.5]	1.05	0.033	1.05	0.034	1.06	0.007
		>0.5 vs. [−1, 0.5]	1.04	0.170	1.03	0.308	1.05	0.067
1−12	All whole day	<−0.5 vs. [−0.5, 0.5]	1.05	0.008	1.04	0.031	1.06	0.004
		>0.5 vs. [−0.5, 0.5]	1.05	0.054	1.04	0.130	1.07	0.013
4−10	All whole day	<−0.5 vs. ≥−0.5	1.04	0.129	1.04	0.168	1.04	0.098
11−3	All whole day	<−1 vs. [−1, 0.5]	1.06	0.064	1.06	0.094	1.07	0.045
		>0.5 vs. [−1, 0.5]	1.10	0.026	1.09	0.036	1.08	0.052
1−12	All 14:00–21:59	<−0.5 vs. [−0.5, 0.5]	1.09	0.003	1.05	0.084	1.05	1.119
		>0.5 vs. [−0.5, 0.5]	1.09	0.019	1.06	0.154	1.06	0.110
11−3	All 14:00–21:59	<−1 vs. [−1, 0.5]	1.13	0.012	1.05	0.320	1.02	0.638
		>0.5 vs. [−1, 0.5]	1.15	0.023	1.11	0.081	1.10	0.144
11−3	All 14:00–21:59	<−0.5 vs. [−0.5, 0.5]	1.14	0.008	1.06	0.282	1.05	0.287
		>0.5 vs. [−0.5, 0.5]	1.18	0.009	1.13	0.064	1.11	0.099
1−12	All 22:00–7:59	<−0.5 vs. ≥−0.5	1.08	0.019	1.12	0.001	1.11	0.004
4−10	All 22:00–7:59	<−0.5 vs. ≥−0.5	1.11	0.017	1.14	0.005	1.13	0.009
1−12	Age ≤ 65 whole day	>0.5 vs. ≤ 0.5	1.12	0.004	1.03	0.157	1.09	0.026
4−10	Age ≤ 65 whole day	>0.5 vs. ≤ 0.5	1.11	0.042	1.04	0.455	1.09	0.076
4−10	Age ≤ 65 whole day	NAOI [a]	1.06	0.017	1.05	0.059	1.03	0.243
11−3	Age ≤ 65 whole day	<−0.5 vs. [−0.5, 0.5]	1.06	0.265	1.06	0.232	1.11	0.029
		>0.5 vs. [−0.5, 0.5]	1.13	0.024	1.10	0.088	1.11	0.064
1−12	Age > 65 whole day	<−0.5 vs. ≥−0.5	1.06	0.010	1.04	0.105	1.04	0.103
4−10	Age > 65 whole day	<−0.5 vs. ≥−0.5	1.08	0.013	1.07	0.048	1.08	0.030
4−10	Age > 65 whole day	NAOI [a]	0.97	0.065	0.97	0.111	0.97	0.152

[a] RR per increase 1.

No significant associations were found between the NAO index and EACs for EABP during the period of 8:00–13:59. In the afternoon-evening, a significant impact of a lower and a higher NAOI was observed only on the day of the call. The stronger impact was observed during the colder months (Figure 2, Table 1). At night until early morning, only the negative NAO (NAOI < −0.5) was associated with an increased number of EACs; the stronger impact was observed during the period of April–October (Figure 2).

For subjects, aged ≤65 years, during the period of April–October, a higher NAOI (>0.5) was associated with an increased risk of EACs. Also, a positive association between the daily NAOI and the daily EACs was found (Table 1). During the colder period, the risk of EACs was associated both with a lower and a higher NAOI (Table 1). For the elderly subjects, during the period of April–October, an increased risk of EACs was associated with a negative NAO (NAOI < −0.5). Also, the association between the NAOI and the risk of EACs tended to be negative (RR = 0.97, p = 0.065 on the same day) (Table 1).

For sensitivity analysis, we randomly divided the sample into five similar size parts and assessed regression coefficients β and its standard deviations for NAOI > 0.5 and NAOI < −0.5 with a lag of 2 days. After, we calculated the mean values of β and it confidence interval. For the daily number of EACs, the correspondence values for RRs ($\exp(\beta)$) were, respectively, 1.052 (1.0005, 1.1061) and 1.068 (1.022, 1.115).

4. Discussion

In our study, a non-linear association between the NAO index and unfavourable health events were determined for the first time. According to the results of our study, an increased risk of the daily EACs for EABP was associated both with a positive (>0.5) and a negative (<−1 or <−0.5) values of the NAO index, adjusting the impact of month and weather variables. Such effect may be explained by the impact of the NAO on weather pattern and changes in air pollution level.

Days of NAOI > 0.5 were found to be associated with an increase in the risk of EAC for all and younger subjects during the colder period. According to the analysis of multiannual data [17], a positive NAO associated with a higher T during the period of November–March. Also, the reverse thermal effects of the positive NAO phase during March of 2009–2011 was observed: BP on days of NAOI > 0.5 was significantly higher as compared to days of NAOI ≤ 0.5 and the mean daily air temperature (−0.23 °C) were the lowest during the period of 2000–2012. The 48.7% of days of NAOI > 0.5 during the colder period was in March. March is the first month of spring associated with vitamin deficiency, flu outbreaks and fatigue as well as reduced physical activity after the winter season, therefore at that time, the human body was most vulnerable. Therefore, the impact of the cold on days of NAOI > 0.5 in March may lead to an increased risk of EACs. So, during the studied period, on 61.6% days mean air temperature was lower than −1 °C during the period of January–February. According to the results of our previous studies [15,18], regression models of EACs for EABP for all and aged ≤65 years subjects not included the variable of air temperature lower than −1 °C; only for the elderly, variables reflecting the impact of the colder air included in the model for the risk of EACs during the whole day and during 14:00–21:59. It is possible that only negative T concomitant weather pattern on days of the positive NAO are risky for younger subjects.

According to our research, an increase in the risk of EACs was associated with a negative NAO during the colder period, especially with NAOI < −0.5 for the period of 14:00–21:59. During colder months of our study, the mean daily T and BP were significantly lower on days of NAOI < −0.5 as compared to days of NAOI between −0.5 and 0.5. It is probable that the complex impact of both lower T and BP may be explained by an increase in the risk of EACs during the days of NAOI < −0.5 in the colder period.

During the colder period, a negative impact of both a positive and a negative NAOI on human health may be caused by an increased exposure to air pollutant transfer from other regions of Europe and Northern America. During winter months for NAO+ the transport of pollutants from North

America to Europe is enhanced and the tracer plume is moving towards high latitudes of Europe and reaches it in 8–10 days after tracer emissions [6]. During high NAO phases enhanced northward NO_2 and black carbon transport is observed [6]. The differences between NAO+ and NAO− pollutants air concentrations are more expressed in winter. The concentrations of $PM_{2.5}$ and PM_{10} during high NAO+ are up to 10 and 20 μgm^{-3} respectively higher than during NAO− and represent variations of up to 20–40% between NAO phases [6]. During winter, a negative NAO associated with a higher concentration of gaseous tracers and water-soluble aerosols emitted from Europe [6,7] in the Baltic countries. The winter-spring peak of aerosol concentration over the north-eastern Europe usually is recorded [9]. Concentrations of nitrogen oxides (NO and NO_2) were negatively related to the NAOI in the city of Gothenburg, west Sweden during the winter months for the period 1997–2006 [19]. In our multivariate model, concentrations of ozone, PM_{10} and CO were included but we did not adjust for the impact of other pollutants.

According to our results, during the period of April-October, an increase in the risk of EACs was associated with a negative NAO in the elderly and with a higher NAOI in the aged ≤65 years subjects. In mid-latitudes ~55° N, 20–30° E, during the period of June–September, negative NAO phases are associated with a stronger westerly wind flow, a higher precipitations level, a higher cloudiness and a lower air temperature [4,8]. A positive NAO during summer associated with a lower precipitation level, weaker winds and a higher temperature. During the warmer months, NAOI < −0.5 on the previous day was associated with a higher WS and a lower BP during the period of our study. The supressed precipitation amount during the summer leads to higher concentrations of aerosols in the atmosphere. Therefore, positive NAO phases favour increased aerosol concentrations in northern Europe regions during summer [8]. In summer, the region of North Poland and Lithuania is also very affected and the high difference between aerosol concentration for NAO+ and NAO− are determined. The similar differences were recorded not only to particulate matter but also for salt, dust, SO_4, NO_3 and NH_4 concentrations [8]. The other meteorological parameters related to the NAO and atmospheric pollution are wind speed, temperature and atmosphere oxidative capacity. During the periods of the weaker winds associated with the NAO+ events favour the increase of particulate matter in polluted regions such as large cities or entire industrial regions. It is probable that the complex of meteorological conditions during days of NAOI < −0.5 had an additional negative effect associated with an increase in the risk of EACs in the elderly and an increase in the air pollution level during a positive NAO had a negative impact on the younger subjects. Studies on the associations between physical activity in the elderly and weather conditions in Europe showed that physical activity decreased significantly with increasing wind speed, precipitation and humidity [20], a shorter day length and duration of sunshine, a high precipitation amount and a lower maximum temperature [21,22]. These weather conditions are associated with a negative NAO excluding winter months—therefore, it can be assumed that NAOI < −0.5 are associated with fewer physical activity opportunities for the elderly, who are likely to be stressed. This can explain the fact that days of NAOI < −0.5 increase the risk of EACs in the elderly during the warmer period. It is probable that the impact of weather pattern was stronger as the impact of air pollutants in the elderly, whereas for younger the impact of weather pattern during a negative NAO was slight.

According to literature, both positive and negative NAO phases are associated with worse cardiovascular outcomes. Messner et al. [14] analysed associations between daily Arctic Oscillation (AO) indices, which are close to the NAO and the incidence of and mortality from acute myocardial infarction (AMI) in northern Sweden. This study established a negative impact of a positive AO with a lag of 3 days: an increase in the AO index bringing warmer weather over Scandinavia was associated with an increase in the incidence of and mortality from AMI. Statistically significant inverse associations between the climate index (which represents winters with a strong negative phase of the NAO) and the level of IHD mortality were found in England [13].

According to our results, the impact of the NAOI as well as weather variables was not identical for different times of the day. No significant impact of the NAOI was observed in the morning until the early afternoon. During this period of the day, the impact of weather pattern was stronger as compared to other periods of the day [15] and perhaps because the additional impact of the NAOI not observed. The stronger impact of NAOI < −0.5 and NAOI > 0.5 was found in the afternoon until the evening, especially during the colder period. It is likely that in the afternoon or several hours before, people were more exposed to environment—a lower air temperature or higher precipitations and cloudiness during the warmer period and a higher pollution levels during NAOI > 0.5. At night-early in the morning, the increased risk of EACs on days of NAOI < −0.5 during the warmer period may be explained by an increased stress after the day with poorer weather conditions—a lower T and BP and a higher precipitations and cloudiness.

5. Conclusions

An increased risk of the daily emergency ambulance calls for elevated arterial blood pressure was associated both with a positive (>0.5) and a negative (<−1 or <−0.5) values of the NAO index, adjusting the impact of month and weather variables. The different impact of the NAO was found during the periods of November–March and April–October. The impact of the NAO index was not identical for different times of the day. No significant impact of the NAO index was observed during the period of 8:00–13:59. The stronger impact of NAOI < −0.5 and NAOI > 0.5 was found in the afternoon until the evening, especially during the colder period.

Author Contributions: J.V. conceived the idea, performed statistical analyses, interpreted the results and was the lead writer. P.D. was responsible for the medical data and read and approved the final manuscript. J.Z. was responsible for the analysis of NAO and this phenomenon impact on climate and air pollution. A.B. assisted with the writing of the manuscript and revised the manuscript. All authors read and approved the final version of the manuscript.

Funding: This research did not receive any specific grant from funding agencies in the public, commercial, or not-for-profit sectors.

Acknowledgments: We acknowledge the contribution of Kaunas city ambulances involved in the registration of cardiovascular emergency admissions and the formation of the computer database.

Conflicts of Interest: The authors declare no conflict of interest.

References

1. Hurrell, J.W.; Kushnir, Y.; Ottersen, G.; Visbeck, M. An overview of the North Atlantic Oscillation, in the North Atlantic Oscillation: Climate Significance and Environmental Impact. *Geophys. Monogr. Ser.* **2003**, *134*, 279. [CrossRef]

2. Bardin, M.Yu.; Polonsky, A.B. North Atlantic Oscillation and Synoptic Variability in the European-Atlantic Region in Winter. *Izv. Atmos. Ocean. Phys.* **2005**, *41*, 127–136.

3. Hurrell, J.W.; Deser, C. North Atlantic climate variability: The role of the North Atlantic Oscillation. *J. Mar. Syst.* **2010**, *79*, 231–244. [CrossRef]

4. Blade, I.; Liebmann, B.; Fortuny, D.; Jan van Oldenborgh, G. Observed and simulated impacts of the summer NAO in Europe: Implications for projected drying in the Mediterranean region. *Clim. Dyn.* **2012**, *39*, 709–727. [CrossRef]

5. Mickevicius, V.; Bukantis, A. Effects of the North Atlantic Oscillation on Lithuanian climate. *Water Manag. Eng.* **2013**, *42*, 5–13. (In Lithuanian)

6. Eckhardt, S.; Stohl, A.; Beirle, S.; Spichtinger, N.; James, P.; Forster, C.; Junker, C.; Wagner, T.; Platt, U.; Jennings, S.G. The North Atlantic Oscillation controls air pollution transport to the Arctic. *Atmos. Chem. Phys.* **2003**, *3*, 1769–1778. [CrossRef]

7. Christoudias, T.; Pozzer, A.; Lelieveld, J. Influence of the North Atlantic Oscillation on air pollution transport. *Atmos. Chem. Phys.* **2012**, *12*, 869–877. [CrossRef]

8. Jerez, S.; Jimenez-Guerrero, P.; Montávez, J.P.; Trigo, R.M. Impact of the North Atlantic Oscillation on European aerosol ground levels through local processes: A seasonal model-based assessment using fixed anthropogenic emissions. *Atmos. Chem. Phys.* **2013**, *13*, 11195–11207. [CrossRef]

9. Byčenkienė, S.; Ulevičius, V.; Prokopčiuk, N.; Jasinevičienė, D. Observations of the aerosol particle number concentration in the marine boundary layer over the south-eastern Baltic Sea. *Oceanologia* **2013**, *55*, 573–597. [CrossRef]

10. Plauškaite, K.; Špirkauskaitė, N.; Byčenkienė, S.; Kecorius, S.; Jasinevičienė, D.; Petelski, T.; Zielinski, T.; Andriejauskienė, J.; Barisevičiūtė, R.; Garbaras, A.; et al. Characterization of aerosol particles over the southern and South-Eastern Baltic Sea. *Mar. Chem.* **2017**, *190*, 13–27. [CrossRef]

11. Morabito, M.; Crisci, A.; Grifoni, D.; Orlandini, S.; Cecchi, L.; Bacci, L.; Modesti, P.A.; Gensini, G.F.; Maracchi, G. Winter air-mass-based synoptic climatological approach and hospital admissions for myocardial infarction in Florence, Italy. *Environ. Res.* **2006**, *102*, 52–60. [CrossRef] [PubMed]

12. Morabito, M.; Crisci, A.; Orlandini, S.; Maracchi, G.; Gensini, G.F.; Modesti, P.A. A synoptic approach to weather conditions discloses a relationship with ambulatory blood pressure in hypertensives. *Am. J. Hypertens.* **2008**, *21*, 748–752. [CrossRef] [PubMed]

13. McGregor, G.R. Winter North Atlantic Oscillation, temperature and ischaemic heart disease mortality in three English counties. *Int. J. Biometeorol.* **2005**, *49*, 197–204. [CrossRef] [PubMed]

14. Messner, T.; Lundberg, V.; Wikström, B. The Arctic Oscillation and incidence of acute myocardial infarction. *J. Int. Med.* **2003**, *253*, 666–670. [CrossRef]

15. Vencloviene, J.; Braziene, A.; Dobozinskas, P. Short-Term Changes in Weather and Space Weather Conditions and Emergency Ambulance Calls for Elevated Arterial Blood Pressure. *Atmosphere* **2018**, *9*, 114. [CrossRef]

16. Vencloviene, J.; Braziene, A.; Dedele, A.; Lopatiene, K.; Dobozinskas, P. Associations of short-term exposure to ambient air pollutants with emergency ambulance calls for the exacerbation of essential arterial hypertension. *Int. J. Environ. Health Res.* **2017**, *27*, 509–524. [CrossRef] [PubMed]

17. Bukantis, A.; Bartkeviciene, G. Thermal effects of the North Atlantic Oscillation on the cold period of the year in Lithuania. *Clim. Res.* **2005**, *28*, 221–228. [CrossRef]

18. Vencloviene, J.; Babarskiene, R.M.; Dobozinskas, P.; Sakalyte, G.; Lopatiene, K.; Mikelionis, N. Effects of weather and heliophysical conditions on emergency ambulance calls for elevated arterial blood pressure. *Int. J. Environ. Res. Public Health* **2015**, *12*, 2622–2638. [CrossRef] [PubMed]

19. Grundstrom, M.; Linderholm, H.W.; Klinberg, J.; Pleijel, H. Urban NO_2 and NO pollution in relation to the North Atlantic Oscillation NAO. *Atmos. Environ.* **2011**, *45*, 883–888. [CrossRef]

20. Klenk, J.; Büchele, G.; Rapp, K.; Franke, S.; Peter, R. Walking on sunshine: Effect of weather conditions on physical activity in older people. *J. Epidemiol. Community Health* **2012**, *66*, 474–476. [CrossRef] [PubMed]

21. Sumukadas, D.; Witham, M.; Struthers, A.; McMurdo, M. Day length and weather conditions profoundly affect physical activity levels in older functionally impaired people. *J. Epidemiol. Community Health* **2009**, *63*, 305–309. [CrossRef] [PubMed]

22. Wu, Y.T.; Luben, R.; Wareham, N.; Griffin, S.; Jones, A.P. Weather, day length and physical activity in older adults: Cross-sectional results from the European Prospective Investigation into Cancer and Nutrition (EPIC) Norfolk Cohort. *PLoS ONE* **2017**, *12*, e0177767. [CrossRef] [PubMed]

Article

Modelling Productivity Loss from Heat Stress

Keith Dear

School of Public Health, The University of Adelaide, Adelaide SA 5005, Australia; keith.dear@adelaide.edu.au; Tel.: +61-403-475-160

Received: 29 May 2018; Accepted: 18 July 2018; Published: 22 July 2018

Abstract: Workers exposed to high ambient temperatures, either indoors or out, work slower. The few studies that have measured this loss of productivity show a degree of consistency across widely varying settings. I develop a class of 5-parameter probability models that express productivity as a function of environmental heat and show how the method of fitting can be adapted according to the completeness of the data available. As well as modelling the mean output, it is important to also consider variation between workers, and the model presented here achieves this. The method is illustrated using three previously published datasets from different industries and work environments.

Keywords: workplace; heat stress; productivity loss; beta distribution

---

1. Introduction

There is growing interest in the extent to which productivity is reduced in workers exposed to high ambient temperatures and high humidity, and the related impacts of climate change [1–3]. A recent survey of the economic costs of extreme weather noted that "labour productivity loss [is] expected to make a substantial contribution to the overall economic impacts associated with a warmer climate" [4]. In the latest climate change impact assessment related to human health concerns [5], the heat effects on human performance and work capacity were given high visibility. The importance of considering humidity, as well as temperature, was stressed in a projection of labour capacity out to the year 2200, which suggested a global reduction to a 40% capacity in the peak months under unmitigated climate change [6].

Besides the likely economic implications [7–9], there are also important health considerations. In poor communities, individual workers may be faced with the dilemma of [10] whether to reduce their work output and suffer a loss of family income or to continue to work in conditions of unsafe heat stress, risking personal injury or even death [11,12]. When the heat stress is temporary, physiological heat acclimatization plays a limited protective role [12], but in more continuous heat exposure situations, behavioural or technological acclimatization will be of great importance, including applying air-cooling systems. However, for many jobs, air-cooling is difficult or impossible to apply, and the limitation of climate change impacts will depend on reducing the extent of the global climate change [5]. The many ways in which climate change will impact worker safety and health are surveyed in Schulte et al. (2016) [13].

Absolute productivity, that is, the hourly output of some industrial or agricultural product, is peculiar to each work environment, so we are concerned rather with the relative productivity. Nevertheless, data will most often be available in terms of the actual output, so here we provide a class of models suitable for modelling such data and capturing the relative output as a discrete component of the model. Despite the importance, both economic and epidemiological, in how worker productivity is reduced through heat stress, few published reports of heat-related productivity loss from actual field observations of workplaces are so far available, and the very few datasets can be used for quantitative analysis [14–16]. It is therefore important to make the best possible use of the meagre data that are available, and the method presented here provides a method for analysing such data and comparing the results between disparate settings.

In this paper, I propose a new model for the output of an individual worker in a particular work environment. The model has two components, each with its own assumptions. First, we suppose that each worker has a maximum potential output under ideal conditions, and model this theoretical maximum as being normally distributed across the population of workers in that industry. Secondly, we suppose that in any work environment, an individual's actual output will be reduced by environmental stress to something less than their potential maximum. This proportional reduction will itself vary between workers, and we model it using beta distributions that vary with the environment. A worker's actual output under given conditions is then the product of their potential maximum output and the proportion that they can maintain, or choose to maintain, in those conditions.

An important feature of this model is that it captures the inter-individual variation in the reduction of output. It is workers in the tails of this distribution that are at the greatest risk to their health and/or income. We turn next to the details of this aspect of the model.

2. Experiments

In a given work environment, including an ideal environment, workers will vary in their relative output, so a probabilistic model is required to capture the mean relative productivity and the variation around the mean. We define ideal conditions as those that maximise the probability of achieving a 100% relative output. Less ideal conditions will reduce this probability, perhaps to zero if no workers can maintain full output, and will increase the probability of lower levels of productivity, potentially to no output at all, if work becomes impossible for at least some workers.

A suitable family of distributions to express variation in a proportion or percentage is the two-parameter beta family, usually parameterised as

$$f_X(x; \alpha, \beta) = (x^{(\alpha-1)} (1 - x)^{(\beta-1)})/B(\alpha, \beta) \; ; x \in [0, 1], \alpha > 0, \beta > 0 \tag{1}$$

This distribution has a mean $\mu = \alpha/(\alpha + \beta)$ and variance

$$\sigma^2 = \alpha\beta/[(\alpha + \beta)^2(\alpha + \beta + 1)] = \mu(1 - \mu)/(\alpha + \beta + 1) \tag{2}$$

A convenient reparameterisation for modelling purposes is directly in terms of μ and a scale parameter $\theta = \sqrt{\alpha\beta}$, from which the usual parameters can be recovered as $\alpha = \theta\phi$, $\beta = \theta/\phi$ where $\phi^2 = \mu/(1 - \mu)$.

Usually, we will model μ to express how it varies with the work conditions. Here we model μ as a logistic function of heat stress T, and estimate θ as a constant. If a particular parametric model implies a mean output of μ, maximising the log-likelihood of the data with respect to the parameters that determine μ in each environment, and also with respect to θ, provides the maximum likelihood estimates of those parameters.

We let the potential maximum output for individual i under ideal conditions following a normal distribution, $Z_i \sim N(\psi, \tau^2)$, and express the proportional reduction in the work environment j as $P_{ij} \sim B(\mu(T_j), \theta)$. Here, T_j represents the heat stress of that environment on some scale such as ordinary ambient temperature, or a more specialised heat stress measure such as effective temperature ET or wet bulb globe temperature, WBGT.

The observed output is $Y = Z \times P$. Since the mean of Z is ψ and that of P is μ, the mean of Y is approximately $E[Y] \approx \psi\mu$. A given level of output might represent a small proportion P of a large potential Z, or vice-versa, so the model, as so far described, will be poorly conditioned in terms of ψ and μ, and a further model constraint is required. Published reports often describe an optimal environment under which heat stress is essentially absent. For example, Wyndham (1969) [14] assumed no loss of productivity until the ambient heat (expressed as natural wet bulb temperature, Tnw) exceeded 27.7 °C. It seems reasonable, then, to assume that the beta distribution representing reduced productivity in such an environment has its mode at $P = 1$, approaching a degenerate distribution in which all workers deliver 100% output. Under the conventional Beta(α, β) parameterisation, such distributions are the subfamily in which $\beta = 1$. This sub-family can be specified through a logistic model in the form:

$$\text{logit } \mu(T) = 2\ln(\theta) + b(T - T_0) \tag{3}$$

Then $\mu(T_0) = \theta^2/(1 + \theta^2)$, $\alpha(T_0) = \theta^2$, and $\beta(T_0) = 1$ as required. The overall model now has four parameters: the mean ψ and variance τ^2 of the maximum potential output; the (negative) regression parameter b that relates the mean relative output μ to the environmental stress measure T; and the beta-distribution scale parameter θ, which also determines the intercept of the regression. Sometimes, T_0 can also be estimated from the data, but this may be unstable in small datasets, and then better results are achieved by setting T_0 from external considerations such as local knowledge or expert opinion. Both approaches are applied in the examples below.

As well as modelling the mean output through a logistic regression equation, which may include many covariates, the model also flexibly captures between-worker variance in output, through two parameters τ^2 and θ. The first measures the inter-worker variation of output in ideal, comfortable conditions, while θ determines how the variance changes with temperature. A large θ implies that the variance reduces in proportion with the mean output, as assumed by the original author in the first example below [14]. A small θ, on the other hand, implies that the variance is maintained, or even increased, as adverse conditions cause the output to fall, which is the case in which any given setting will in itself be a research question of interest, to be answered through the analysis of relevant field data.

3. Results

Little research has been published reporting either experimental or observational data on occupational productivity loss. We present analyses of three papers, dating from 1969, 1992 and 2013, which report various types of data as discussed below.

3.1. Example 1: Wyndham 1969 [14], Gold-Mining in South Africa

Wyndham [14] presents fitted curves, but no data, of percentage performance (P) against natural wet bulb temperature (Tnw) for three wind speeds: 100, 400, and 800 ft/min (approximately 0.5, 2 and 4 ms^{-1}). This temperature metric will typically (except in extremely humid conditions) take lower values than other scales such as Effective Temperature (ET), or WBGT. Wyndham's curves can be expressed as

$$\log_{10} P = 2 - b\left(r^{(T_{nw}-27.7)} - 1\right), \ T_{nw} \geq 27.7 \tag{4}$$

This ensures $P = 100\%$ when $T_{nw} = 27.7\,^\circ\text{C}$, which is the complement of how Wyndham defined the vertical axis of his graphs ("percentage falloff in productivity from the level at 27.7 $^\circ$C wet bulb temperature"). We recovered Wyndham's fit by digitising values from the published figures and fitting this equation to them by least squares (on the P scale). We found that a common value of $r = 1.880$ provides an essentially perfect fit (data not shown), with $b = 0.00460$, 0.00245, and 0.00165, respectively, for wind speeds of 100, 400, and 800 ft/min. The exponential model chosen by Wyndham can be closely approximated by our preferred logistic model. Figure 1 shows logistic equations fitted by least squares to the same digitised values, with Wyndham's exponential curves for comparison.

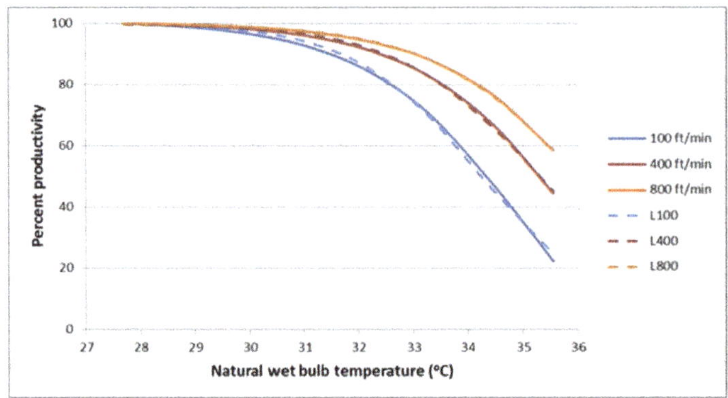

Figure 1. The Wyndham's exponential model (solid lines) and our logistic curves (dashed lines).

The estimated logistic curves, shown as dashed lines in Figure 1, are

- (100 ft/min) logit(μ) = 29.4 $-$ 0.858 $\times$ Tnw,
- (400 ft/min) logit(μ) = 27.4 $-$ 0.775 $\times$ Tnw,
- (800 ft/min) logit(μ) = 26.4 $-$ 0.734 $\times$ Tnw.

In terms of the model proposed here, these represent logistic curves for the mean μ of the beta distribution, with slope b that reduces with increasing wind speed. Wyndham gave no actual output data, so we have no direct information about the mean ψ and variance τ^2 of the maximum output.

However, we can infer something about θ. Wyndham adds 78% confidence intervals (± 1.23 SE) to these graphs, stating that "where they do not overlap, there is a 95% chance that there is a real difference between the two curves", presumably on the basis that $(1 - 0.78)^2 \approx 0.05$. Regardless of the statistical logic, when combined with the reported design points, and the sample size of 10 individuals tested in each environment, these intervals furnish information on the between-subject standard deviation, which simulations using nonlinear least squares modelling suggest was about 0.05 on the $\log_{10}$ scale or roughly $\pm 12\%$. Constant variance of log-transformed data, as Wyndham assumed, implies that on the original productivity scale, the SD would be proportional to the mean; that is, the variance of Y falls at high temperatures in proportion to μ, and so has no contribution from the variation in P but only the original between-subject variation in Z. Under our model, this implies a degenerate beta distribution for P with zero variance, represented by an unbounded θ. Unfortunately, Wyndham did not present his original data, from which we might have tested this and estimated θ. Nevertheless, these data provide some support for the general applicability of our logistic model.

The estimated relative output (%) is shown in Table 1 for the range of temperatures in the Wyndham data, assuming a wind speed of 200 ft/min (approximately 1 m·s^{-1}). As noted above, standard deviations for the between-worker variation around these means cannot be estimated from the data published.

Table 1. The summary of examples.

Setting	Wyndham 1969 [14]		Nag 1992 [15]		Sahu 2013 [16]	
Industry	Gold mining		Beedi rolling		Rice harvesting	
Effort level	Heavy, underground		Light, indoors		Moderate, outdoors	
Data	Author's model output		Group output and SE		Group output	
Metric and threshold	Wet bulb, Tnw ($1\,\mathrm{ms}^{-1}$)		Effective temp, ET		WBGT (5th hour)	
	$T_0 = 27.7\,^\circ\mathrm{C}$ (declared)		$T_0 = 19.7\,^\circ\mathrm{C}$ (estimated)		$T_0 = 0\,^\circ\mathrm{C}$ (set here)	
Temperature	**% Output**	**SD**	**% Output**	**SD**	**% Output**	**SD**
25 °C			98	1.6		
26 °C			97	2.0		
27 °C			96	2.4	87	1.4
28 °C	100	N/A	95	2.9	83	1.7
29 °C	99	N/A	93	3.5	78	1.9
30 °C	98	N/A	91	4.2	72	2.2
31 °C	95	N/A	88	4.9	65	2.4
32 °C	90	N/A	85	5.7	58	2.5
33 °C	80	N/A	81	6.6		
34 °C	64	N/A	76	7.4		
35 °C	44	N/A	70	8.2		
36 °C	26	N/A	64	8.8		
37 °C			57	9.3		

3.2. Example 2: Nag 1992 [15], Light Manual Workers in India

Reports of productivity in relation to heat stress do not always provide individual-level data, but may only give summary statistics such as the mean production in each environment and some measure of variation or uncertainty such as the standard error of each reported mean (SE) or the between-worker standard deviation (SD). The model parameters can be estimated from these summary statistics by maximising the sum of the two log-likelihood terms, one for the mean $\bar{x} \sim \mathrm{N}\left(\psi\mu, \mathrm{SE}^2\right)$ and one for the standard deviation SD, where $\upsilon.\mathrm{SD}^2 \sim V.\chi^2_\upsilon$, and $V = \psi^2\sigma^2 + \mu^2\tau^2$. There are $\upsilon = \mathrm{n} - 1$ degrees of freedom for standard deviations estimated from groups of n workers. The likelihood contribution from each experimental environment is then

$$\log L = -\frac{1}{2}\left(\frac{\bar{x} - \psi\mu}{\mathrm{SE}}\right)^2 - \frac{\upsilon}{2}\left(\ln V + \frac{\mathrm{SD}^2}{V}\right) \tag{5}$$

This function is maximised over our model parameters T_0, ψ, τ, b and θ, which jointly determine the mean and standard deviation of production in each environment.

Figure 2 of Nag and Nag (1992) [15] shows the means with error bars of actual production in the first, second, and third hour of 3-hour observation periods, at nine controlled heat stress levels from 26.0 °C to 35.8 °C effective temperature (ET), which is similar to WBGT. Each of the six workers was observed in each environment over nine days, experiencing one environment per day in a different order for each worker. Here, we ignore the first hour as representing a "run-in" period and average the output of the second and third hours. The data used here, digitised and summarised from the published graph, are shown in Table 2. The units are beedi/h.

Table 2. Production data digitised from Figure 2 of Nag and Nag (1992) [15]

ET (°C)	26.0	27.0	28.4	30.2	31.6	33.3	32.1	33.8	35.8
Hour 2	90.0	82.9	79.3	72.9	72.1	72.1	75.7	63.6	55.7
Hour 3	84.3	82.1	75.0	65.7	72.9	82.9	71.4	64.3	52.1
Average	87.1	82.5	77.1	69.3	72.5	77.5	73.6	63.9	53.9

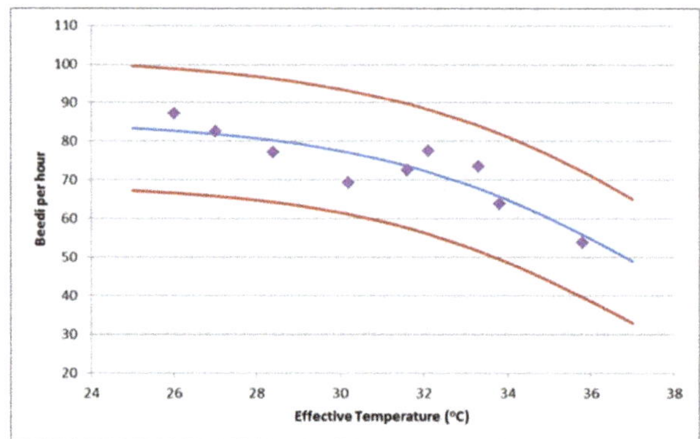

Figure 2. The group data from Nag and Nag (1992) [15]: beedi/h (dots); Fitted marginal distribution of absolute production, with a 90% interval on the individual output T_0 = 19.7 °C, mean ψ = 85.4, SDτ = 9.97, slope b = −0.285, scale θ = 13.6.

To estimate the between-worker variation, we can measure the width of the published error bars, averaged where distinguishable on the original graph, which is approximately 8 beedi/h. Assuming these are standard-error bars, the standard error of the means is about 4, which implies (since n = 6) a between-subject standard deviation in each hour of $4\sqrt{6} \approx 9.8$ beedi/h. The within-subject correlation between hours is unknown, but a correlation close to 1 seems likely in the workers' output in two consecutive hours, in which case the standard deviation of the individual output averaged across two hours remains 9.8 beedi/h.

We next use a logistic model for μ, and here we are able to estimate the optimal-conditions parameter T_0 from the data using the maximum likelihood. Thus, we write logit(μ) = 2ln(θ) + b(T − T_0), where θ, b and T_0 are all parameters to be estimated. We also estimate the two parameters of the N(ψ, τ^2) distribution of maximum potential output. The parameter estimates are T_0 = 19.7 °C, ψ = 85.4, τ = 9.97, b = −0.285, and θ = 13.626. Thus, the optimal temperature is about 20 degrees Celsius, under which conditions workers produce about 85 ± 10 beedi/h. Production falls to about 65% when the temperature rises to 36 degrees, but relative production at that temperature varies between workers from about 50% to 80%. At T = 20 °C effective temperature, the relative production follows a Beta(178, 1.04) distribution, with a mean of 0.994 and a 90% probability interval (0.983, 0.9997), so that all workers are then, in effect, fully productive.

3.3. Example 3: Sahu 2013 [16], Rice Harvesters in India

Sahu et al. (2013) [16] report the number of bundles of rice gathered over several days, by groups of 18 workers in the first and fifth hours of each day. Figure 3 shows the data: productivity strongly declines with temperature, and is lower for any given temperature towards the end of the working day, presumably due to tiredness and perhaps dehydration. By the end of a hot day, work proceeds at only about 60% of the rate at the start of a cool day.

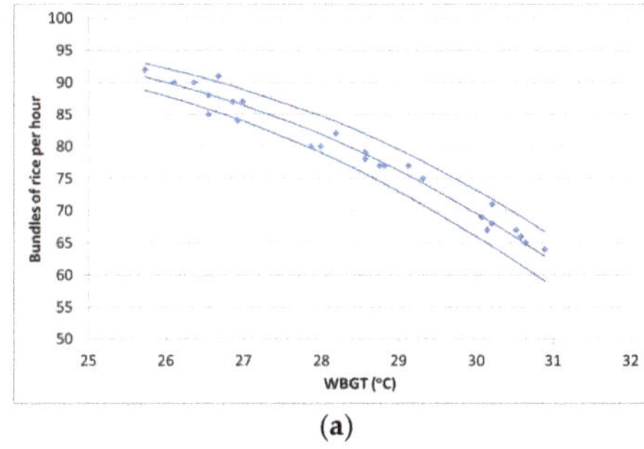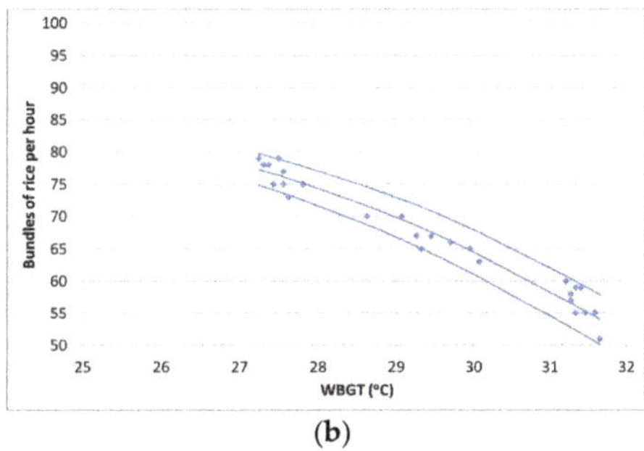

(a)   (b)

Figure 3. The data from Sahu (2013) [16]: hourly rice production in bundles/h (dots); fitted model and approximate 90% confidence intervals (lines). (**a**) First hour: $T_0 = 0\,°C$, mean $\psi = 100.0$, $SD\tau = 1.04$, slope $b = -0.339$, scale $\theta = 245$; (**b**) Fifth hour: $T_0 = 0\,°C$, mean $\psi = 90.2$, $SD\tau = 0.545$, slope $b = -0.316$, scale $\theta = 183$.

Individual data are not available, so we model the output of a group as the dependent variable. There is no evidence of a threshold temperature even at the "coolest" conditions observed (which were nevertheless very hot), so we can reasonably set T_0 to any sufficiently low value, well below the observed range. Figure 3a shows the fitted model for hour 1 and Figure 3b, for hour 5, using $T_0 = 0\,°C$ on the WBGT scale.

These results agree well with those reported in Sahu et al. (2013) [16] as regards to the reduction in the mean output per degree increase in WBGT. Their linear model showed a mean reduction of 5.42 bee*di/h* per degree in the first hour, 5.14 in the fifth, where our model, illustrated in Figure 3, shows an average reduction of about 5.4 in the first hour, 5.3 in the fifth, but falling more steeply at higher temperatures. To these results, we can now add information about the between-worker variation in this reduction, finding that the variation itself strongly increases with the temperature, as shown in the last column of Table 1.

Estimating the parameters by maximum likelihood requires the marginal probability distribution of the product of the two random variables Z and P, which is not analytically tractable. The estimation may be done through approximate numerical quadrature, integrating over the distribution of Z using Simpson's rule. The calculations were done using the "solver" utility of Microsoft Excel. An approximate variance for the product is available as $Var(Y) \approx \psi^2\sigma^2 + \mu^2\tau^2$, and this can be used to generate probability intervals for individual output at a given temperature, as shown in Figure 3.

4. Discussion

Assessing how worker productivity is impacted by hot working environments is important and urgent. There is as yet very little data, and none of it is individual data, but given the importance of the topic, more data should and doubtless will be generated. A class of models is needed for investigating such data.

There is interest in two aspects—health and economics—and models should ideally allow exploration of both. From a health perspective, individual risk matters. The model can also be of value as input to the four-stage SOBANE strategy (Screening, OBservation, ANalysis, Expertise) for the prevention and control of thermal problems in the workplace, specifically at the third analysis stage [17,18]: if the data show a great variation in the output under heat stress, there is more of a cause for concern about the health of the workers. This does indeed appear to be the case for those two of our three examples in which we are able to assess it. Both the Nag data and the Sahu data suggest that under heat stress the degree to which workers maintain their individual output becomes increasingly variable. This, in turn, suggests that a proportion of the workforce may high experience levels of stress, not fully apparent from looking only at the mean output of groups of workers.

Economically, we are interested in the overall output of an industry, so the mean is the key aspect. Assessing the cost-effectiveness of preventive measures to protect health clearly involves knowing the productivity cost of inaction [19].

We have presented here a class of models, the logistic-beta models, that have the following features: (a) they can be fitted to individual or groups data; (b) they capture well the patterns seen so far, in three very diverse settings; (c) they yield estimates of both mean output and variation about the mean, and how these vary with ambient heat, however measured.

The model can be applied to any measure of environmental stress that impacts worker productivity. Our three examples used three different measures: Tnw, ET and WBGT, respectively. While all three attempt to capture mainly heat and humidity, they are not directly comparable, and many other measures might be used [12,20].

Our work has some limitations. There appears to be very little available, published data. One of our sources [14] provided only a fitted model, with little indication of how the data varied around it. The other two provided work output from groups of workers, not from individuals. In one case [15], the threshold, i.e., the temperature below which full productivity is obtained, was estimable from the data. On the other [16], the temperature range represented in the data was too narrow for the threshold to be identified and the output appeared still to be rising strongly at the lower limit of the temperature range (Figure 3). The threshold was therefore set to zero degrees, at which value the shape of the fitted curve subjectively fitted the data well, though values as high as 10 degrees were also consistent.

The model is appropriate for workplaces where the measurable output is generated, such as factories and fields. Heat stress may limit worker efficiency and threaten worker health in other settings also, such as in the boiler rooms of ships [21].

The logistic-beta model is somewhat complicated, and care is needed in using general-purpose software. There is, therefore, a need for the method to be implemented in special-purpose software, such as a Stata program, SAS macro or R function. This work is in progress.

5. Conclusions

I have proposed a new model for individual or grouped worker productivity and illustrated it using three sets of data from very different settings. The model is widely applicable to such data and can be fitted to datasets with different properties and levels of completeness using readily available software (Microsoft Excel). This is the first time a model has been proposed that is capable of quantitatively capturing, estimating and displaying not just how average worker productive declines in hot conditions, but also how between-worker variation is affected, a consideration of great epidemiologic importance.

The model might be used as input to assessing (a) health risk to workers and (b) economic risk to industries under climate change, as extreme work environments arise and become increasingly frequent and severe.

Funding: This research received no external funding.

Acknowledgments: I thank Professor Tord Kjellstrom for collecting and sharing the three datasets, and for helpful comments on early drafts, and the referees for their comments which substantially improved the paper.

Conflicts of Interest: The author declares no conflict of interest.

References

1. Kjellstrom, T.; Kovats, S.; Lloyd, S.J.; Holt, T.; Tol, R.S.J. The direct impact of climate change on regional labour productivity. *Arch. Environ. Occup. Health* **2009**, *64*, 217–227. [CrossRef] [PubMed]
2. Hsiang, S.M. Temperatures and cyclones strongly associated with economic production in the Caribbean and Central America. *Proc. Natl. Acad. Sci.* **2010**, *107*, 15367–15372. [CrossRef] [PubMed]

3. Climate change and Labour: impacts of heat in the workplace. Issue paper. Geneva, CVF Secretariat, UNDP. Available online: http://www.ilo.org/wcmsp5/groups/public/---ed_emp/---gjp/documents/publication/wcms_476194.pdf (accessed on 28 April 2016).
4. Schmitt, L.H.M.; Graham, H.M.; White, P.C.L. Economic evaluations of the health impacts of weather-related extreme events: A scoping review. *Int. J. Environ. Res. Public Health* **2016**, *13*, 1105. [CrossRef] [PubMed]
5. Smith, K.R.; Woodward, A.; Campbell-Lendrum, D.; Chadee, D.D.; Honda, Y.; Liu, Q.; Olwoch, J.M.; Revich, B.; Sauerborn, R. 2014: Human health: Impacts, adaptation, and co-benefits. In *Climate Change 2014: Impacts, Adaptation, and Vulnerability. Part A: Global and Sectoral Aspects. Contribution of Working Group II to the Fifth Assessment Report of the Intergovernmental Panel on Climate Change*; Cambridge University Press: Cambridge, UK; New York, NY, USA; pp. 709–754.
6. Dunne, J.P.; Stouffer, R.J.; John, J.G. Reductions in labour capacity from heat stress under climate warming. *Nat. Clim. Change* **2013**, *3*, 563–566. [CrossRef]
7. DARA. Climate Vulnerability Monitor 2012. Madrid: Fundacion DARA Internacional, 2012. Available online: http://daraint.org/climate-vulnerability-monitor/climate-vulnerability-monitor-2012/ (accessed on 14 August 2016).
8. Kopp, R.; Hsiang, S.; Rasmussen, D.J.; Mastrandrea, M.; Jina, A.; Rising, J.; Muir-Wood, R.; Wilson, P.; Delgado, M.; Mohan, S.; et al. American Climate Prospectus: Economic risks in the United States. New York: Rhodium Group, 2014. Available online: http://rhg.com/wp-content/uploads/2014/06/RHG_AmericanClimateProspectus_June2014_LowRes1.pdf (accessed on July 18 2014).
9. Kjellstrom, T. Impact of climate conditions on occupational health and related economic losses: A new feature of global and urban health in the context of climate change. *Asia Pac. J. Public Health* **2016**, *28*, 28S–37S. [CrossRef] [PubMed]
10. Venugopal, V.; Chinnadurai, J.; Lucas, R.; Wishwanathan, V.; Rajiva, A.; Kjellstrom, T. The social implications of occupational heat stress on migrant workers engaged in public construction: A case study from southern India. *Int. J. Constr. Environ.* **2016**, *7*, 25–36. [CrossRef]
11. Centers for Disease Control and Prevention. Heat-related deaths among crop workers—United States, 1992–2006. *Morb. Mortal. Wkly. Rep.* **2008**, *57*, 649–653.
12. Parsons, K. *Human thermal environment: The effects of hot, moderate and cold temperatures on human health, comfort and performance*, 3rd ed.; CRC Press: New York, NY, USA, 2014.
13. Schulte, P.A.; Bhattacharya, A.; Butler, C.R.; Chun, H.K.; Jacklitsch, B.; Jacobs, T.; Kiefer, M.; Lincoln, J.; Pendergrass, S.; Shire, J.; et al. Advancing the framework for considering the effects of climate change on worker safety and health. *J. Occup. Environ. Hyg.* **2016**, *13*, 847–865. [CrossRef] [PubMed]
14. Wyndham, C.H. Adaptation to heat and cold. *Environ. Res.* **1969**, *2*, 442–469. [CrossRef]
15. Nag, A.; Nag, P.K. Heat stress of women doing manipulative work. *Am. Ind. Hyg. Assoc. J.* **1992**, *53*, 751–756. [CrossRef] [PubMed]
16. Sahu, S.; Sett, M.; Kjellstrom, T. Heat Exposure, Cardiovascular Stress and Work Productivity in Rice Harvesters in India: Implications for a Climate Change Future. *Ind. Health* **2013**, *51*, 424–431. [CrossRef] [PubMed]
17. Malchaire, J.B.M.; Gebhardt, H.J.; Piette, A. Strategy for evaluation and prevention of risk due to work in thermal environments. *Ann. Occup. Hyg.* **1999**, *43*, 367–376. [CrossRef] [PubMed]
18. Malchaire, J.B.M. Occupational heat stress assessment by the predicted heat strain model. *Ind. Health* **2006**, *44*, 380–387. [CrossRef] [PubMed]
19. Parsons, K. Maintaining health, comfort and productivity in heat waves. *Glob. Health Action* **2009**, *2*, 2057. [CrossRef] [PubMed]
20. Epstein, Y.; Moran, D.S. Thermal comfort and the heat stress indices. *Ind. Health* **2006**, *44*, 388–398. [CrossRef] [PubMed]
21. Palella, B.I.; Quaranta, F.; Riccio, G. On the management and prevention of heat stress for crews onboard ships. *Ocean Eng.* **2016**, *112*, 277–286. [CrossRef]

 *atmosphere*

Article

Making Administrative Systems Adaptive to Emerging Climate Change-Related Health Effects: Case of Estonia

Kati Orru [1,*], Mari Tillmann [2], Kristie L. Ebi [3] and Hans Orru [4,5]

[1] Institute of Social Sciences, University of Tartu, Lossi 36, 51003 Tartu, Estonia
[2] Estonian Police and Border Guard Board, Pärnu mnt 139, 15060 Tallinn, Estonia; mari.tillmann@politsei.ee
[3] Department of Global Health, University of Washington, Seattle, WA 98105, USA; krisebi@uw.edu
[4] Institute of Family Medicine and Public Health, University of Tartu, Ravila 19, 50411 Tartu, Estonia; hans.orru@ut.ee
[5] Department of Public Health and Clinical Medicine, Umea University, 901 87 Umea, Sweden
* Correspondence: kati.orru@ut.ee; Tel.: +372-737-5188

Received: 30 April 2018; Accepted: 5 June 2018; Published: 9 June 2018

Abstract: To facilitate resilience to a changing climate, it is necessary to go beyond quantitative studies and take an in-depth look at the functioning of health systems and the variety of drivers shaping its effectiveness. We clarify the factors determining the effectiveness of the Estonian health system in assessing and managing the health risks of climate change. Document analyses, expert interviews with key informants from health systems whose responsibilities are relevant to climate change, and analysis of a population-based survey conducted in 2015, indicate that the health effects of climate change have not been mainstreamed into policy. Therefore, many of the potential synergistic effects of combining information on health systems, environment, and vulnerable populations remain unexploited. The limited uptake of the issue of climate change-related health risks may be attributed to the lack of experience with managing extreme weather events; limited understanding of how to incorporate projections of longer-term health risks into policies and plans; unclear divisions of responsibility; and market liberal state approaches. Minority groups and urban dwellers are placing strong pressure on the health system to address climate change-related risks, likely due to their lower levels of perceived control over their physical wellbeing. The results have implications for national, community, and individual resilience in upper-middle income countries in Eastern Europe.

Keywords: health systems; climate adaptation; health infrastructure; rescue services; Northern Europe

---

1. Introduction

It is expected that the health risks of climate change will increase globally, with increases in morbidity and mortality from selected climate-sensitive health outcomes, putting additional pressures on health systems [1,2]. The effectiveness of adaptation is influenced by a country's physical location and exposure to climate shocks, levels of socioeconomic development, and healthcare capacity to prepare for and manage climate-related shocks [3]. Tailoring effective adaptation measures requires a clear understanding of which factors led to today's preparedness to manage climate change. Researchers have been urged to pay more attention to the role of the nation state in climate governance, as these actors have the legitimacy and resources to develop long-term visions, stimulate and oversee local approaches, and carry forward adaptation programs, e.g., [4]. In their analysis of health adaptation initiatives in ten Organization for Economic Co-operation and Development (OECD) countries, Austin et al. concluded that national governments play a key role in health adaptation to climate change, but there are competing views on what responsibilities and obligations this will—or should—include [5]. One view suggests increasing investments in existing public health infrastructure to manage the projected health risks [6]. The second argues climate change will likely affect health by, for instance, destabilising supporting systems or threatening infrastructure, thus necessitating new and innovative responses. Suggested factors for successful adaptation include adopting legislation, ensuring interdepartmental coordination, and increasing self-governance [7–9]. We take an in-depth look at health and rescue systems in Estonia, because they will play a key role in climate adaptation policies in connection with the EU climate change strategy [10].

Informed by the research on risk governance [11,12], we analyse the health and rescue system as a control system for achieving policy goals; assess factors shaping those goals; and assess the factors shaping the process of information-gathering and implementation of protective measures, such as early warning systems or emergency preparations. We examine the extent to which outside (e.g., issue salience) and inside (e.g., institutional capacities and rules) drivers of the health systems shape the success of climate adaptation. We apply the World Health Organization's Operational Framework for Building Climate Resilient Health Systems [13] to understand how the quality of leadership and governance, health workforce, health information systems, medical technologies, service delivery, and climate and health financing shape the quality, efficiency, equity, accountability, and resilience of the health systems [14].

1.1. Climate Change Adaptation of Health Systems

We adopt a case-study framework of adaptation of health systems (Figure 1). For analytic purposes, the framework distinguishes between the processes of elaborating policy programmes, monitoring, and information-gathering, and enforcing protective measures. The framework recognises significant overlaps and feedback loops between these responsibilities of the Estonian health system. Effective and timely responses depend on effective **policies, strategies, and action plans** crafted by political leaders to manage climate-sensitive health outcomes. The WHO Operational Framework specifies that in order to continuously deliver health services throughout climate-related events, governments should modify emergency management plans and incorporate up-to-date risk assessments. In line with the European Union's climate change adaptation strategy [15], most European countries have developed national strategies, but few countries have implemented regulations and operational plans for authorities and healthcare systems to manage health-related issues.

Monitoring and gathering information of health and health system conditions provides information for timely responses and behaviour change, but also for setting new goals and amending legislation. The WHO Operational Framework [13] identifies three key areas that health information systems must develop: vulnerability, capacity, and adaptation assessments; risk monitoring; and climate and health research. Risk monitoring systems provide timely, detailed information on current and future environmental conditions that may affect health and the ability of health systems to provide services. In the context of climate change, hurdles for new interventions may arise in the form of evidence provision; climate will continue to change over coming decades, with uncertainties associated with the rate and magnitude [16,17]. The tendency of turning a "blind eye" towards policy problems by limiting investments in surveillance and monitoring may work to maintain the status quo in state investments [18].

As for timely **responses and protective measures**, monitoring systems coupled with communication networks (early warning systems) can be designed to alert members of the public when environmental hazards may affect their health. However, existing health safety programs excessively focus on providing information [19] without considering the need to change motivation and supportive infrastructure, living arrangements, and skills. For example, climate change can affect the ability of healthcare systems to provide services during extreme weather events because of infrastructure damage and medical supply disruptions [13].

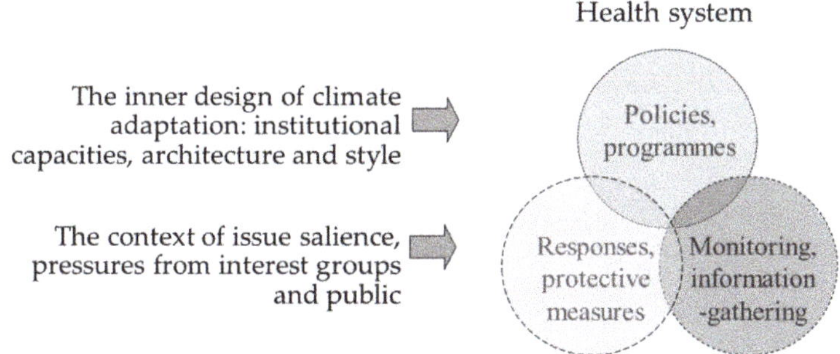

Figure 1. Determinants of climate adaptation of health systems.

1.2. Factors Influencing Adaptation of Policy Programmes, Monitoring Systems, and Response Measures

Effectively adapting to the health risks of climate change includes internal factors related to administering health systems, such as management structure and capability, and external factors, including the pressure of interest groups and the public. The **internal functioning** of health systems is often characterised by administrative cultures of reactivity and lack of long-term future orientation. The administrative focus is tightly related to the risk calculus that politicians make; this includes issues such as the visibility of the problem (which may be obvious to experts but far less so to the general public and politicians) and the avoidability of blame (for events that lead to damage) [20,21]. By contrast, the notion of institutional adaptive capacity [22] highlights the proactive provision of means and information to enable social actors to anticipate possible futures and take preventive measures.

Administrative capacity is the health system's economic aptitude, technical preparedness, and competence and sufficiency of human resources; these factors may influence the development and application of adaptation policy [23]. At lower level administrative units, larger municipalities are likely to have more human and financial resources to direct to adaptation [24,25]. Maintaining a healthy and effective practitioner workforce, products, and technologies are critical challenges for health systems [13]. Beyond human and financial resources, institutions should have the authority to generate political and legal incentives for actors to change [26].

The administrative architecture is crucial in the design of adaptation policy, as multilevel institutions and networks are needed for the process to be successful [27]. Developing effective adaptation plans requires coordination and collaboration between health ministries and other government agencies and non-governmental partners, to ensure that the actions undertaken foster positive health outcomes [13]. Policy coherence literatures (see Tosun and Lang, 2017 [28]) highlight the demanding task of integrating institutions, particularly in cases with an increasing number of affected interests, such as health [29] or security [30].

The most obvious **contextual driver** of political and administrative action is the effect of extreme events spurring increased awareness and policy innovation [18]. In general, climate change is difficult to understand and psychologically distant for lay people and political and administrative actors [31]. However, according to the policy-window hypothesis, following a disaster, the political climate may be conducive to legal, economic, and social change that can begin to reduce structural vulnerabilities [32].

Based on strong interests, non-governmental organisations, entrepreneurs, or the scientific community can influence health risk governance [12]. Robust climate and health research agendas should expand and improve the quality of knowledge [33]. However, the scarcity of scientific evidence of the emergence of climate change-related health impacts may challenge the ability to develop effective adaptation options [34].

In countries of the former Soviet Union, the role of non-governmental organisations in developing and applying regulatory regimes has been modest [18]. This may change as national groups advocating a certain policy or action gain prominence when an issue is placed on the political agenda and reaches a certain level of salience [35].

International organisations pose crucial pressures; the adoption of climate policies in the OECD countries was influenced by learning from international organisations [36]. The issue of climate change entered the political agendas of Central and Eastern European countries through the EU [37,38]. Exemplary neighbouring states are urging other EU countries to tighten their mitigation policies [36,39].

In the context of climate change, this study takes a comprehensive look at external pressures and factors characterising the internal working of the Estonian health system. We first present empirical data and methods that are used to validate our expectations. We then discuss background information about Estonia's health systems' adaptation to climate change. In subsequent sections, we discuss how the dominant ideas, administrative structures and capacities, and windows of opportunity have shaped the current situation with respect to adaptation politics.

2. Methods

Estonia lies in the North-Eastern corner of Europe. Out of a population of 1.3 million, 35.5% are nationalities other than Estonian, including Russians (25%). One million people (77.2% of the population) live in urban areas [40]. Income disparities are relatively large at 33.2% [40]. The health inequalities among different age, education, and ethnicity groups are considerably larger in Estonia than in other Nordic countries [41]. For example, life expectancy among Estonians is 78.5 years, and non-Estonians 76.5, and total life expectancy is 77.6 years in Estonia, compared with 81.1 years in Finland [42].

For analysing the health systems' amenability to internal and contextual pressures, Estonian legislative acts, directives, strategic documents, such as activity plans, and their implementation were examined. In addition, 21 semi-structured interviews were conducted with key experts connected to health system adaptation in Estonia, ranging from policy designers to officials to scientists to social workers (Appendix A). The interviews clarified the expert's observations and experiences with the functioning of the Estonian health system and the actors and processes shaping its effectiveness. The interviews addressed the questions of the capability of the health and rescue system, operability of implemented measures, and the importance of various factors in shaping health system adaptation (Appendix B). The interviews were conducted in the spring of 2015 when compiling an adaptation strategy for the reducing the health risks of climate change and when health impact assessments were in their infancy [43].

To address public salience as an external driver of adaptation governance, we use a survey on environmental health risk perception and coping conducted in Estonia in 2015 of persons aged 18–75 years, stratified by age, sex, and geographical location [44]. The survey invited 2207 participants (administered by IBP Saar Poll), of which 1000 agreed (45.3% response rate). We used a semi-structured questionnaire constructed to assess climate change-related issues, such as perceived exposure to extreme weather events, demand for state support for coping with extreme events, beliefs about state institutions' efficacy in taking care of the healthfulness of the environment, and concerns about health risks from the environment (Appendix C). Additionally, the instrument had entries for respondents' demographic data and self-rated health status.

We used a logistic regression analysis to estimate which factors were associated with perceived needs for measures for coping with the health risks of climate change. For statistical modelling of covariates of perceived need, we collapsed perceived need ratings into dichotomous groups by combining the answers high to total agreement (scores 4–5) into a group perceiving need, and the other categories as not. We recoded worry about health risk to self and family arising from the environment, as scores 1 and 2 = group 1, score 3 = group 2, and scores 4 and 5 = group 3.

3. Results

Despite the relatively milder impacts of climate change on Northern and Eastern Europe under the more pessimistic RCP8.5 emission pathway, up to 1000 additional premature deaths due to heatwaves are projected annually by the end of the 21st century [45]. In northern Europe, the number of heatwave days are projected to increase by 2.4 times during the warmest months [33]. Also, the likelihood of forest fires could increase, and pollen seasons will be prolonged, with an increase in the number of new allergens [46,47]. Climate change could affect the spread of leishmaniosis, hantavirus, tick-borne encephalitis, dengue fever, and tularaemia in the region [48]. Among other European areas, the northern European region is projected to experience significant increases in future strong storms (intensity of high-impact wind speed and extreme cyclone frequency) [49]. An increase in the frequency of extreme weather, storms, and torrential rain may bring increased risks of physical danger. Climate change-related increases in the number of glazed frost days and prolonged periods of heatwaves threaten the elderly and chronically ill. These projected risks pose a direct challenge not only to medical care and rescue capacity, but also to social security services and informal support networks.

3.1. Adapting Health Systems to Climate Change

3.1.1. Planning Responses to Climate Change-Related Health Risks

Managing climate change-related health risks are divided between several ministries and their subdivisions in Estonia (Table 1). The Ministry of Social Affairs coordinates environmental health politics following the National Health Plan 2009–2020 [50], without explicitly considering the health risks associated with climate change. The Ministry of Environment Environmental Strategy 2030 and National Environmental Action Plan recognise the need to increase understanding of health risks, including through increasing awareness about health risks in environment and educational specialists and in the population, and in those responsible for managing emergency situations [51,52].

Table 1. Key institutions and responsibilities in policymaking, monitoring, and protective measures in relation to climate change adaptation.

Key Institutions	Policies and Programmes	Monitoring and Information-Gathering	Responses and Protective Measures	Addressing Climate Change?
Ministry of Social Affairs, Health Board, National Institute for Health Development	National Health Plan 2020	Health Board: monitoring and prevention of infectious diseases, vector transmitted diseases, including cyanotoxins in bathing water	Health Board: Information on web on prevention of infectious diseases, vector transmitted diseases.	No
Ministry of Environment, Weather Service	Environmental Strategy 2030	Weather Service [53–55]: European forecast model HIRLAM, Weather Alert System Meteoalarm, Baltic Sea forecast models HIROMB; UV radiation; pollen content for allergens	Weather Service: forecasts and warnings on web: Estonian Weather Service, Air Quality Management System [56].	Included in environment monitoring
Ministry of Interior	Emergency Act [57], Internal Security Development Plan 2015–2020	Rescue Board in collaboration with Health Board conducts risk analysis for emergency situations including floods, extensive forest fires, extremely cold and warm weather	Emergency plans for large-scale forest or landscape fires; storms and floods in densely populated areas; epidemics. Instructions on behaviour during emergencies on Crisis web	Included in rescue capacity

Emergency situations that potentially affect health, including extremely hot or cold weather, storms, large-scale landscape or forest fires, floods in densely populated areas, and epidemics, are regulated by the Emergency Act [57]. According to the Internal Security Development Plan 2015–2020 [58], the goal is to ensure preparedness for disasters resulting from extreme weather conditions caused by climate change, and to develop procedures for disaster management.

Baseline research for a more comprehensive Climate Strategy was initiated in 2015 to follow the EU Climate Adaptation Directive [15]. Interviews found there was not a perceived need to deal with adaptation separately. The respondents found that healthcare, environmental health, and rescue services are sufficiently regulated by legislative acts and strategic documents. An official of the Ministry of the Interior (29 April 2015): *"Everything is in its place legislatively, and if anything needed to be changed it is rather the organisation of work."*

3.1.2. Gathering Information and Monitoring the Situation

Different organisations monitor and forecast weather conditions. The Estonian Weather Service uses several European forecasting models and alert systems (Table 1) [53–55] and measures UV radiation. Measuring the quality of drinking and bathing water, spread of infectious diseases, and vector-transmitted diseases is the responsibility of the Health Board [59]. Not addressed are the spread of potentially infectious agents, vectors, airborne mould, and changes in indoor environments associated with climate change.

The Environmental Health Research Centre under the Estonian Health Board was active during the period 2013–2015, with the help of EU structural funding. The goals of the Centre were to collect data from different state agencies to analyse the associations between morbidity and environmental factors; to conduct risk assessments and health impact analyses; and to develop evidence-based policy proposals concerning environmental health [60]. Due to the of lack of resources, the work of the Centre on comprehensive risk and vulnerability analyses has stopped.

The Estonian Rescue Board, in collaboration with Health Board [61], conducts risk analyses for emergency situations (Table 1). However, most of the analyses do not consider climate change when assessing the probability and impact of events. The Internal Security Development Plan 2015–2020 [58] identified, as a problem, the lack of a common environment for analyses of the sustainability of essential services, and of phenomena and events with impacts on the living environment; together, these make the timely prediction of needs impossible. An official from the Ministry of the Interior stated in an interview the need to make the compilation of risk analyses more efficient (29 April 2015): *"We would like to create a situation where the risk analysis is more frequent, based on the needs, in other words, when there is a need for a more frequent analysis than after every two years, then it is done . . . "*

3.1.3. Protective Measures and Responses

Emergency plans specify administration structures during emergency situations, the tasks of participating institutions, and procedures for information exchange and for notifying the public in case of large-scale forest or landscape fires, storms, floods in densely populated areas, and epidemics [62–65]; heatwaves are excluded. Under the coordination of Health Board, health service providers are expected to deliver health services to those affected [59,61,64,65]. The weak capacity of the Health Board to manage emergency situations was mentioned by a representative of the Rescue Board (7 June 2015): *"Only the police and rescue services have a real operational capacity currently in Estonia, the rest who should run the situation, like the Health Board or Environmental Board, only do it declaratively since they have no capacity, experience or structure to speak of today."* Risk analyses compiled by the Health Board [66] noted that microclimates in hospital wards and work premises are likely to deteriorate as a consequence of warmer weather, because most hospitals lack air conditioners. An official from the Ministry of the Interior (29 April 2015) and one from the Health Board's medical bureau (25 May 2015) noted that hot weather plans as not being necessary because: *"We do not have such extreme conditions as a result of which hospitals should start rearranging their work."* Family physicians, hospitals, and ambulance services operate under private law, and their responsibilities and preparedness competences in case of emergency events, like extreme weather, have not been analysed nor regulated [62].

According the Internal Security Development Plan 2015–2020 [58], plans for managing emergency situations are not tested on a regular basis, as required by law (RT I, 2009, 39, 262). For example, during a national exercise "Snowstorm 2010" that took place in the January of 2010, Padaorg valley was mentioned as an area of greatest risk. In the end of the same year, when snowstorm Monica struck Estonia, hundreds of people were left to the elements in Padaorg. Although according to the law, the organisation of exercises and training should be the responsibility of the ministry that is responsible for the emergency, the reality is that national exercises have only been organised by the Ministry of Interior [67].

The interviews also revealed a need to address the issue of "grassroots level applications" to create possibilities for people to protect their health in everyday life. In the case of more frequent extreme weather in the future, the most vulnerable, including the elderly, the chronically ill, or mentally disabled, require special attention from social care and primary healthcare providers. Although social care professionals possess vital information on the location and specific needs of the most vulnerable, they have not been engaged in the process of adaptation (interview with an expert in social care 4 March 2015).

Response measures targeted at individuals during a crisis mainly focus on information-sharing on the web about health risks and preventative measures and instructions on behaviour, and do not consider practical arrangements or infrastructure (Table 1). Web-based information-sharing is limited because not all vulnerable groups have access. According to a Health Board official (7 May 2015), recommendations for the population, including risk groups, in the case of extreme hot weather, are published on the Health Board's website; however, there was no money for published brochures. An official from the Ministry of Environment (21 May 2015): *"What needs to be improved is the notification of the public, early warning of the climate change health risks."* A family physician (7 July 2015) criticised the whole information-based approach for adverse impact prevention: *"Let me give you an example, even if we have info on the radio on the dangerously high UV-factor, people still go to the beach to get tanned."*

3.2. Explanations for the Current Situation with Respect to Health Adaptation

There are questions about how the situation evolved in which goals concerning climate change health effects have been set only in the fields of environment and internal security; why monitoring climate parameters are fragmentary and do not cover all the important factors for assessing health effects; why the quality of risk analyses are disputed; why rescue and private health service providers and social care cooperation procedures have not been developed; and why exercises to increase the capacity to cope with extreme weather conditions are irregular and not very effective.

3.2.1. The Inner Functioning of the State Apparatus Influences the Adaptation Activities

The absence of an integrated approach to climate adaptation can be explained by the nature of the **administrative culture** of the institutions involved in the Estonian health system. On the one hand, adaptation policies are influenced by an understanding of the necessity of state intervention (or the lack thereof) and, on the other hand, the manner of making decisions and their transparency. A prevalent viewpoint is that individuals can adapt to climate change because it occurs over long time periods, so preparedness only needs to be ensured for extreme changes. As stated by a climate expert (29 April 2015): *"We need the support of the state first and foremost when health effects have occurred as a result of short and abrupt changes. For certain, human body has the ability to adapt to long-term changes."* Likewise, the majority of the interviewees found that state regulations are sufficient, and that increasing the readiness of people to cope with changing weather patterns is all that is needed. As one official from the Ministry of the Interior said (29 April 2015): *"The readiness of people to cope has decreased. This should be taught from early on."*

In coping with extreme weather events, transparency in decision-making was called into question. Interviews pointed out that although shaping policies is open to different parties, decisions in the area of health and rescue are made by a small closed group. The officials pointed out that sometimes government decisions come without explanation as to what was the motive or evidence behind a decision. As a specialist in environmental health from the Ministry of Social Affairs (4 May 2015) said: *"Today such decisions are made in politics the background of which is not known even to us—no matter have the scientists or experts been involved or not."* There was no evidence supporting the hypothesis that there is a change in the climate-related norms of the political actors in Estonia or for growing support for climate adaptation policy.

The administrative architecture is important for monitoring climate change-related health effects and assessing the readiness to cope with emergencies; these are lagging due to the dispersal of responsibilities across institutions. A specialist in public health discussed the challenge of not having an owner of the issue of the health risks of climate change: *"Climate health effects, as environmental health in general, is a highly cross-sectoral phenomenon and nobody really wants to claim the problems and take responsibility for the area."* A specialist in environmental health in the Ministry of Social Affairs (4 May 2015) brought out problems in distributing resources in connection with the application of measures and executing control: *"The Ministry of the Environment has the money and the Ministry of Social Affairs has the problems, however, the ministry will not allocate money to solve the problems. The fragmentation of the area between the ministries would not even be a problem if there was clarity as to which ministry should be paying for solving the problem."*

Distributing responsibilities between ministries with insufficient cooperation between them results in problems when dealing with emergency situations associated with climate change, as was pointed out by a representative of the Rescue Board (7 June 2015): *"Each ministry has their own plans for coping in different situations, but there is no general plan."* A representative of the Rescue Board (18 June 2015) said: *"Currently, the Ministry of the Interior is the coordinator of crises situations and other ministries do not want to come along in this matter. The Ministry of the Interior does not have the means to force other ministries either. This is why such freewheeling exists. Since the state has not given any direct guidelines the departments just see how they can manage."* A family physician (7 May 2015) commented on the ability of healthcare to react in case of emergency situations: *"Everybody is doing what comes to mind. There is no uniformly controlling agency to whom to turn to. The tasks and who precisely is responsible for what has not been clearly determined. This is a very topical problem."* A representative of Rescue Board said: *"It would be complicated to get the hospitals and doctors behind one table, to map their resources and situation, because they are private bodies. One can only assume that when something happens that would exceed the capacity of a local hospital, other hospitals would accept the patients; however, there is actually no certainty in this matter."* The issue of silos of primary healthcare, emergency situations, and social care was also stressed by a city medical officer (12 June 2015): *"Now when the family physicians are no longer part of local governments and the communication takes place through Ministry of Social Affairs, the health care part is distancing itself from the local governments' health adaptation issues."*

One of the factors influencing the **administrative capability** to cope with climate change on the state level is the scarcity of the necessary expertise. A separate programme for environmental health to address adaptation issues was not considered necessary. There is almost no continuing medical training in environmental health for physicians. Climate change-related health risks are an unknown territory for social care professionals. As an expert in social work (4 March 2015) explained: *"A social worker or caretaker does not have the knowledge about what kind of weather conditions present danger and how to prepare a ward. The current activities are all a creative effort on the spot."*

Lack of financing has inhibited monitoring, integrating databases to assess health effects and rescue necessities, and diversifying information systems. Creating a common pool of resources and synergies by coordination of information systems is resisted for a paradoxical reason mentioned by an official from the Ministry of the Interior (29 April 2015): *"We are, nevertheless, a poor country. There is a fight in departments for money and for each individual."* Poor preparedness of medical infrastructure and staff are explained by a lack of financing and prioritisation. The interviewed medical workers stressed the inadequacy of human resources to cope with climate-related disease outbreaks. As a family physician (7 May 2015) explained: *"As for today, we don't have extra resources in case of an outbreak."* At the same time, some officials claimed there are insufficient human resources to ensure the capability to deal with climate change-related health risks. An official from the Ministry of the Interior (29 April 2015) said: *"Climate change is currently highly unlikely and the effect climate change has does not exceed the usual capability of the hospitals; we do not treat this as a likely emergency right now."* This also explains the insufficient investment into cooling systems for hospitals and care homes.

3.2.2. Adaptation Activities Influenced by the External Pressure Factors of the State Apparatus

The lack of **public interest** may reduce the pressure on health systems to address climate change-related health risks. An Eurobarometer survey [68] confirms that, in comparison with the rest of Europe, Estonia has the fewest numbers of people who consider climate change as a very serious problem. Analyses of the population-based survey focused on people with strong beliefs that state measures are necessary to reduce the health risks from climate change (respondents scores 4–5 on a five-point-scale, ranging from strong disagreement (1) to strong agreement (5), accounting for 30% (301) of respondents). There are significantly ($p < 0.05$) more women, ethnicities other than Estonians, and urban residents with high interest in adaptation measures (Appendix D). The level of interest does not differ significantly by age, educational group, or level of individual self-rated health.

We identified factors explaining high interest in measures for reducing the health risks of climate change using logistic regression. The base model predicting demand for measures included gender, age, education, home language, and self-rated health status. Table 2 shows that Estonian-speakers had a 0.61 (95% CI 0.43–0.86) lower odds of requesting measures than individuals speaking Russian or other languages. When adding the possible predictors of demand to the base model one-by-one, no effect was seen between the trust in institutional efficacy in taking care of the healthfulness of the environment and demand for measures. Compared to individuals with high worry, individuals with low worry had 0.57 (0.37–0.88), and individuals with medium worry had 0.58 (0.38–0.88) lower odds of demanding measures. The odds of demanding measures increased by 1.24 (1.06–1.45) when the perceived exposure to extreme weather events increased by one degree on a five-point-scale.

Table 2. Association between factors and the belief in need for measures to reduce the health risks from climate change in Estonia, odds ratios (95% CI).

Base Model	Sig.	Exp(B)
Gender, man (ref woman)	0.79	0.96 (0.68–1.34)
Age, 20–34 (ref 55–75)	0.19	0.73 (0.46–1.17)
Age, 35–54 (ref 55–75)	0.10	0.71 (0.48–1.06)
Education, primary (ref higher)	0.84	1.08 (0.50–2.33)
Education, secondary (ref higher)	0.57	0.90 (0.62–1.30)
Home language, Estonian (ref Russian or other)	0.01	0.61 (0.43–0.86)
Self-rated health status	0.71	0.96 (0.79–1.18)
Social and psychological factors *		
Belief in state institutions taking care of the healthfulness of living environment	0.52	0.94 (0.79–1.18)
Worry about health effects from environment on personal and family health		
Low (ref high worry)	0.01	0.57 (0.37–0.88)
Medium (ref high worry)	0.01	0.58 (0.38–0.88)
Exposure to extreme weather events	0.01	1.24 (1.06–1.45)

* Logistic regression was adjusted for gender, age, education, home language, and self-assessed health status.

The opinion that the health risks of climate change are a comparatively less important concern for Estonian society is also reflected in the (lack of) political debate and administrative prioritisation. This was brought out in key expert interviews. All the respondents started the interview with a recognition that they do not believe that climate in Estonia will start changing considerably, and that there is not enough evidence to confirm that climate change is occurring. For example, an official from the Ministry of the Interior (29 April 2015) said: *"The numbers do not show that the climate would change dramatically in Estonia; we have no facts to support that today."*

One of the most important factors that has kept public and political salience low is the insufficiency of corroborating scientific evidence that climate change affects health. Up to 2015, there were very few studies in Estonia on the interaction between climate change, environmental factors, and health indicators. Limited governmental resources have forced the state to set priorities; environmental health is not one of them. A public health specialist (8 May 2015) emphasised: *"Unfortunately, environmental health is not a priority and therefore there is no state funding available."* Paradoxically, the officials consider the role of scientific research important in making decisions. As a specialist in environmental health from the Ministry of Environment (4 May 2015) said: *"We do nothing just because of gut feeling. We will not contribute the resources if there are no prognoses or studies made. At least when a change has been initiated by us, there must always be a scientific explanation behind it."*

The shortage of expertise in this area is also suggested by the fact that Estonia does not train specialists in environmental health, and in medical school, the topic is covered in only one subject. A public health specialist (8 May 2015) stated: *"We have many specialists on environment in different universities in Estonia, but in none of them has a course on environmental health in which climate change and its actual health effects would be clarified for the future decision makers."* The National Health Plan 2009–2020 [38] includes measures for training of health effect assessment specialists and for collecting environmental health related knowledge. So far, there are no resources to support this.

The key experts do not see commercial interests as having an impact on the design or implementation of health adaptation policies. Increasing introductions of viruses could interest new product development in pharmaceutical companies. For example, tick-borne encephalitis can be prevented by vaccination. According to a family physician (7 May 2015), it is simply a matter of time until the pharmaceutical companies *"get wind of it"* and start gathering support for the national vaccination programme.

3.2.3. International Pressures

The lack of public interest may reduce the state-level policies that are often influenced by international agreements or the experiences of neighbouring countries. A key factor in the development of Estonian health adaptation policies is European Union regulations. Therefore, developing a national strategy for health adaptation is based on the EU strategy on adaptation to climate change [43]. When developing policies in the area of safety, according to an official from the Ministry of the Interior, the examples of Sweden, Finland, and Germany were taken into account (official from the Ministry of the Interior 20 April 2015). However, the progressive examples of Sweden and Finland were not considered.

In addition to the need to develop an adaptation strategy, the interviewees considered the influence of the EU to be an important example of the demand to finance and monitor environmental factors. Thus, a public health specialist (8 May 2015) emphasised: *"The European Union has a huge influence in improving for example environmental monitoring, since as a condition of the accession to the EU, Estonia has had to introduce several amendments to the laws and that effect is only positive, since the efficiency of monitoring is on the increase as a result of it."* For example, the purpose of the "Report on the risks of flooding" is to map flood-prone areas in Estonia using observations according to Articles 4 and 5 of the EU Floods Directive [69].

4. Discussion

Our research clarified the factors determining the preparedness of the health system in Estonia to manage the health risks of climate change. The study was conducted in 2015, when health adaptation had not been legally defined. The material analysed was used as background material for compiling the strategy "Climate Change Adaption Development Plan until 2030" that was ratified by the Estonian parliament in 2017 [70]. A "Climate Change Adaption Development Plan's Action Plan for 2017–2020" was subsequently agreed upon. However, the resources dedicated for the action plan were a magnitude smaller than what was proposed [71]. Therefore, many of the problems raised in the study remain to be addressed. Our analyses take an in-depth look at one country, Estonia. The lessons learned could be of relevance for other countries with similar Eastern European backgrounds. In some Eastern

European countries (e.g., Poland), climate change adaptation and mitigation are taken seriously at the city level [25]. However, Eastern European countries, including Estonia, fall far behind Finland, Sweden, and the United Kingdom in adaptation planning and implementation.

The policy innovation literature, e.g., [4,72] offers some clues as to the circumstances under which adaptation policies emerge, e.g., bringing out the role of leadership and the state's participation in international organisations. However, the literature stops short in giving an integrated look at the drivers behind implementation, monitoring, and protective responses. We took a broader look at the health systems functioning and the variety of drivers shaping its effectiveness and equality in provision. An in-depth case study allowed exploration of the mechanisms behind the adoption and implementation of adaptation policies, including the broader perspective of health systems, particularly prevention, healthcare provision, and rescue services.

The WHO Operational Framework foresees that for effective policymaking activities, policymakers and other stakeholders require accurate, timely information on health and health system conditions. However, up until the initiation of baseline studies for a climate adaptation strategy pursuant to the EU Directive [15], there had been no studies on climate change and related health systems functioning. Our analyses confirm the findings by Massey et al. [23] that external drivers, primarily, the centrality of EU is required for the diffusion of adaptation policies across Europe. EU institutions are important driving forces for gathering information, especially in improving monitoring and early warning systems.

It has been argued that the "EU values" on climate change are not internalised by the political actors in Central and Eastern Europe [25]. The analyses in this paper highlight several key impediments for why health systems have not fully adopted these ideas. One is the **style of regulation**. The market liberal state approach characteristic of environmental health governance in Estonia can dissociate responsibilities with the "lean state" rhetoric [18]. The overruling perspective among policymakers is that if climate change occurs at all, then people should be able to cope with the long-term effects by themselves. Such a perspective can be explained by limited economic capacity and the lack of evidence. This dissociation of responsibilities is understandable as the evidence pointing to the need for intervention is scarce, and longer-term planning based on projections of the health risks of climate change is limited. The tendency of turning a "blind eye" towards policy problems by limiting investments in monitoring and inspection also occur in other environmental health issues [18]. We may assume that health systems in other small countries with dispersed human, financial, and political resources may lack the resources necessary for adaptive capacity [22] to deal with what is still often considered as an avant-garde issue of climate change risks.

Even if the capability to cope during acute situations exists, there is no certainty as to the extent to which family physicians and hospitals operating under the private law would be ready for extreme events. Maintaining an informed and effective practitioner workforce is a critical challenge for health systems grappling with extreme events [13], however, no contingency plan has been elaborated for hospitals. Limited preparation and the state approach of minimal interference has important implications for the safety of the most vulnerable population groups, including elderly and minority groups.

Our results highlight the significance of the **regulatory architecture and the related allocation of responsibilities**. Climate change-related health risks are highly cross-sectoral, which means that other state departments are expected to assume primary responsibility. In extreme weather conditions, healthcare systems and rescue services will indeed cope within the limits of their competences, but in case of more complex issues, prevention and timely responses are inhibited by the lack of cooperation between rescue and social services, and healthcare providers. Also, for information-gathering, monitoring of some parameters or vulnerable groups is organised separately, fragmented between institutions. Fragmentation inhibits the integrated assessment of climate change-related health and related risks and vulnerabilities. Furthermore, lack of coordination and integration impedes accessing a range of possible solutions. The process of mainstreaming a policy issue (also called policy integration [73]) through the integration of adaptation policy and measures into ongoing national sectorial planning is expected to increase policy coherence, minimise contradictory policies, and capture the opportunities for synergistic efforts in terms of increased adaptive capacity [74]. However, developing cross-sectorial policy goals becomes more demanding with an increase in the number of affected institutions and interests [75,76]. A solution addressing the institutional and organisational dimension of the problem would be to appoint a responsible institution that could facilitate policy integration in-between existing policy pillars. Attempts at mainstreaming are also more likely when the issue receives political attention [24]. The EU Climate Adaptation Strategy has been an important force to increase the political salience of the topic. However, this has yet to translate into actual better coordination between the institutions.

Our analysis shows the **societal and institutional silencing** of the health risks of climate change. Institutional responses constitute important influences on people's perceptions [77], and the implementation of adaptation measures require some degree of demand from citizens [78,79]. The low level of worries and attention to climate change adaptation measures can be explained through the limited "pool of worries" hypothesis, and the tendencies of attenuation of risk in countries preoccupied with socioeconomic instability. With the increase in income levels, the East of Europe countries are expected to increase their concern with respect to climate change [80,81]. Although there is an increase in levels of concern in Eastern member states, including Estonia, instead of these countries "catching up" with the levels of concern of southern and northern EU countries, the Eurobarometer studies show an increase in the relative difference between the average levels of concern in these groups of countries [68].

Low level of engagement with adaptation has been attributed to a lack of knowledge of climate change and related adaptation issues [82,83]. Although experience with weather-related phenomena may give rise to a perceived need for adaptation action [84–86], Estonia has yet to experience any major weather extremes (except for a major heatwave in 2010 [87]. This may have impeded awareness of the risks amongst the public. The low demand for health adaptation may also be explained by the general population being used to relatively extreme weather over four seasons (a few hot weeks are warmly welcomed). Further, the geographical size of the country may play a role, as the small size of the country decreases the chances of anything happening, leaving an impression that climate-related events that happen in other, even close-by countries, do not affect our safety and wellbeing. It has been argued that extraordinary storms and heatwaves can result in a focus on the current emergency, leaving the long-standing risks unattended [20]. However, the Estonian case indicates that increased experience with attention-grabbing extreme events in the northeastern parts of Europe could bring to the fore the need to consider long-term climate trends, and risks for health and health systems.

The occurrence of extreme weather events, coupled with research projecting that these are likely to increase in frequency and severity, resulting in increasing costs in the future, motivates richer states [23]. In the case of Estonia, an upper-middle income country, the lack of public and political salience means research of climate change-related health effects is less of a priority. The associated lack of evidence, in turn, may inhibit raising public awareness and efforts to establish health adaptation goals and implement policies and programs, including increasing individual and community-based capacities to respond to emergencies. Recent projections indicate substantial increases in summer temperatures by the end of the century over central and Eastern Europe [88], along with projected increases in pollution levels (near-surface ozone and aerosol particles) [89]. Health risks will increase under these projections, unless proactive adaptation is undertaken.

Ethnic minorities (mainly Russian-speaking) and city-dwellers are more worried about climate change, and express higher demand for protective measures. Their concerns can be explained by their social status, and the perceived level of control a person has over one's physical or psychological wellbeing. As shown also in the literature on risk perception among minority groups [90], the demand for measures is high among individuals with weakened positions, and who, in general, may have restricted ability to influence decision-making. This has further implications for managing the vulnerability of minority groups. Their perceived inhibited access to informal and institutional support may be an indication of actual vulnerability, but also a lack of awareness of support networks. As a novel finding, respondents in urban areas expressed higher demand for measures for coping with the health risks of climate change. A possible explanation for the urban dwellers' higher demand for protective measures could be attributed to their dependence on urban expert systems and infrastructures that may make city dwellers feel more vulnerable, with a general sense of higher exposure to extreme weather events. In rural areas, individuals have lower expectations from the state, being situated further away from expert systems; they have maintained skills to cope with current extreme weather events, but may not be prepared for future, more extreme events.

5. Conclusions

The study shows the significance of the EU pressures for aligning the health systems of a small country, like Estonia, with the climate adaptation goals of building resilience to future increases in, e.g., projected storminess and length of heatwaves. However, the effect of these external pressures remains short-lived, when the political salience of climate change and the related political will is low to mainstream climate change into policies, and to invest scarce resources in adaptation policy programmes, monitoring, and protective responses. Although efforts are being made in emergency preparedness and rescue, which, as a co-benefit, increase the ability to cope with the health risks of climate-related extreme events, the state has limited avenues for pressuring the private domains of primary care and hospitals. There are significant opportunities to gain synergistic benefits from conducting risk and vulnerability analyses, and from building community resilience through mainstreaming climate change over related policy fields, including rescue services, health, environment, social care, and even education. In a situation where there is lack of issue ownership, policy integration and mainstreaming could be facilitated by appointing a responsible institution. Institutional responses resonate with people's perception and the demand for adaptation measures. As characteristic of a small health system, the shortage of regionally specific scientific assessments and lack of pressure from other organised interest groups, attenuate the social and political urgency for adaptation. Nevertheless, growing experience with extreme weather events, particularly among increasing urban and minority populations who are detached from traditional coping strategies, may increase demand for the provision of state support for health adaptation.

Author Contributions: K.O. conceived the research questions and design; conducted the analysis and did most of the writing of the manuscript; M.T. conducted expert interviews and participated in their analysis; K.L.E. provided advice on the data analysis and helped in writing the manuscript; H.O. conducted the population-based survey and contributed to writing the manuscript.

Funding: K.O.'s work on the preparation of this article was supported by the HEALTHDOX project (MSHUH14155) financed by NORFACE Programme. H.O.'s work on the preparation of this article was supported by the Estonian Ministry of Education and Research grant IUT34-17.

Conflicts of Interest: The authors declare no conflict of interest.

Appendix A

Interviewed experts:

1. Expert on social work (4 March 2015)
2. Local government official (10 March 2015)
3. Tartu City Government official (17 March 2015)
4. Environmental health scientist (20 April 2015)
5. Official from the ministry of the Interior (29 April 2015)
6. Climate expert (29 April 2015)
7. Environmental health specialist, Ministry of Social Affairs (4 May 2015)
8. Official from the Estonian Environmental Research Centre (4 May 2015)

9. Family physician (member of the managing board, Association of Family Physicians) (7 May 2015)
10. Health Board official (7 May 2015)
11. Public health specialist (8 May 2015)
12. Official from the Ministry of the Environment (21 May 2015)
13. Representative of a hospital (member of managing board) (4 June 2015)
14. City medical officer (12 June 2015)
15. Representative of volunteer rescue service providers (16 June 2015)
16. Representative of Rescue Board (Lääne päästekeskus) (18 June 2015)
17. Official from the Health System Development Department, Ministry of Social Affairs (19 June 2015)
18. Chief specialist from the Environmental Health Research Centre, Health Board (27 May 2015)
19. Official from the Health Board's medical bureau (25 May 2015)
20. Official from Pärnu City Government (26 August 2015)
21. Official from Government Office (27 August 2015)

Appendix B

Interview guide:

- What do you think about the climate change and its health effects?

  ○ Health risks in the world/Estonia
  ○ Future trends

- How would you assess health systems functioning to cope with CC effects on health?

  ○ Sufficiency of regulations, programmes
  ○ Monitoring and information gathering (prognoses, risk analyses)
  ○ Protective responses, prevention

- What could be the key drivers/impediments on (a) tailoring programmes, measures, and (b) their implementation for CC health adaptation, (c) gathering information? Assess the role of ...

  ○ yourself as an expert; public salience, NGOs, private companies, scientific groups, international organisations
  ○ state officials, their competences and resources, regulatory style (and size), architecture (dispersion of responsibilities).

Appendix C

Table A1. Questionnaire items on key variables.

1	Personal Exposure	No Exposure			Extreme Exposure	
	Please rate on a scale of 1–5 your exposure to extreme weather events	1	2	3	4	5
2	**Personal worry**	No exposure			Extreme exposure	
	In general, how worried are you about the health risks posed to you and your family by your residential environment?	1	2	3	4	5
3	**Belief in need for measures**	No exposure			Extreme exposure	
	Measures against climate change related risks are urgently needed	1	2	3	4	5
4	**Belief in institutional efficiency**	No exposure			Extreme exposure	
	I trust that the authorities will take care of the healthfulness of my living environment	1	2	3	4	5

Appendix D

Table A2. Low and high demand for state intervention to mitigate climate change CC health effects per demographic and self-rated health, and group comparisons.

	Low Demand, n (%)	High Demand, n (%)	Chi^2	*p* Value
Gender			3.11	0.08
Male	320 (72.7)	120 (27.3)		
Female	377 (67.6)	301 (32.4)		
Age			2.47	0.29
Age, 20–34	110 (67.9)	52 (32.1)		
Age, 35–54	211 (69.2)	94 (30.8)		
Age, 55–75	109 (62.3)	66 (37.7)		
Education			0.03	0.98
primary	23 (65.7)	12 (34.3)		
secondary	262 (67.2)	128 (32.8)		
higher	143 (66.8)	71 (33.2)		
Home language			13.86	0.00
Estonian	496 (73.6)	178 (26.4)		
Russian or other	201 (62.0)	123 (38.0)		
Place of residence			9.93	0.00
Urban	473 (66.9)	234 (33.1)		
Rural	224 (77)	67 (23)		
Self-rated health status			8.69	0.12
Very good	52 (71.2)	21 (28.8)		
Good	299 (72.7)	112 (27.3)		
Average	277 (67.6)	133 (32.4)		
Bad	59 (69.4)	26 (30.6)		
Very bad	6 (75.0)	2 (25.0)		

References

1. Sellers, S.; Ebi, K.L. Climate change and health under the shared socioeconomic pathway framework. *Int. J. Environ. Res. Public Health* **2017**, *15*, 3. [CrossRef] [PubMed]
2. Watts, N.; Adger, W.N.; Agnolucci, P.; Blackstock, J.; Byass, P.; Cai, W.; Chaytor, S.; Colbourn, T.; Collins, M.; Cooper, A.; et al. Health and climate change: Policy responses to protect public health. *Lancet* **2015**, *386*, 1861–1914. [CrossRef]
3. Smith, K.R.; Woodward, A.; Campbell-Lendrum, D.; Chadee, D.D.; Honda, Y.; Liu, Q.; Olwoch, J.; Revich, B.; Sauerborn, R. Human health: Impacts, adaptation, and co-benefits. In *Climate Change 2014: Impacts, Adaptation, and Vulnerability. Part A: Global and Sectoral Aspects. Contribution of Working Group II to the Fifth Assessment Report of the Intergovernmental Panel on Climate Change*; Field, C.B., Barros, V.R., Dokken, D.J., Mach, K.J., Mastrandrea, M.D., Bilir, T.E., Chatterjee, M., Ebi, K.L., Estrada, Y.O., Genova, R.C., et al., Eds.; Cambridge University Press: Cambridge, UK, 2014; pp. 709–754.
4. Jordan, A.; Huitema, D. Innovations in climate policy: The politics of invention, diffusion, and evaluation. *Environ. Politics* **2014**, *23*, 715–734. [CrossRef]
5. Austin, S.E.; Biesbroek, R.; Berrang-Ford, L.; Ford, J.D.; Parker, S.; Fleury, M.D. Public health adaptation to climate change in oecd countries. *Int. J. Environ. Res. Public Health* **2016**, *13*, 889. [CrossRef] [PubMed]
6. Hess, J.J.; McDowell, J.Z.; Luber, G. Integrating climate change adaptation into public health practice: Using adaptive management to increase adaptive capacity and build resilience. *Environ. Health Perspect.* **2012**, *120*, 171–179. [CrossRef] [PubMed]
7. Bowen, K.J.; Ebi, K.L. Governing the health risks of climate change: Towards multi-sector responses. *Curr. Opin. Environ. Sustain.* **2015**, *12*, 80–85. [CrossRef]

8. Amundsen, H.; Berglund, F.; Westskog, H. Overcoming barriers to climate change adaptation—A question of multilevel governance? *Environ. Plan. C Gov. Policy* **2010**, *28*, 276–289. [CrossRef]

9. Austin, S.E.; Ford, J.D.; Berrang-Ford, L.; Araos, M.; Parker, S.; Fleury, M.D. Public health adaptation to climate change in canadian jurisdictions. *Int. J. Environ. Res. Public Health* **2015**, *12*, 623–651. [CrossRef] [PubMed]

10. Roose, A. Climate changes—How to adapt? *Publ. Inst. Geogr. Univ. Tartu* **2015**, *112*, 6–19.

11. Rothstein, H.; Demeritt, D.; Paul, R.; Beaussier, A.-L.; Wesseling, M.; Howard, M.; de Haan, M.; Borraz, O.; Huber, M.; Bouder, F. Varieties of risk regulation in europe: Coordination, complementarity and occupational safety in capitalist welfare states. *Socio-Econom. Rev.* **2017**, mwx029. [CrossRef]

12. Hood, C.; Rothstein, H.; Baldwin, R. *The Government of Risk: Understanding Risk Regulation Regimes*; Oxford University Press: New York, NY, USA, 2001.

13. World Health Organization. *Operational Framework for Building Climate Resilient Health Systems*; World Health Organization: Geneva, Switzerland, 2015.

14. Kieny, M.P.; Bekedam, H.; Dovlo, D.; Fitzgerald, J.; Habicht, J.; Harrison, G.; Kluge, H.; Lin, V.; Menabde, N.; Mirza, Z.; et al. Strengthening health systems for universal health coverage and sustainable development. *Bull. World Health Organ.* **2017**, *95*, 537–539. [CrossRef] [PubMed]

15. Commission of the European Communities. *White Paper–Adapting to Climate Change: Towards a European Framework for Action*; COM: Brussels, Belgium, 2009.

16. Hildén, M. Evaluation, assessment, and policy innovation: Exploring the links in relation to emissions trading. *Environ. Politics* **2014**, *23*, 839–859. [CrossRef]

17. Hildén, M.; Jordan, A.; Rayner, T. Climate policy innovation: Developing an evaluation perspective. *Environ. Politics* **2014**, *23*, 884–905. [CrossRef]

18. Orru, K.; Rothstein, H. Not 'dead letters', just 'blind eyes': The europeanisation of drinking water risk regulation in estonia and lithuania. *Environ. Plan. A Econ. Space* **2015**, *47*, 356–372. [CrossRef]

19. Vihalemm, T.; Keller, M.; Kiisel, M. *From Intervention to Social Change: A Guide to Reshaping Everyday Practices*; Ashgate: Surrey, UK, 2015.

20. Knaggård, Å. What do policy-makers do with scientific uncertainty? The incremental character of swedish climate change policy-making. *Policy Stud.* **2014**, *35*, 22–39. [CrossRef]

21. Howlett, M. Why are policy innovations rare and so often negative? Blame avoidance and problem denial in climate change policy-making. *Glob. Environ. Chang.* **2014**, *29*, 395–403. [CrossRef]

22. Gupta, J.; Termeer, C.; Klostermann, J.; Meijerink, S.; van den Brink, M.; Jong, P.; Nooteboom, S.; Bergsma, E. The adaptive capacity wheel: A method to assess the inherent characteristics of institutions to enable the adaptive capacity of society. *Environ. Sci. Policy* **2010**, *13*, 459–471. [CrossRef]

23. Massey, D.; Chaboyer, W.; Aitken, L. Nurses' perceptions of accessing a medical emergency team: A qualitative study. *Aust. Crit. Care Off. J. Confed. Aust. Crit. Care Nurses* **2014**, *27*, 133–138. [CrossRef] [PubMed]

24. Rauken, T.; Mydske, P.K.; Winsvold, M. Mainstreaming climate change adaptation at the local level. *Local Environ.* **2015**, *20*, 408–423. [CrossRef]

25. Reckien, D.; Salvia, M.; Heidrich, O.; Church, J.M.; Pietrapertosa, F.; De Gregorio-Hurtado, S.; D'Alonzo, V.; Foley, A.; Simoes, S.G.; Krkoška Lorencová, E.; et al. How are cities planning to respond to climate change? Assessment of local climate plans from 885 cities in the eu-28. *J. Clean. Prod.* **2018**, *191*, 207–219. [CrossRef]

26. Gupta, S.; Harnisch, J.; Barua, D.C.; Chingambo, L.; Frankel, P.; Vazquez, R.J.G.; Gomez-Echeverri, L.; Haites, E.; Huang, Y.; Kopp, R.; et al. Cross-cutting investment and finance issues. In *IPCC Working Group III AR5.*; Cambridge University Press: Cambridge, UK, 2014.

27. Bowen, K.J.; Ebi, K.; Friel, S.; McMichael, A.J. A multi-layered governance framework for incorporating social science insights into adapting to the health impacts of climate change. *Glob. Health Action* **2013**, *6*, 21820. [CrossRef] [PubMed]

28. Tosun, J.; Lang, A. Policy integration: Mapping the different concepts. *Policy Stud.* **2017**, *38*, 553–570. [CrossRef]

29. Ollila, E. Health in all policies: From rhetoric to action. *Scand. J. Public Health* **2011**, *39*, 11–18. [CrossRef] [PubMed]

30. May, P.J.; Jochim, A.E.; Sapotichne, J. Constructing homeland security: An anemic policy regime. *Policy Stud. J.* **2011**, *39*, 285–307. [CrossRef]

31. Spence, A.; Venables, D.; Pidgeon, N.; Poortinga, W.; Demski, C. *Public Perceptions of Climate Change and Energy Futures in Britain*; Technical Report; School of Psychology: Cardiff, UK, 2010.
32. Kingdon, J.W. *Agendas, Alternatives, and Public Policies*; Longman: Boston, MA, USA, 2011.
33. Portier, C.; Thigpen Tart, K.; Carter, S.; Dilworth, C.; Grambsch, A.; Gohlke, J.; Hess, J.; Howard, S.; Luber, G.; Lutz, J.; et al. *A Human Health Perspective on Climate Change: A Report Outlining the Research Needs on the Human Health Effects of Climate Change*; Environmental Health Perspectives/National Institute of Environmental Health Science: Durham, NC, USA, 2010.
34. Wardekker, J.A.; de Jong, A.; van Bree, L.; Turkenburg, W.C.; van der Sluijs, J.P. Health risks of climate change: An assessment of uncertainties and its implications for adaptation policies. *Environ. Health* **2012**, *11*, 67. [CrossRef] [PubMed]
35. Braun, M. Eu climate norms in east-central europe. *JCMS J. Common Mark. Stud.* **2014**, *52*, 445–460. [CrossRef]
36. Biesenbender, S.; Tosun, J. Domestic politics and the diffusion of international policy innovations: How does accommodation happen? *Glob. Environ. Chang.* **2014**, *29*, 424–433. [CrossRef]
37. Skjærseth, J.B.; Wettestad, J. Is eu enlargement bad for environmental policy? Confronting gloomy expectations with evidence. *Int. Environ. Agreem. Politics Law Econ.* **2007**, *7*, 263–280. [CrossRef]
38. Felix, C.; Marcus, H.; James, R.; Kacper, S. Challenging the european climate debate: Can universal climate justice and economics be reconciled with particularistic politics? *Glob. Policy* **2014**, *5*, 6–14.
39. Tompkins, E.L.; Amundsen, H. Perceptions of the effectiveness of the united nations framework convention on climate change in advancing national action on climate change. *Environ. Sci. Policy* **2008**, *11*, 1–13. [CrossRef]
40. World Bank. Rural Population. Indicators. Available online: https://data.worldbank.org/indicator/SP.RUR.TOTL.ZS (accessed on 24 May 2018).
41. Reile, R.; Helakorpi, S.; Klumbiene, J.; Tekkel, M.; Leinsalu, M. The recent economic recession and self-rated health in estonia, lithuania and finland: A comparative cross-sectional study in 2004–2010. *J. Epidemiol. Community Health* **2014**, *68*, 1072–1079. [CrossRef] [PubMed]
42. World Health Organization. Global Health Observatory Data Repository. Life Expectancy. Available online: http://apps.who.int/gho/data/view.main.SDG2016LEXv?lang=en (accessed on 24 May 2018).
43. Ministry of Environment. *Proposal for Compiling Action Plan for Coping with Climate Change Till 2030*; Ministry of Environment: Tallinn, Estonia, 2015.
44. Orru, K.; Hendrikson, R.; Nordlund, A.; Nutt, N.; Veeber, T.; Orru, H. *Keskkonnatervis: Arusaamine Riskidest ja Motivatsioon Tervisemõjude Vähendamiseks. Keskkonnatervise Uuringute Keskus*; Health Board: Tallinn, Estonia, 2015.
45. Orru, H.; Lanki, T.; Forsberg, B.; Saava, A.; Åström, D.O.; Indermitte, E.; Orru, K.; Åström, K.; Rekker, K.; Tillmann, K.; et al. Tervis. In *Kliimamuutuste Mõjude Hindamine ja Kohanemismeetmete Väljatöötamine Planeeringute, Maakasutuse, Inimtervise ja Päästevõimekuse Teemas (kati)*; Roose, A., Ed.; Tartu Ülikool: Tartu, Estonia, 2015; pp. 162–252.
46. Giorgi, F.; Torma, C. Climate Variability and Change Over Europe. In *Connections to Pollen Concentrations*; ICTP: Trieste, Italy, 2015.
47. Hamaoui-Laguel, L.; Vautard, R.; Liu, L.; Solmon, F.; Viovy, N.; Khvorostyanov, D.; Essl, F.; Chuine, I.; Colette, A.; Semenov, M.A.; et al. . Effects of climate change and seed dispersal on airborne ragweed pollen loads in europe. *Nat. Clim. Chang.* **2015**, *5*, 766. [CrossRef]
48. Semenza, J.C.; Suk, J.E.; Estevez, V.; Ebi, K.L.; Lindgren, E. Mapping climate change vulnerabilities to infectious diseases in europe. *Environ. Health Perspect.* **2012**, *120*, 385–392. [CrossRef] [PubMed]
49. Mölter, T.; Schindler, D.; Albrecht, A.; Kohnle, U. Review on the projections of future storminess over the north atlantic european region. *Atmosphere* **2016**, *7*, 60. [CrossRef]
50. Ministry of Social Affairs. *Action Plan of National Health Plan 2009–2020*; Ministry of Social Affairs: Tallinn, Estonia, 2008.
51. Ministry of Environment. *Estonian Environmental Strategy Until 2030*; Ministry of Environment: Tallinn, Estonia, 2007.
52. Ministry of Environment. *Estonian Environmental Action Plan 2007–2013*; Ministry of Environment: Tallinn, Estonia, 2007.
53. Estonian State Weather Service. Sõnastik. Available online: http://www.ilmateenistus.ee/ilmatarkus/sonastik/#hiromb (accessed on 6 June 2018).

54. Estonian State Weather Service. Sunshine. Available online: https://www.ilmateenistus.ee/kliima/climate-maps/sunshine/?lang=en (accessed on 6 June 2018).
55. Estonian State Weather Service. *High-Resolution Operational Model for the Baltic Sea, Hiromb*; Estonian State Weather Service: Tallinn, Estonia, 2016.
56. Air Quality Management System. Seirevõrgustik. Available online: http://airviro.klab.ee/ (accessed on 8 June 2018).
57. Ministry of Interior. Emergency act. In *RT I, 2009, 39, 262*; Ministry of Interior: Tallinn, Estonia, 2009.
58. Ministry of Interior. *Development Plan for Interior Security 2015–2020*; Ministry of Interior: Tallinn, Estonia, 2015.
59. Ministry of Social Affairs. Public health act. In *RT I, 1995, 57, 978*; Ministry of Social Affairs: Tallinn, Estonia, 1995.
60. Health Board. Centre for Environmental Health Studies. Available online: http://www.terviseamet.ee/keskkonnatervis/keskkonnatervise-uuringute-keskus.html (accessed on 6 June 2018).
61. Ministry of Interior. *The List of Emergencies According to Which a Risk Analysis and Solution Plan Will Be Composed and the Identification of the Institutions Responsible for Risk Analysis and Emergency*; Ministry of Interior: Tallinn, Estonia, 2013.
62. Ministry of Interior. *Plan. for Solving Emergencies Due to Storm*; Ministry of Interior: Tallinn, Estonia, 2013.
63. Ministry of Interior. *Plan. for Solving Emergencies Due to Flooding in Densely Populated Areas*; Ministry of Interior: Tallinn, Estonia, 2013.
64. Ministry of Interior. *Risk Analysis for Emergency Situations in 2013*; Ministry of Interior: Tallinn, Estonia, 2013.
65. Ministry of Social Affairs. *Development Plan for the Primary Care in 2009–2015*; Ministry of Social Affairs: Tallinn, Estonia, 2009.
66. Health Board. *Risk Analysis for Emergency Situation Related to Epidemics*; Health Board: Tallinn, Estonia, 2013.
67. Ministry of Interior. Law on changing the law on emergency situations. In *RT I, 2017, 1*; Ministry of Interior: Tallinn, Estonia, 2017.
68. European Commission. *Special Eurobarometer 459. Report. Climate Change*; TNS Opinion & Social: Brussels, Belgium, 2017.
69. Ministry of Environment. *Report on Assessment of the Flood Risks*; Ministry of Environment: Tallinn, Estonia, 2011.
70. Ministry of Environment. *Climate Change Adaption Development Plan Until 2030*; Ministry of Environment: Tallinn, Estonia, 2017.
71. Uustal, T. *Development Plan for Adapting to the Climate Change in 2030*; Proposal; Estonian Environmental Research Centre: Tallinn, Estonia, 2016.
72. Bulkeley, H.; Jordan, A. Transnational environmental governance: New findings and emerging research agendas. *Environ. Plan. C Gov. Policy* **2012**, *30*, 556–570. [CrossRef]
73. Brouwer, S.; Rayner, T.; Huitema, D. Mainstreaming climate policy: The case of climate adaptation and the implementation of eu water policy. *Environ. Plan. C Gov. Policy* **2013**, *31*, 134–153. [CrossRef]
74. Bouwer, L.M.; Aerts, J.C. Financing climate change adaptation. *Disasters* **2006**, *30*, 49–63. [CrossRef] [PubMed]
75. Andrew, J.; Andrea, L. Environmental policy integration: A state of the art review. *Environ.Policy Gov.* **2010**, *20*, 147–158.
76. Peters, B.G. *Pursuing Horizontal Management: The Politics of Public Sector Coordination*; University Press of Kansas: Laurence, KS, USA, 2015.
77. Pidgeon, N.; Kasperson, R.E.; Slovic, P. *The Social Amplification of Risk*; Cambridge University Press: Cambridge, UK, 2003.
78. Dietz, T.; Stern, P.C. *Public Participation in Environmental Assessment and Decision Making*; National Academies Press: Washington, DC, USA, 2008.
79. Shwom, R.; Bidwell, D.; Dan, A.; Dietz, T. Understanding u.S. Public support for domestic climate change policies. *Glob. Environ. Chang.* **2010**, *20*, 472–482. [CrossRef]
80. Shum, R.Y. Effects of economic recession and local weather on climate change attitudes. *Clim. Policy* **2012**, *12*, 38–49. [CrossRef]
81. Scruggs, L.; Benegal, S. Declining public concern about climate change: Can we blame the great recession? *Glob. Environ. Chang.* **2012**, *22*, 505–515. [CrossRef]
82. Milfont, T.L. The interplay between knowledge, perceived efficacy, and concern about global warming and climate change: A one-year longitudinal study. *Risk Anal. Off. Publ. Soc. Risk Anal.* **2012**, *32*, 1003–1020. [CrossRef] [PubMed]

83. Kellstedt, P.M.; Zahran, S.; Vedlitz, A. Personal efficacy, the information environment, and attitudes toward global warming and climate change in the united states. *Risk Anal. Off. Publ. Soc. Risk Anal.* **2008**, *28*, 113–126. [CrossRef] [PubMed]

84. Demski, C.; Capstick, S.; Pidgeon, N.; Sposato, R.G.; Spence, A. Experience of extreme weather affects climate change mitigation and adaptation responses. *Clim. Chang.* **2017**, *140*, 149–164. [CrossRef]

85. Reser, J.P.; Bradley, G.L.; Ellul, M.C. Encountering climate change: 'Seeing' is more than 'believing'. *Wiley Interdiscip. Rev. Clim. Chang.* **2014**, *5*, 521–537. [CrossRef]

86. McDonald, R.I.; Chai, H.Y.; Newell, B.R. Personal experience and the 'psychological distance' of climate change: An integrative review. *J. Environ. Psychol.* **2015**, *44*, 109–118. [CrossRef]

87. Barriopedro, D.; Fischer, E.M.; Luterbacher, J.; Trigo, R.M.; Garcia-Herrera, R. The hot summer of 2010: Redrawing the temperature record map of europe. *Science* **2011**, *332*, 220–224. [CrossRef] [PubMed]

88. Anders, I.; Stagl, J.; Auer, I.; Pavlik, D. Climate change in central and eastern europe. In *Managing Protected Areas in Central and Eastern Europe under Climate Change*; Ranow, S., Neubert, M., Eds.; Springer: Dordrecht, The Netherlands, 2014; pp. 17–30.

89. Orru, H.; Ebi, K.L.; Forsberg, B. The interplay of climate change and air pollution on health. *Curr. Environ. Health Rep.* **2017**, *4*, 504–513. [CrossRef] [PubMed]

90. Macias, T. Environmental risk perception among race and ethnic groups in the united states. *Ethnicities* **2015**, *16*, 111–129. [CrossRef]

 *atmosphere*

Article

Cold Waves in Poznań (Poland) and Thermal Conditions in the City during Selected Cold Waves

Arkadiusz M. Tomczyk *, Marek Półrolniczak and Leszek Kolendowicz

Department of Climatology, Adam Mickiewicz University in Poznań, B. Krygowskiego 10,
61–680 Poznań, Poland; marekpol@amu.edu.pl (M.P.); leszko@amu.edu.pl (L.K.)
* Correspondence: atomczyk@amu.edu.pl

Received: 27 April 2018; Accepted: 25 May 2018; Published: 28 May 2018

Abstract: The objective of the paper was to characterize the occurrence of cold days and cold waves in Poznań in the years 1966/67–2015/16, as well as to characterize thermal conditions in the city during selected cold waves in the years 2008/09–2015/16. The study was based on daily data on maximum and minimum air temperature for station Poznań-Ławica from the years 1966/67–2015/16 and daily air temperature values from eight measurement points located in the territory of the city in different types of land use from the years 2008/08–2015/16. In addition, to characterize thermal conditions during selected days forming cold waves, satellite images were used, on the basis of which the land surface temperature (LST) was calculated. A cold day was defined as a day with daily maximum temperature (Tmax) below the value of 5th annual percentile of Tmax, and a cold wave was defined as at least five consecutive cold days. The study showed an increase in Tmax in winter, which translated to a decrease in the number of cold days over the last 50 years, although the changes were not statistically significant. Thermal conditions in the city showed high variability in the winter season and during the analyzed cold waves.

Keywords: air temperature; cold days; cold waves; climate change; urban area types; Poznań; Poland

---

1. Introduction

The currently observed climate warming is unquestionable and evident among others through an increase in mean global air temperature [1]. The result is a decrease in the number of cold days and cold waves, although the changes are not as intensive as in the case of an increase in the frequency of hot days in the summer period. Current research conducted in Poland also points to a decrease in the number of cold days [2–4], although more evident changes are observed in the case of frequency of occurrence of days with strong frosts [5,6]. Despite the observed changes, during numerous winter seasons, several-day periods with very low air temperature have been recorded in Poland in recent years, for example in January 2016 and 2017 and February 2018 [7–9].

The meteorological conditions of the city and suburban areas are considerably different because of transformation of the environment. Due to a small contribution of natural plant surfaces, numerous vertical surfaces, as well as human activity, substantial amounts of heat are accumulated within the city during the day. They are then released to the atmosphere causing their slower cooling than in the surrounding areas [10]. Therefore, air temperature in the city is usually higher than in suburban areas.

In recent years, teledetection data have been applied in research on thermal conditions in urban areas increasingly frequently. An example of such research can be the analysis of relations between surface temperature and air temperature in the central part of Japan [11]. Voogt and Oke [12] emphasized that the urban heat island observed from thermal remote sensing data is the surface urban heat island (SUHI). Analyses of the urban heat island based on teledetection data have been conducted among others in Brno [13], Kraków [14], Madrid [15], and Poznań [16]. Moreover, satellite teledetection data are increasingly frequently used for the analysis of lake surface temperature [17,18] and dates of the beginning of the vegetative season [19].

The study of the temperature field in urban areas at a resolution of 30 × 30 m allows distinguishing part of the city exposed to very high [20] and low values of air temperature. These thermal conditions are the cause of thermal stress both in the summer and winter seasons. In this study, the obtained maps allow to compare the different types of land cover according to Urban Atlas in terms of their thermal conditions. This paper constitutes a continuation of research on thermal conditions in the city during extreme air temperatures [20]. In the context of forecasts of the UN [21] showing an increase in the number of population living in cities, it is desirable to thoroughly investigate thermal conditions in the city both in the case of high and low temperatures. Taking this into consideration, the objective of the paper is:

1. characteristics of occurrence of cold days and cold waves in Poznań in the years 1966/67–2015/16;
2. characteristics of occurrence of cold days and cold waves in the city in the years 2008/09–2015/16;
3. analysis of thermal conditions in the city during selected days of cold waves based on satellite images.

2. Research Area, Data and Methods

The study area is the city of Poznań. It is the largest city in the Wielkopolskie Voivodship, inhabited by 545.7 thousand people. In terms of the size of population, Poznań is the fifth city in Poland, and eighth in terms of surface area. The city is in three physiographic mesoregions, namely: Poznańskie Lakeland, Wrzesińska Plain, and Poznański Warta River Gorge. The three areas are parts of the Wielkopolskie Lakeland [22,23].

The paper is based on two sets of data concerning air temperature. The first set covered daily maximum (Tmax) and minimum (Tmin) air temperature values for the station in Poznań (Poznań-Ławica) from the multi-annual period 1966–2016 provided by the Institute of Meteorology and Water Management–National Research Institute. The second data set includes daily air temperature values from the years 2008–2016 for eight measurement points located in Poznań in different types of land use (Table 1, Figure 1). The data were obtained from the resources of the Department of Climatology of the University of Adam Mickiewicz in Poznań. At the one of the measurement points, namely Collegium Geographicum, season 2008/2009 was excluded from the analysis due to the lack of a complete sequence of data.

Table 1. Location and characteristics of measurement points in Poznań.

No.	Location	Latitude (N)	Longitude (E)	Distance from the City Center–Piekary (km)/Direction	Land Cover
1.	Piekary	52°24′19.96″	16°55′3960″	0.0/-	Industrial, commercial, public, military, and private units
2.	Collegium Minus	52°24′31.13″	16°54′53.22″	0.9/W	Industrial, commercial, public, military, and private units
3.	Słoneczna	52°23′40.27″	16°52′29.69″	3.9/W	Green urban areas
4.	Rusa	52°23′29.65″	16°59′0.75″	4.0/SE	Discontinuous Dense Urban Fabric
5.	Dębina	52°21′19.88″	16°54′46.14″	5.7/S	Green urban areas
6.	Ławica	52°24′59.46″	16°50′4.71″	6.9/W	Airports
7.	Collegium Geographicum	52°27′46.80″	16°56′28.92″	6.7/N	Industrial, commercial, public, military, and private units
8.	Strzeszyn	52°27′15.14″	16°50′50.79″	7.7/NW	Discontinuous Dense Urban Fabric
9.	Świerczewo	52°22′11.90″	16°53′56.11″	4.5/S	Continuous urban fabric

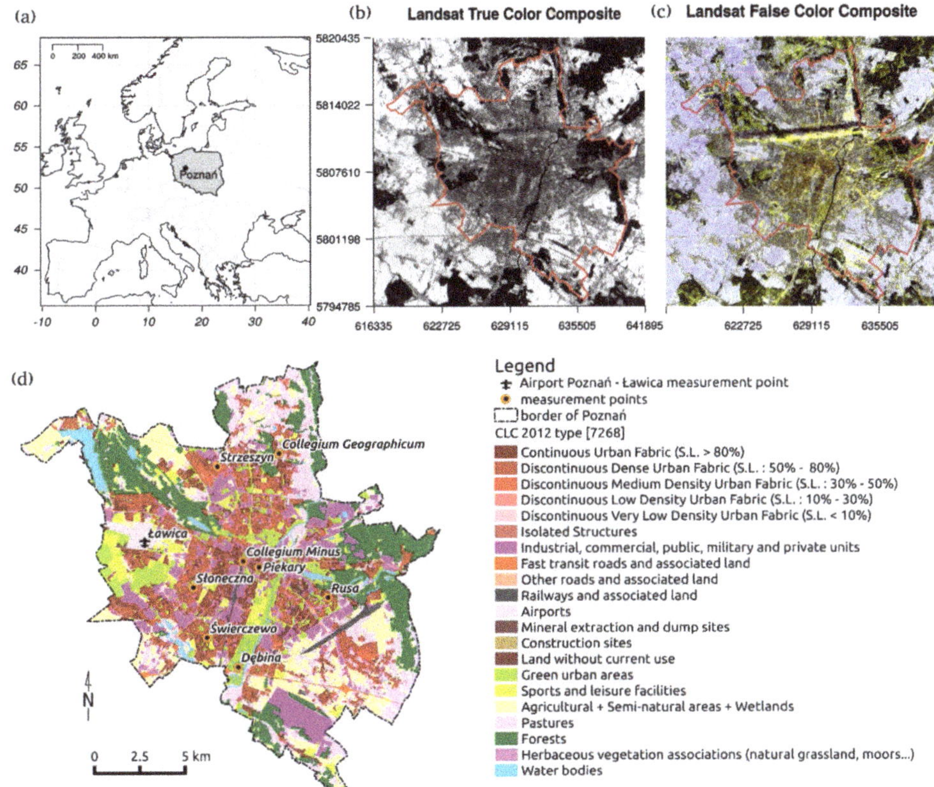

Figure 1. Location of Poznań (**a**) as well as Landsat 5 (thermal) 8-bit greyscale image of the study area (**b**) and image composite of three infrared bands of the ETM+ radiometer RGB = 654 (**c**) (acquisition date: 9 January 2010) and location of measurement points against type of land use (**d**) (Urban Atlas 2012).

The first data set provided the basis for the calculation of mean Tmax for particular winter seasons (December–February). Cold days and cold waves were also identified. Then, for the aforementioned characteristics, the direction of changes and statistical significance of such changes was determined. This involved the application of a non-parametric Mann-Kendall test, detecting a trend in time series. The strength of tendencies of characteristics was determined based on the non-parametric Sen's method [24]. Next, based on threshold values from station Poznań-Ławica, cold days were designated at all measurement points in the years 2008/09–2015/16.

A cold day was defined as a day with maximum daily air temperature below 5th annual percentile of Tmax (from 1966–2015), and a cold wave as a continuous sequence of at least five such days.

For detailed determination of thermal conditions in the city, land surface temperature (LST) was calculated with the application of two satellite images registered during days included in a cold wave. The analysis of satellite images is intended to illustrate the differentiation of thermal conditions of the city at a resolution of 30 × 30 m. The used images were registered by satellite LANDSAT-5 TM on 9 February 2010 and 21 February 2011 at 9:34 UTC. A sample of a thermal image showing the study area in shades of grey is presented in Figure 1b (darker pixels refer to colder objects and brighter ones to warmer objects). Figure 1c is much easier to interpret by the human eye. It can be obtained by mixing three infrared bands in RGB composite (Red = 6 band, Green = 5 band, Blue = 4 band). In this image, the interpretation is easier and more intuitive because warmer objects are in yellow, the coldest in dark grey, blue, and violet. Figure 1c clearly shows the trail of smoke and its shadow covering some of the pixels that were masked before any calculations.

The calculation of basic LST statistics for each type of land use required applying a layer with information on the range of each of the types on the LST raster map for the purpose of reading the LST value for overlapping pixels. Due to this, the Urban Atlas [25] vector layer was rasterized to a raster layer with a size of a pixel corresponding to a surface of 10×10 m in the field. Moreover, due to the presence of a smoke streak and the shadow of the streak coming out of the chimney of the heat and power plant Karolin right outside the eastern boundary of Poznań, some of the pixels were masked before performing the calculations.

A detailed description of the performed calculations and used methods and tools was presented in a paper concerning thermal conditions in the city during selected heat waves [20].

The last stage of work involved the determination of weather conditions based on weather forecast from Daily Meteorological Bulletin of the Institute of Meteorology and Water management—National Research Institute. The presented weather maps were obtained from websites of Met Office.

3. Results

3.1. Tmax in the Winter Season in the Years 1966/67–2015/16 (Poznań-Ławica)

The average Tmax in winter (December-February) in the years 1966/67-2015/16 in Poznań amounted to 2.3 °C. The lowest average Tmax was recorded in the season 1969/70. It amounted to −2.8 °C (Figure 2). Equally cold seasons occurred in the years 1984/85, 1995/96, 1978/79, 1986/87, as well as 2009/10 and 2010/11. The highest average Tmax, amounting to 6.0 °C, was recorded in seasons 1989/90 and 2006/07. Moreover, warm winters with an average considerably above norm occurred in the years 1997/98, 2015/16, 1988/89, 1974/75, and 2007/08. Standard deviation for the given multi-annual period amounted to 2.2 °C. An increase in Tmax at a level of 0.36 °C was recorded in the analyzed years, although the changes were not statistically significant.

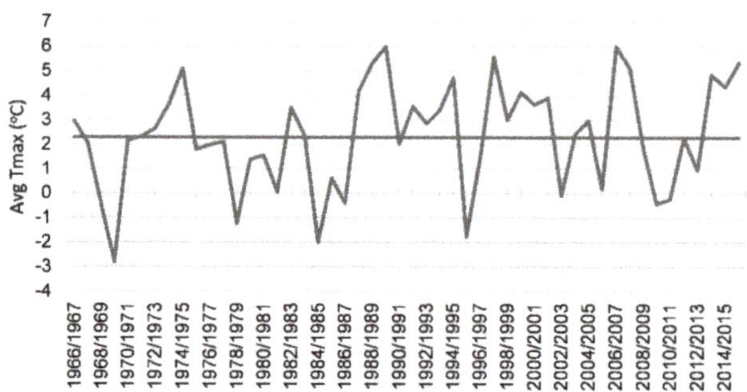

Figure 2. Course of the average Tmax in the winter season with long-term average value in Poznań-Ławica in the years 1966–2015.

3.2. Cold Days and Cold Waves in the Years 1966–2015 (Poznań-Ławica)

Cold days (i.e., days with a maximum temperature below −1.6 °C) in Poznań occurred for 18 days in a year on average. In particular seasons, their number was largely variable, as suggested by the value of standard deviation amounting to 12.6. The number of cold days changes from 1 day (1974/75) to 50 days (1969/70) (Figure 3). A decrease in the number of cold days was recorded in the analyzed period, although the changes were not statistically significant. The analyzed days occurred from November to March, although their highest number was recorded from December to February with a maximum in January (43.9% of all recorded days). The first cold day was observed the earliest on

4 November 1980, and the latest on 23 March 2013. According to the above data, the potential period of occurrence of cold days in Poznań amounted to 140 days.

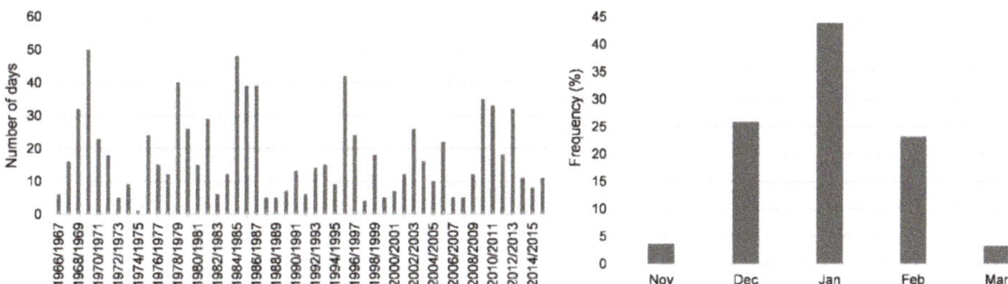

Figure 3. Number of cold days in the period 1966/67–2015/16 and the frequency of occurrence of cold days in months.

A total of 59 cold waves were observed in the analyzed multi-annual period. They lasted for a total of 551 days (Table 2). The lowest number of cold waves was recorded in the years 1996/97–2005/06 (8 cases), and the highest in the years 1976/77–1985/86 (16 cases). Waves of 5 and 7 days were the most frequent, respectively 14 and 12. The longest cold wave occurred in winter 1984/85 and lasted for 27 days from 26 December to 21 January. An equally long cold wave was recorded in the season 1996/97. It lasted for 24 days (from 20 December to 12 January). The average length of cold waves in the analyzed multi-annual period amounted to 9.3 day, and in particular decades it varied from 7.9 (1986/87–1995/96) to 10 days (1996/97–2005/06). Cold waves occurred from November to March and were the most frequent in January (52.5% of all waves). The earliest cold wave in the analyzed multi-annual period was observed in 1993, from 19 to 23 November, and the latest from 1 to 7 March 1987.

Table 2. Characteristics of cold waves in the period 1966/67–2015/16 in Poznań-Ławica.

Years	Number of Cold Waves	Total Duration of Cold Waves (Days)	Average Length (Days)	Average T_{max} (°C)	Average T_{min} (°C)
1966/67–1975/76	12	113	9.4	−5.6	−11.8
1976/77–1985/86	16	158	9.9	−5.0	−11.4
1986/87–1995/96	11	87	7.9	−5.5	−11.9
1996/97–2005/06	8	80	10.0	−5.5	−11.2
2006/07–2015/16	12	113	9.4	−5.7	−11.4
1966/67–2015/16	**59**	**551**	**9.3**	**−5.5**	**−11.5**

The average Tmax during the analyzed cold waves amounted to −5.5 °C, and Tmin −11.5 °C. The lowest average Tmax and Tmin was recorded during the cold wave in 1987 (7–22.01). They amounted to −10.6 °C and −18.3 °C, respectively. The highest average Tmax and Tmin were recorded during the wave in 1978. They equaled −2.9 °C and −5.5 °C, respectively. During the longest cold wave, i.e., a wave lasting from 26 December 1984 to 21 January 1985, the average Tmax amounted to −7.0 °C, and average Tmin −12.8 °C.

3.3. Tmax and Cold Days in the City in the Years 2008/09–2015/16

In the years 2008/09–2015/16 in the winter season, the average Tmax in the city varied from 2.2 °C in Słoneczna to 2.6 °C in Collegium Minus and Piekary. In all measurement points, the coolest season occurred in winter 2009/10, and the average Tmax ranged from −0.5 °C in Collegium Geographicum to 0.0 °C in Collegium Minus (Figure 4). An equally cold winter occurred in 2010/11. The warmest season was observed in winter 2015/16, when the average Tmax varied from 5.0 °C in Słoneczna and Strzeszyn to 5.4 °C in Collegium Minus, Piekary, and Ławica.

The total number of cold days in the city ranged from 141 days in Piekary and 142 days in Collegium Minus to 160 days in Ławica. The highest number of such days was recorded during the coldest seasons, i.e., in winter 2009/10 (Figure 4). Their number varied from 33 in Collegium Minus to 36 days in Rusa and Collegium Geographicum. Cold days occurred equally frequently in winter 2010/11. They were recorded the most seldom in season 2014/15, when their number varied from 7 to 8 days. In the warmest season, from 8 cold days were recorded in Collegium Minus and Piekary to 11 cold days in Ławica.

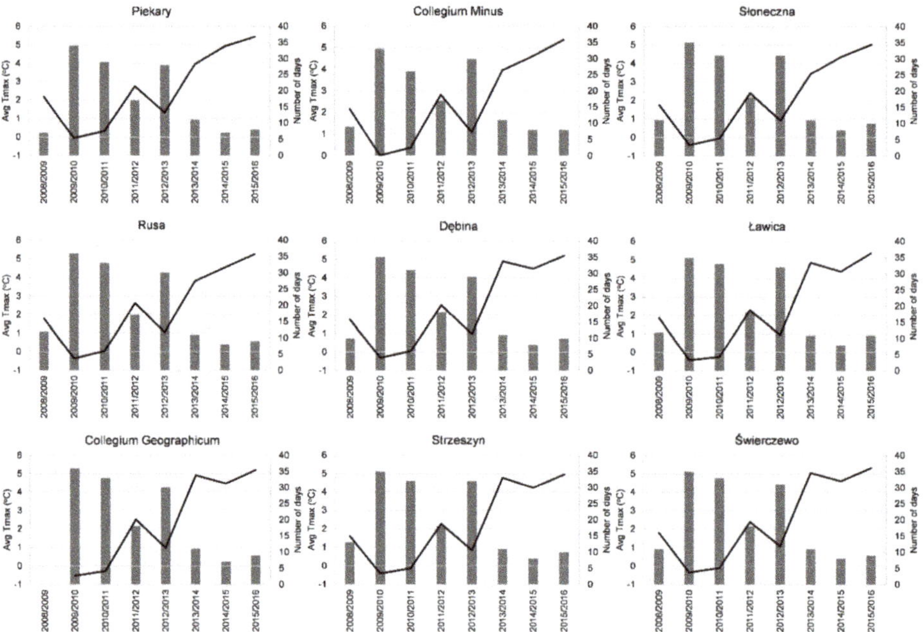

Figure 4. Average Tmax (black solid line) in the winter season and number of cold days (grey bars) in 2008/09–2015/16 in Poznań.

3.4. Cold Waves of 2010 and 2011

3.4.1. Tmax in the Cold Waves

The cold wave in the entire area of the city began on 7 and lasted until 11 February (Figure 5a). The average Tmax during the wave ranged from −4.3 °C in Collegium Geographicum to −3.3 °C in Collegium Minus and −3.5 °C in Piekary. In most of measurement points, the lowest Tmax was recorded on 9 February. It varied from −5.2 °C in Collegium Geographicum and Słoneczna to −3.3 °C in Collegium Minus.

The beginning and ending of the second analyzed cold wave was variable over the area of the city. It began the earliest in Collegium Geographicum and in Rusa and Słonecznej—on 14 February, one day later in Ławica, Strzeszyn, and Świerczew, and on the subsequent day in Dębina (Figure 5b). In Collegium Minus and Piekary, the beginning of the wave was recorded only on 19 February. The discussed cold wave ended on 26 February, and one day earlier in Collegium Minus. The lowest Tmax over the entire area of the city was recorded on 22 February. It varied from −9.4 °C in Ławica to −6.7 °C in Collegium Minus.

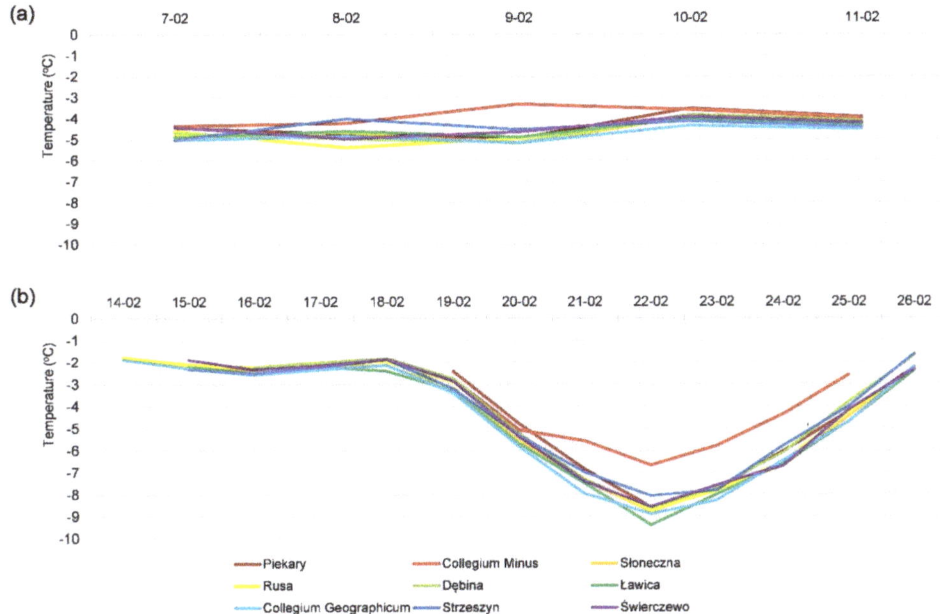

Figure 5. Tmax during the cold wave of 7–11 February 2010 (**a**) and 14–26 February 2011 (**b**).

3.4.2. Daily air temperature on 9 February 2010 and 21 February 2011

On 9 February, the daily course of air temperature throughout the city was approximate (Figure 6a). The lowest air temperature was observed at 6–7:30 UTC, and the highest at 12–14:30 UTC. Minimum temperature in Poznań ranged from $-15.8\,°C$ in Collegium Geographicum to $-11.1\,°C$ in Słoneczna, and maximum from $-5.2\,°C$ in Collegium Geographicum and Słoneczna to $-3.3\,°C$ in Collegium Minus. During the day, differences in air temperature between particular measurement points were considerably lower than in night-morning hours. The smallest differences (approximately $1.1\,°C$) occurred around midnight, and the highest in morning hours with a maximum at 6:30 UTC ($5.0\,°C$). During the analyzed day, air temperature amplitude varied from $5.9\,°C$ in Słoneczna to $11.0\,°C$ in Strzeszyn.

On 21 February 2011, similarly as on 9 February 2010, the daily course of air temperature was approximate in all measurement points (Figure 6b). The lowest temperature was observed at 6:30–7:30 UTC, and the highest at 12–13:30 UTC. Minimum temperature ranged from $-16.7\,°C$ in Ławica to $-14.6\,°C$ in Piekary, and maximum from $-8.2\,°C$ in Collegium Geographicum to $-5.6\,°C$ in Collegium Minus. The smallest differences in air temperature between particular points were observed in morning hours with the minimum at 8:30 UTC ($1.1\,°C$), and the highest in evening hours with a maximum at 19 UTC ($4.3\,°C$). The daily air temperature amplitude in particular points ranged from $7.5\,°C$ in Słoneczna to $9.3\,°C$ in Collegium Minus.

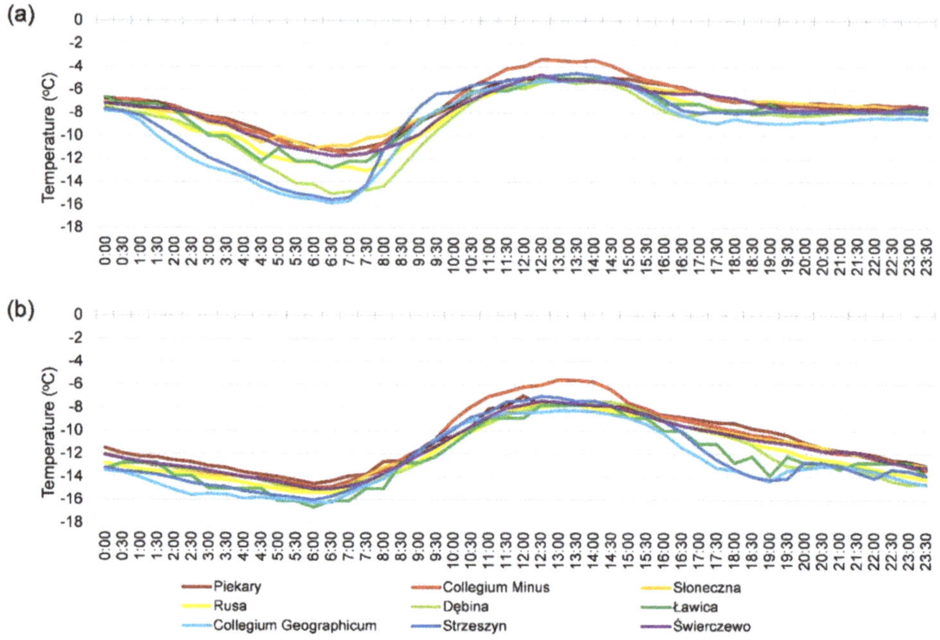

Figure 6. Daily course of air temperature in Poznań on 9 February 2010 (**a**) and 21 February 2011 (**b**).

3.4.3. Land Surface Temperature and Land Use

On 9 February 2010 at 9:34 UTC over the study area, considerably high variability of the value of the LST field occurred because of variable land cover and land use (Figure 7a). The highest values of the LST median (-3.2 °C) are characteristic of continuous urban fabric and discontinuous dense urban fabric (-3.5 °C). For both types of land use, statistically significant medians are different from the medians calculated for all the remaining types, directly suggesting evident distinctiveness of the LST field in the areas. The lowest values of the LST median occur in the following areas: agricultural + semi-natural areas and wetlands as well as pastures (-6.4 °C, -6.6 °C, respectively) (Figure 8a).

21 February 2011 was cooler than the day described above. Similarly, as previously, the highest values of the LST median (excluding water areas) occurred in urbanized areas to the greatest degree, i.e., continuous urban fabric (-5.4 °C) and discontinuous dense urban fabric (-5.5 °C) (Figure 7b). Also, in this case, LST medians were statistically significantly different from medians for the remaining types of land use. The lowest values of the LST median (-7.4 °C) occurred in the following areas: agricultural + semi-natural areas, wetlands, and pastures (Figure 8b). On that day, a higher value of the LST median (-4.3 °C) over water areas in comparison to medians in the remaining types of land use draws attention. The comparison of values of medians for water areas from 21 February 2011 and 9 February 2010 shows a slight difference in LST (0.7 °C), whereas for the remaining types of land use the differences are considerably higher and amount to almost 2.0 °C. The direct cause of such a situation is presumably substantially higher heat capacity of water areas in comparison to land areas and their resulting slower response to changes in air temperature.

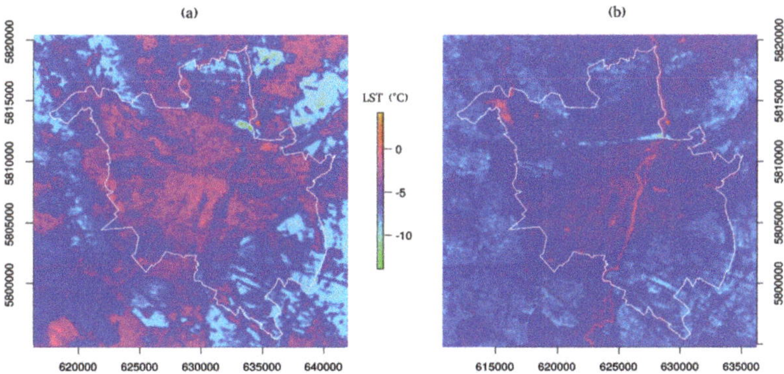

Figure 7. Land surface temperature pattern in Poznań ((**a**)—9 February 2010, (**b**)—21 February 2011).

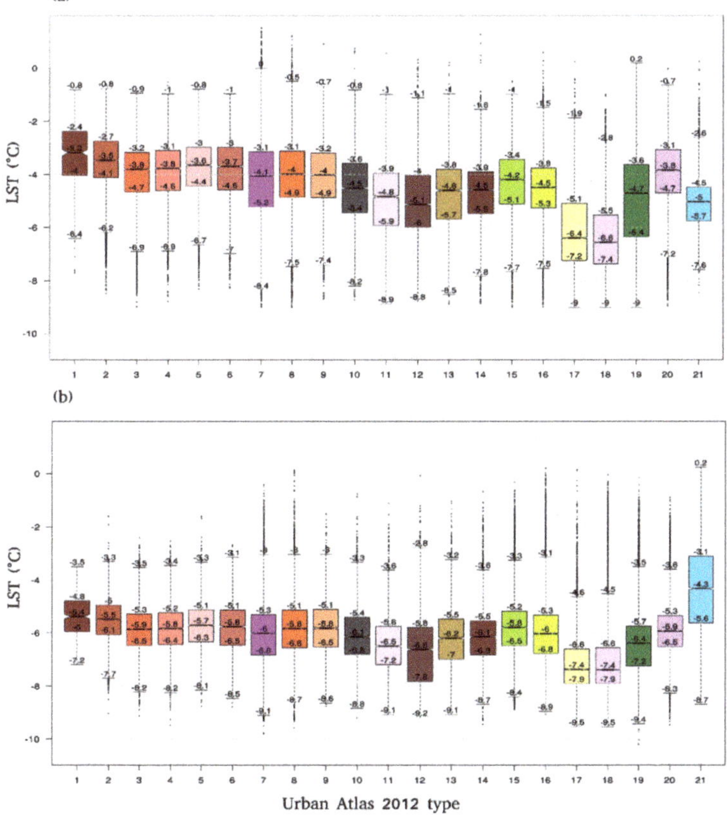

Figure 8. Statistic of land surface temperature in Poznań ((**a**)—9 February 2010, (**b**)—21 February 2011) based on Landsat images according to Urban Atlas 2012 types (colors and order of types according to legend in Figure 1). On the boxplot, the middle values denote medians, the box extends to the Q1 (first quartile) and Q3 (third quartile), and the whiskers show the range (99.3%): the upper whisker shows Q3 + 1.5 × IQR (the interquartile range), the lower shows Q1 − 1.5 × IQR. The notches extend to ±1.58 IQR/sqrt(n) and the dots represent outlier.

3.4.4. Weather situation on 9 February 2010 and 21 February 2011

On 9 February, Poland remained under the influence of a retreating high from over Russia, gradually replaced by a low from over Scandinavia (Figure 9a). Old polar-marine air flew in over Poland. On that day, the weather was cloudy with clear spells and local clearings. Maximum temperature varied from −7.0 °C in the east and center of the country to −1.0 °C on the coast. The wind was weak, from eastern directions. In Poznań, the wind was north and north-east, and its average daily speed was 6.8 km/h.

On 21 February, Poland remained within the range of an extensive high pressure system with the center over Russia, providing for advection of arctic air masses (Figure 9b). Over a major part of the country, the weather was cloudy or slightly cloudy, the cloudiness grew to moderate only locally. Maximum temperature varied from approximately −12.0 °C in north-east to −4.0 °C on the coast. The wind was weak, northern, and north-eastern. In Poznań, the wind was north and north-east, and its average daily speed was 9.4 km/h.

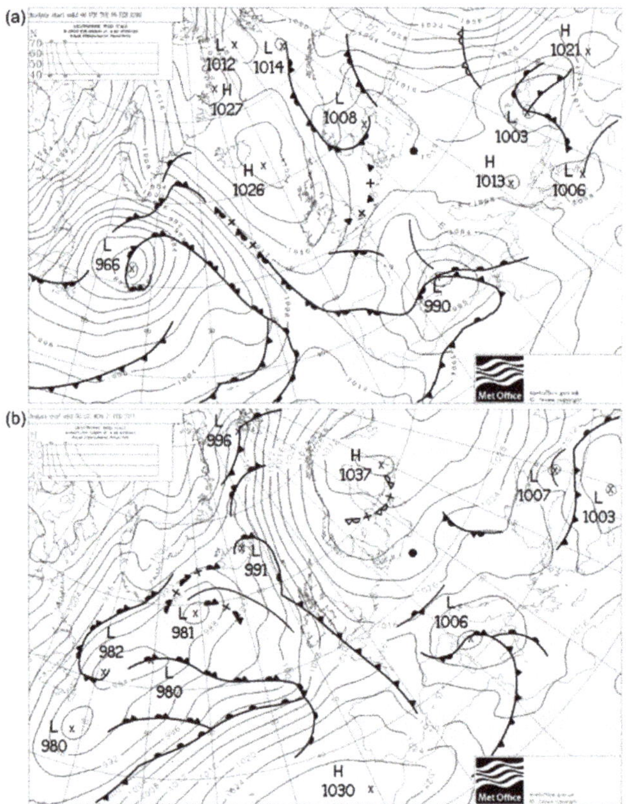

Figure 9. Synoptic maps on 9 February 2010 (**a**) and 21 February 2011 (**b**). Source: Met Office.

4. Discussion and Summary

The study showed an increase in Tmax in the winter season, although the recorded changes were not statistically significant. Earlier studies also determined the lack of statistically significant changes in air temperature in winter in many regions of Poland [26,27], simultaneously pointing to the strongest warming in spring and summer [4,26,27]. The consequence of the observed warming is a decrease in the number of cold days and cold waves. A similar direction of changes in Poland and Central Europe was also pointed out by other authors [4,28–31]. In the studied years, cold waves of 5 and 7 days occurred the most frequently. The longest cold wave lasted for 27 days and occurred in winter 1984/85. The aforementioned wave also occurred in other regions of the country and was similarly included to the longest among others in south-east Poland [32].

The analysis of point measurements in the city showed high differentiation of thermal conditions in winter and during the analyzed cold waves. In particular measurement points, for the analyzed cold days, the differences varied from 1.1 to 5.0 °C. A similar range of differences was also observed during selected hot days, with a maximum of 5.2 °C [20]. The coldest areas include semi-natural urban green areas, sports and recreation areas, as well as forests and fallow arable land. The warmest areas include strongly transformed areas, i.e., areas of continuous urban fabric and discontinuous dense urban fabric and industrial areas.

Research on the differentiation of urban conditions during a cold wave by means of satellite images showed a considerable effect of particular urban structures on LST. On both days, irrespective of the average air temperature for each of them, the effect of the type of land on LST was very similar, as suggested by the structure and proportions of LST for each type of land. Considerably higher and statistically significantly different values of the median as well as values of the lower and upper quartile of LST concern the direct city center, i.e., areas of continuous urban fabric and discontinuous dense urban fabric with density according to the Urban Atlas at a level of 30–80%. The warming effect of the city on LST is also evident in areas with a lower density of urban fabric. In this case, very similar and evidently distinctive values of statistics of LST temperatures occurred on the first of the analyzed days (warmer) in the categories of discontinuous urban fabric (medium, low, and very low). On this day, another group probably affecting LST includes areas of industrial fast transit and other roads with associated land. The remaining types of land use have a considerably lower effect on LST in the city, and the lowest effect concerns types marked as pastures and agricultural, semi-natural areas and wetlands. On the second of the analyzed days (cooler), statistics of LST values show a similar effect of particular types of land use except for lower importance of areas of discontinuous urban fabric (medium, low and very low), as suggested by the equalization of values of LST statistics with values for industrial fast transit and other roads with associated land. On both days, the smallest range of difference of LST medians for water areas draws attention. It suggests a considerably higher inertia of their response to changes in air temperature. An approximate distribution of the coldest and warmest areas in the city was presented by Majkowska et al. [16] analyzing the urban heat island in Poznań, and Walawender [14] analysing the urban heat island in Kraków based on satellite images. According to the expectations and earlier results of urban climate research, areas with a higher density of buildings (a greater number of inhabitants and thus more released anthropogenic heat) are usually warmer than other areas with a lower density of buildings.

Differences in air temperature between particular points are largely determined by weather conditions. During cloudless and windless weather, natural surfaces are cooled faster than artificial ones [10]. This results in higher variability of air temperature in the city.

Author Contributions: A.M.T. has inspired and carried out the research as well as prepared the manuscript. M.P. developed satellite images and complemented the description of the results. L.K. assisted with the results and discussion.

Acknowledgments: This work was supported by the Polish National Science Centre under grant number: UMO–2017/24/C/ST10/00109.

References

1. IPCC. *Climate Change 2013: The Physical Science Basis. Contribution of Working Group I to the Fifth Assessment Report of the Intergovernmental Panel in Climate Change*; Cambridge University Press: Cambridge, UK, 2013.
2. Matuszko, D.; Piotrowicz, K. Long-term variability of meteotropic situations in Kraków. *Przegląd Geograficzny* **2012**, *84*, 413–422.
3. Tomczyk, A.M. Impact of macro-scale circulation types on the occurrence of frosty days in Poland. *Bull. Geogr. Phys. Geogr. Ser.* **2015**, *9*, 55–65.
4. Owczarek, M.; Filipiak, J. Contemporary changes of thermal conditions in Poland, 1951–2015. *Bull. Geogr. Phys. Geogr. Ser.* **2016**, *10*, 31–50.
5. Bielec-Bąkowska, Z.; Łupikasza, E. Frosty, freezing and severe freezing days and their synoptic implications in Małopolska, southern Poland, 1951–2000. *Bull. Geogr. Phys. Geogr. Ser.* **2009**, *1*, 39–62.
6. Bielec-Bąkowska, Z.; Piotrowicz, K. Extreme temperatures in Poland 1951–2006. *Pr. Geograficzne* **2013**, *132*, 59–98.
7. Informacja Tygodniowa, Zagrożenia—Skutki—Ocena; Rządowe Centrum Bezpieczeństwa; 2016. Available online: http://rcb.gov.pl/wp-content/uploads/BIULETYN-ANALITYCZNY-nr-14.pdf (accessed on 15 January 2018).
8. Informacja Tygodniowa, Zagrożenia—Skutki—Ocena; Rządowe Centrum Bezpieczeństwa; 2017. Available online: http://rcb.gov.pl/wp-content/uploads/BIULETYN-ANALITYCZNY-nr-18.pdf (accessed on 15 January 2018).
9. Ostrzeżenia Meteorologiczne. Available online: http://www.pogodynka.pl/ostrzezenia (accessed on 27 Frebuary 2018).
10. Błażejczyk, K.; Kuchcik, M.; Milewski, P.; Dudek, W.; Kręcisz, B.; Błażejczyk, A.; Szmyd, J.; Degórska, B.; Pałczyński, C. *Urban Heat Island in Warsaw—Climatic and Urban Conditionings*; Wyd. Akademickie SEDNO Warszawa: Warsaw, Poland, 2014; p. 177. ISBN 978-83-7963-018-9.
11. Kawashima, S.; Ishida, T.; Minomura, M.; Miwa, T. Relations between surface temperature and air temperature on a local scale during winter nights. *J. Appl. Meteorol. Climatol.* **2000**, *39*, 1570–1579.
12. Voogt, J.A.; Oke, T.R. Thermal remote sensing of urban climates. *Remote Sens. Environ.* **2003**, *86*, 370–384.
13. Dobrovolny, P. The surface urban heat island in the city of Brno (Czech Republic) derived from land surface temperatures and selected reasons for its spatial variability. *Theor. Appl. Climatol.* **2013**, *112*, 89–98.
14. Walawender, J. Application of LANDSAT satellite data and GIS techniques for estimation of thermal conditions in urban area (using an example of Krakow agglomeration). *Pr. Geograficzne* **2009**, *122*, 81–98.
15. Fabrizi, R.; De Santis, A.; Gomez, A. Satellite and ground-based sensors for the urban heat island analysis in the city of Madrid. *Urban Remote Sens. Event* **2011**, 349–352.
16. Majkowska, A.; Kolendowicz, L.; Półrolniczak, M.; Hauke, J.; Czernecki, B. The urban heat island in the city of Poznań as derived from Landsat 5 TM. *Theor. Appl. Climatol.* **2017**, *128*, 769–783. [CrossRef]
17. Politi, E.; Cutler, M.E.J.; Rowan, J.S. Using the NOAA advanced very highresolution radiometer to characterize temporal and spatial trends in water temperature of large European lakes. *Remote Sens. Environ.* **2012**, *126*, 1–11. [CrossRef]
18. Ptak, M.; Choiński, A.; Piekarczyk, J.; Pryłowski, T. Applying Landsat Satellite Thermal Images in the Analysis of Polish Lake Temperatures. *Pol. J. Environ. Stud.* **2017**, *26*, 2159–2165. [CrossRef]
19. Bartoszek, K.; Siłuch, M.; Bednarczyk, P. Characteristics of the onset of the growing season in Poland based on the application of remotely sensed data in the context of weather conditions and land cover types. *Eur. J. Remote Sens.* **2015**, *48*, 327–344. [CrossRef]
20. Półrolniczak, M.; Tomczyk, A.M.; Kolendowicz, L. Thermal Conditions in the City of Poznań (Poland) during Selected Heat Waves. *Atmosphere* **2018**, *9*, 11. [CrossRef]
21. United Nations. World Population Prospects 2017. Available online: https://esa.un.org/unpd/wpp/Publications/Files/WPP2017_KeyFindings.pdf (accessed on 5 November 2017).
22. Główny Urząd Statystyczny. Rocznik Statystyczny Rzeczypospolitej Polskiej 2015. Available online: https://stat.gov.pl/ (accessed on 5 November 2017).
23. Kondracki, J. *Regional Geography of Poland*; Wydawnictwo Naukowe PWN: Warszawa, Poland, 2002; p. 441.

24. Salmi, T.; Maiittii, A.; Anttila, P.; Ruoho-Airola, T.; Amnel, T. *Detecting Trends of Annual Values of Atmospheric Pollutants by the Mann-Kendall Test and Sen's Slope Estimates—The Excel Template Application MAKESENS*; Publications on Air Quality No. 31; Finnish Meteorological Institute: Helsinki, Finland, 2002; pp. 1–35.
25. Urban Atlas 2012. Available online: https://www.eea.europa.eu/data-and-maps/data/urban-atlas (accessed on 5 December 2017).
26. Michalska, B. Tendencies of air temperature changes in Poland. *Prace Stud. Geograficzne* **2011**, *47*, 67–75.
27. Wójcik, R.; Miętus, M. Some features of long-term variability in air temperature in Poland (1951–2010). *Przegląd Geograficzny* **2014**, *86*, 339–364.
28. Wibig, J.; Podstawczyńska, A.; Rzepa, M.; Piotrowski, P. Heatwaves in Poland—Frequency, trends and relationships with atmospheric circulation. *Geogr. Pol.* **2009**, *82*, 33–46. [CrossRef]
29. Lhotka, O.; Kyselý, J. Characterizing joint effects of spatial extent, temperature magnitude and duration of heat waves and cold spells over Central Europe. *Int. J. Climatol.* **2015**, *35*, 1232–1244. [CrossRef]
30. Spinoni, J.; Lakatos, M.; Szentimrey, T.; Bihari, Z.; Szalai, S.; Vogt, J.; Antofie, T. Heat and cold waves trends in the Carpathian Region from 1961 to 2010. *Int. J. Climatol.* **2015**, *35*, 4197–4209. [CrossRef]
31. Migała, K.; Urban, G.; Tomczyński, K. Long-term air temperature variation in the Karkonosze mountains according to atmospheric circulation. *Theor. Appl. Climatol.* **2016**, *125*, 337–351. [CrossRef]
32. Krzyżewska, A.; Wereski, S. Heat waves and frost waves in selected Polish stations against bioclimatic regions background (2000–2010). *Przegląd Geofizyczny* **2014**, *1–2*, 9–109.

Article

Thermal Conditions in the City of Poznań (Poland) during Selected Heat Waves

Marek Półrolniczak, Arkadiusz M. Tomczyk * and Leszek Kolendowicz

Department of Climatology, Adam Mickiewicz University in Poznań, Krygowskiego 10, 61-680 Poznań, Poland; marekpol@amu.edu.pl (M.P.); leszko@amu.edu.pl (L.K.)
* Correspondence: atomczyk@amu.edu.pl; Tel.: +48-061-829-6266

Received: 10 November 2017; Accepted: 3 January 2018; Published: 7 January 2018

Abstract: The aim of the study was to characterise the occurrence of hot days and heat waves in Poznań in the 1966–2015 period, as well as to describe the thermal conditions in the city during selected heat waves between 2008 and 2015. The basis of the study was the daily maximum and minimum air temperature values for Poznań–Ławica station from 1966–2015 and the daily values of air temperature from eight measuring points located in the city in various land types from 2008 to 2015. A hot day was defined as a day with T_{max} above the 95th annual percentile (from 1966 to 2015), while a heat wave was assumed to be at least five consecutive hot days. The research study conducted shows the increase of T_{max}, number of hot days and frequency of heat waves in Poznań over the last 50 years. Across the area of the city (differentiation of urban area types according to Urban Atlas 2012), there was a great diversity of thermal conditions during the heat waves analysed.

Keywords: air temperature; hot days; heat waves; climate change; city; urban area types; Poznań; Poland

---

1. Introduction

Today's climate warming is unmistakable and visible, among other manifestations, in the increase in global average air temperature [1]. The consequence of the aforementioned changes is the increasingly frequent occurrence of hot days and heat waves [2–6]. The waves of 2003, 2006, 2010 and 2015 are to be highlighted amongst the recent extreme heat waves of Poland and Europe. In the future, an increase in the frequency of heat waves and their duration is expected, which will be a consequence of the increase in air temperature [7,8]. It is estimated that in the Poland of 2071–2100, the increase in air temperature in the summer may reach 2.9 °C for the average maximum air temperature and 3.4 °C for the average minimum air temperature [9].

Urban areas are particularly vulnerable to prolonged and intense heat waves [10], which is a consequence of the transformation of the environment. The small proportion of natural vegetation, numerous vertical surfaces and human activity causes significant heat in the city during the day, which is then released into the atmosphere, causing it to cool down more slowly than the surrounding areas [11]. The above factors make bioclimatic conditions more burdensome in the city than in non-urban areas. As Gabriel and Endlicher's [12] studies in Brandenburg and Berlin show, the effects of heat are more pronounced in highly urbanised area. Numerous studies have shown an increase in the number of deaths during heat waves in large cities, including in Warsaw [13], Munich [14] and Paris [15].

In urban climate studies since the end of the 20th century, more and more investigations have been based on remote sensing data. For example, Landsat 5 TM was used by Kawashima et al. [16] to analyse relations between surface temperature and air temperature in the central part of Japan. Voogt and Oke [17] emphasised that the urban heat island observed from thermal remote sensing data is the surface urban heat island (SUHI).

An attempt at air temperature retrieval from Moderate Resolution Imaging Spectroradiometer (MODIS) data was made by Sun et al. [18] for the North China Plane. Taking into consideration land cover categories, vegetation cover, and building density Dobrovolny [19] studied the SUHI in Brno (Czech Republic). Some indicators obtained from remote sensing data were used by Schwarz et al. [20] to analyse differences between urban and agricultural areas. Ptak et al. [21] applied Lansat thermal images in an analysis of lake surface temperature in Poland. In the present study the temperature field on the area of Poznań city was characterised by using Landsat 5 and 8 TM remote sensing data.

Due to the great risk to human health and life associated with the occurrence of heat waves, particularly in highly urbanised areas, the research results presented may be useful for a wide and diverse group of readers. The implementation of this research study is well founded, especially in the face of UN projections [22] of the increase in the number of people living in cities. With this information in mind, the purpose of the article was to:

- define multiannual changes in the occurrence of hot days and heat waves in Poznań in the years 1966–2015;
- characterise thermal conditions in the city during selected heat waves in the years 2008–2015.

2. Research Area, Data and Methods

Poznań is located in western Poland in the area of the Wielkopolska Lakeland. The city's area is 262 km^2 and the population is 545,700 people. Over 58% of the city area is above 80 m a.s.l.; 7% is in the floodplain of the Warta River Valley, and the rest (35%) in upper river terrains. Within the limits of Poznań, there are natural and artificial lakes covering a total of 1.9% of the city area. Developed and urbanised land constitutes 44.6% of the city's area (of which about 28.6% is residential areas, 30.3% of communication and 9.1% of industrial areas), while forested and wooded land constitutes 15.3% [23].

Poznań is located in a temperate transition zone between the sea and the continental climate. The average annual air temperature is 8.3 °C. On average, the coldest month is January with an average temperature of −1.6 °C, and the warmest is July with an average of 18.1 °C. The average annual rainfall is 517 mm, with the lowest observed in February (26 mm) and the highest in July (75 mm). Over the year, winds from the western sector (SW, W, NW) dominate the study area and about 5% of the day is calm. The average wind speed is 4.1 m/s and it ranges from 3.5 m/s in August to 4.8 m/s in March [24].

This article has used daily values of the maximum (T_{max}) and minimum (T_{min}) air temperature for the station in Poznań (Poznań–Ławica) from 1966 to 2015. The data were obtained from the records of the Institute of Meteorology and Water Management–National Research Institute (IMGW). In addition, daily air temperature data for the years 2008–2015 were used from eight measuring points located in the Poznań area on various types of land (Table 1, Figure 1). The analysis excluded the year 2012 due to the lack of complete data for all the measuring points. Data were obtained from the Department of Climatology of Adam Mickiewicz University in Poznań. The air temperature is measured 2 m above ground level with HOBO U23-001 recorders with 30-min resolution and 0.2 °C accuracy.

Table 1. Location and characteristics of measuring points in Poznań.

	Location	Latitude [N]	Longitude [E]	Distance from the City Center–Piekary (km)/Direction	Land Cover
1.	Piekary	52°24′19.96″	16°55′39.60″	0.0/-	Industrial, commercial, public, military and private units
2.	Collegium Minus	52°24′31.13″	16°54′53.22″	0.9/W	Industrial, commercial, public, military and private units
3.	Słoneczna	52°23′40.27″	16°52′29.69″	3.9/W	Green urban areas
4.	Rusa	52°23′29.65″	16°59′0.75″	4.0/SE	Discontinuous Dense Urban Fabric
5.	Dębina	52°21′19.88″	16°54′46.14″	5.7/S	Green urban areas

Table 1. *Cont.*

	Location	Latitude [N]	Longitude [E]	Distance from the City Center–Piekary (km)/Direction	Land Cover
6.	Ławica	52°24′59.46″	16°50′4.71″	6.9/W	Airports
7.	Collegium Geographicum	52°27′46.80″	16°56′28.92″	6.7/N	Industrial, commercial, public, military and private units
8.	Strzeszyn	52°27′15.14″	16°50′50.79″	7.7/NW	Discontinuous Dense Urban Fabric
9.	Świerczewo	52°22′11.90″	16°53′56.11″	4.5/S	Continuous urban fabric

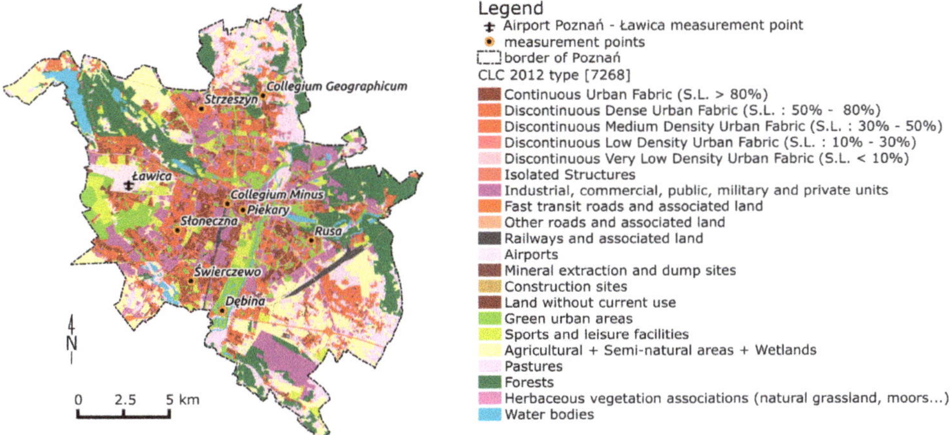

Figure 1. Location of measuring points against the type of land use (Urban Atlas 2012).

Based on the data from the Poznań–Ławica station, the average T_{max} for each summer season (June–August) was calculated and the hot days and heat waves were differentiated. A hot day was defined as a day with T_{max} above the 95th annual percentile (from 1966–2015), and a heat wave was assumed to be at least five consecutive hot days. The aforementioned assumption was based on the definition of an extreme weather event included in IPCC reports [1], according to which a weather phenomenon is so rare within the particular area that it lies within the range of 10th or 90th percentile of an observed probability density function or rarer. It is hence defined as an extreme weather event. The value of the 95th percentile of the air temperature was often used in previous studies on the occurrence of heat waves [25–28].

The next step was to examine changes in the above characteristics in the years 1966–2015 and determine their statistical significance. To this end, the non-parametric Mann–Kendall test was used to detect the trend in the time series. The strength of trends in temperature characteristics and the number of days in the multiannual period were determined by Sen's non-parametric method. Sen's method adopts a linear trend model, f (t) = Qt + B, where Q is an estimator of the linear regression coefficient (trend strength); B is a free term. The calculations were done using the MAKSESENS 1.0 application (freeware) developed by researchers from the Finnish Meteorological Institute [29].

Moreover, with the threshold values from Poznań–Ławica station, hot days and heat waves were distinguished at every measuring point in the city.

Thus, to characterise the thermal conditions in the city accurately, two satellite images were used, i.e., the image recorded by the LANDSAT-5 TM satellite on 12 July 2010 at 9:34 UTC and LANDSAT-8 OLI/TRIS on 11 August 2015 at 9:34 UTC. Both days were characterised by the heat wave occurrence. In addition, on 12 July 2010, there was clear and calm weather with a clearly marked thermal diversification of the urban surface; 11 August 2015 was thus characterised by the passage of warm weather front and less spatial thermal diversification of the city.

The image processing involved the data processing procedure described in detail by Sobrino and Raissouni [30], Jiménez-Muñoz and Sobrino [31], Walawender [32] and Majkowska et al. [33], including separate patterns for LANDSAT-5 TM and LANDSAT-8 OLI/TRIS images. The aforementioned procedure consisted of the following steps:

1. conversion of the values measured by the satellite sensor (in the thermal channel) for each pixel to the energy radiation value;
2. conversion of radiation values to radiation temperature using Planck's law;
3. calculation of the land surface temperature under the Stefan–Boltzmann law, taking into account the differentiation of the emissivity of different surfaces;
4. conversion of the land surface temperature (LST) to the air temperature at 2 m above the surface layer (Tasl) using a linear regression model. This model is used to estimate the air temperature (Tasl) value when the LST value is known:

$$Tasl = a \times LST + b.$$

a, b—linear regression coefficient.

Maps of the air temperature distribution were then constructed for the area analysed for the days and the air temperature course analysed along the designated profiles on 12 July 2010. This day was chosen because of the large variation in air temperature in the city area and the apparent impact of land development on the thermal conditions. Daily sea level pressure (SLP), heights of isobaric surface 500 hPa (z500 hPa) and temperatures on isobaric surface 850 hPa (T850) were used to characterise the weather conditions. The data were obtained from the collection of the National Centre for Environmental Prediction/National Centre for Atmospheric Research (NCEP/NCAR) Reanalysis [34], which are available at the Climate Research Unit resources. Based on the aforementioned data, the maps of SLP, z500 hPa and SLP, z500 hPa and T850 anomalies maps were plotted for the analysed days. In addition, weather maps and comments from the Daily Meteorological Bulletin from IMGW resources were used.

What is more, the Urban Atlas 2012 [35] database for Poland was used. For each type of land use, the average values of LST and their median, first and third quartiles as well as outliers and ranges were calculated. The aforementioned data was used to evaluate the impact of land use on the thermal conditions in the city during the hot days analysed. The following were used in the research: ESRI ArcGiS Desktop 10.5 software with Spatial Analyst, Quantum GIS 2.8 and R software (R Core Team, Vienna, Austria, 2015) and its packages: "raster" [36] and "rgdal" [37] dedicated to the spatial analysis.

3. Results

3.1. T_{max} in the Summer Season in the Years 1966–2015 (Poznań–Ławica)

The average summer season (June–August) T_{max} in the years 1966–2015 was 23.6 °C and ranged from 20.8 °C (1980) to 26.8 °C (1992) (Figure 2). The standard deviation for the given multiannual period was 1.4 °C. A statistically significant ($p < 0.05$) increase in T_{max} was observed in the period analysed, which was 0.35 °C per 10 years. The aforementioned increase was significantly influenced by T_{max} changes at the beginning of the 21st century, when its value in almost all summer seasons was higher than the average for the 1966–2015 period (Figure 2). In the 1966–2000 period, during 19 of the 35 summer seasons, the T_{max} mean was lower than the mean from the multiannual period.

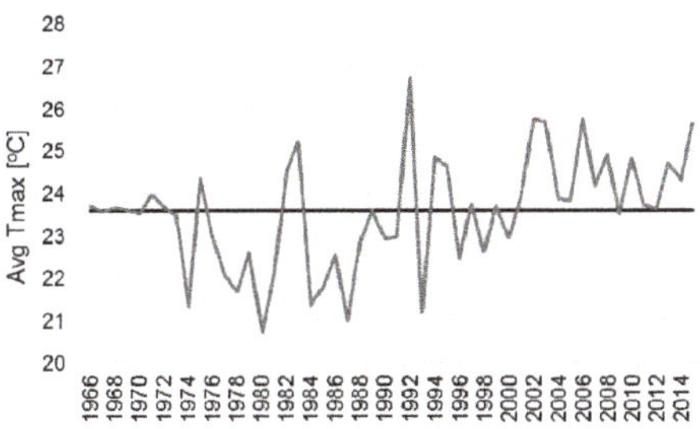

Figure 2. Average T_{max} in the summer season in Poznań–Ławica in the years 1966–2015.

3.2. Hot Days and Heat Waves in the Years 1966–2015 (Poznań–Ławica)

Hot days (i.e., days above the 95th annual percentile—>27.8 °C) in Poznań occurred on average 18 times a year. In individual seasons, the number varied from two days (1980) to as many as 37 days (2006), and the standard deviation was 9.1 days (Figure 3). In the period analysed, there was a statistically significant ($p < 0.05$) increase in the number of hot days (2.9 days/10 years). Days with $T_{max} > 27.8$ °C were observed from April to October, although the largest number were reported in July and August, representing 39.8% and 31.3% of all hot days, respectively. In total, there were seven days in April and only one day in October.

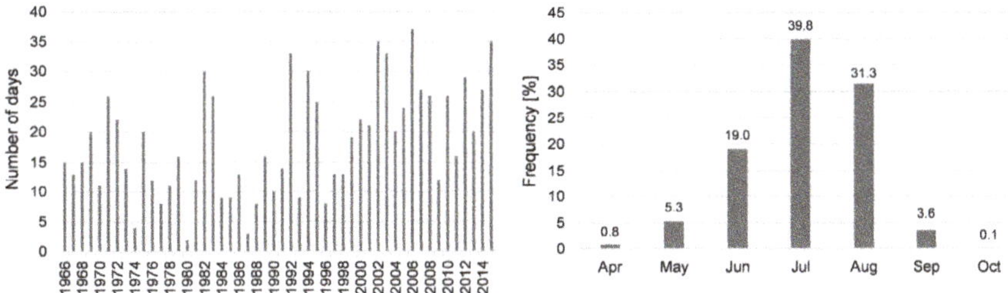

Figure 3. Number of hot days in 1966–2015 and the frequency of hot days' occurrence in certain months.

In the analysed multiannual period, there were 42 heat waves in total, which lasted 310 days (Table 2). Five- and six-day waves were the most common and accounted for, respectively, 34.9% and 20.9% of all heat waves. The longest heat wave was recorded in 1994, which lasted 18 days (from 21 July to 7 August). Long heat waves were observed in 2015 (16 days, from 2 to 17 August), 2006 (14 days, from 18 to 31 July), 1969 and 1971 (12 days—27 July to 3 August and 24 July to 4 August, respectively). The average heat wave length in the multiannual period analysed was 7.4 days, while in the individual years it ranged from six (1996–2005) to 8.8 days (2006–2015). Heat waves occurred between May and August, although most of them (as much as 56%) occurred in July. The earliest recorded heat wave in 1979 was from 31 May to 7 June, while the latest one in 1997 occurred from 22 to 27 August.

Table 2. Characteristics of heat waves in 1966–2015 in Poznań–Ławica.

Years	Number of Heat Waves	Total Duration of Heat Waves (Days)	Average Length (Days)	Average T_{max} (°C)	Average T_{min} (°C)
1966–1975	6	51	8.5	30.7	16.8
1976–1985	9	57	6.3	30.1	16.2
1986–1995	9	63	7.0	30.6	15.0
1996–2005	7	42	6.0	29.9	14.5
2006–2015	11	97	8.8	31.2	17.7
1966–2015	**42**	**310**	**7.4**	**30.5**	**16.2**

The mean T_{max} during the heat waves analysed was 30.5 °C, while T_{min} was 16.2 °C (Table 2). The highest mean T_{max} was found during the heat wave of 1992 (6–11 August), which was 33.9 °C, while the lowest average T_{max} value, which was 28.9 °C, was seen in 1992 (24–28 July), 1997 (22–27 August) and 2001 (26–30 July). During the longest heat wave, i.e., that lasting from 21 July to 7 August, 1994, the average T_{max} was 32.1 °C, while the average T_{min} was 16.1 °C.

3.3. T_{max}, Hot Days and Heat Waves in the City Area in 2008–2015

In 2008–2015, the average T_{max} in the city area was 23.8 °C and changed from 22.3 °C in Collegium Geographicum to 25.1 °C in Piekary. At all points, the coldest season was recorded in 2009. The average T_{max} then changed from 21.6 °C in the Collegium Geographicum to 24.3 °C in Piekary (Figure 4). Also, at the two points (Collegium Geographicum and Słoneczna), the same average T_{max} as in the summer of 2009 was recorded in the 2014 season. On the other hand, at most points, the highest mean T_{max} in the summer occurred in 2015 and ranged from 21.9 °C at the Collegium Geographicum up to 25.8 °C in Strzeszyn and Piekary.

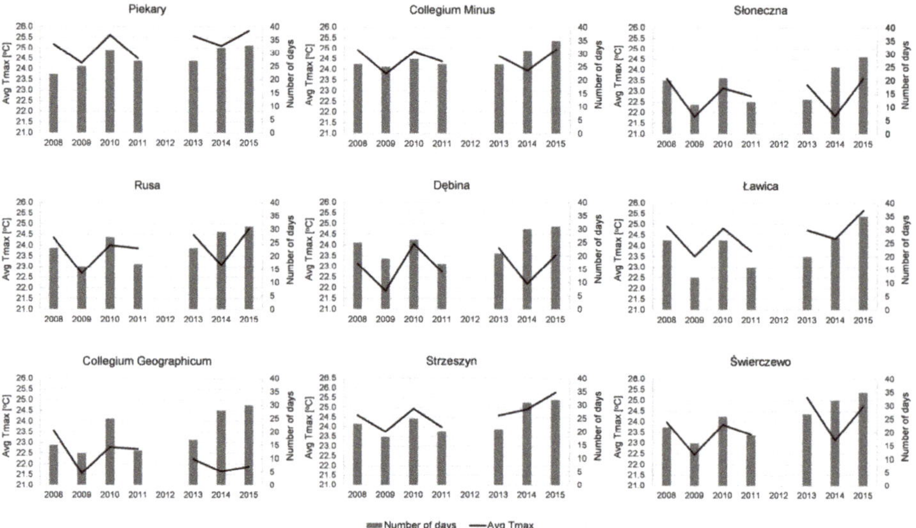

Figure 4. Average T_{max} in the summer season and the number of hot days in 2008–2015 in the city area.

In the studied years, the total number of hot days in the city ranged from 131 days at Słoneczna to 197 days in Collegium Minus and Piekary. Apart from two points (Dębina and Piekary), the least hot days were recorded in the coldest season (summer 2009) and their number ranged from 11 days at Słoneczna to 25 days in Collegium Minus and Piekary (Figure 4). In turn, the hottest days occurred in 2015, and their number changed from 29 days at Słoneczna to 35 days in Collegium Minus, Dębina and Ławica.

A total of 10 heat waves were recorded in the city, of which five heat waves occurred at every measuring point. These were the waves from 2010 (8–17 July), 2014 (4–9 July, 15–23 July, 26–30 July and 2015 (2–18 August). The start and end dates of these waves (except for the wave from 2015) were similar throughout the entire city. In turn, three heat waves occurred in the city area, but they were not recorded in the Poznań–Ławica station. These were the waves of 2008 (30 May–3 June), 2009 (29 June–4 July) and 2011 (3–8 June). Across the city, the longest heat wave was recorded in August 2015. The length of this wave ranged from 10 days (6–15 August) at Słoneczna to 17 days (2–18 August) in Dębina and Collegium Minus.

3.4. Heat Waves of 2010 and 2015

3.4.1. T_{max} in the Heat Waves

The heatwave lasted from 8–17 July 2010. On 8 July, it started in Collegium Minus and Piekary, while for the rest of the area it was on 9 July. Mean T_{max} during the wave ranged from 32.5 °C in Collegium Geographicum and Ławica to 33.7 °C in Dębina. At most points, the highest T_{max} was found on 12 July and fluctuated from 33.1 °C at Słoneczna to 35.4 °C in Collegium Minus (Figure 5a). During the heat wave there was a marked cooling on 14 July, caused by the passage of a warm weather front. Then T_{max} changed from 30.1 °C at Słoneczna to 32.9 °C in Collegium Minus.

In 2015, the start and end of the heatwave varied across the city. At six points, it began on 2 August, while at three points it only started on 6 August. The end of the heat wave was recorded between 15 and 18 August. The mean T_{max} during the wave ranged from 31.9 °C in Ławica to 32.8 °C in Dębina. At all points the highest T_{max} was recorded on August 8th, which ranged from 36.8 °C at Słoneczna to 38.3 °C in Świerczewo (Figure 5b).

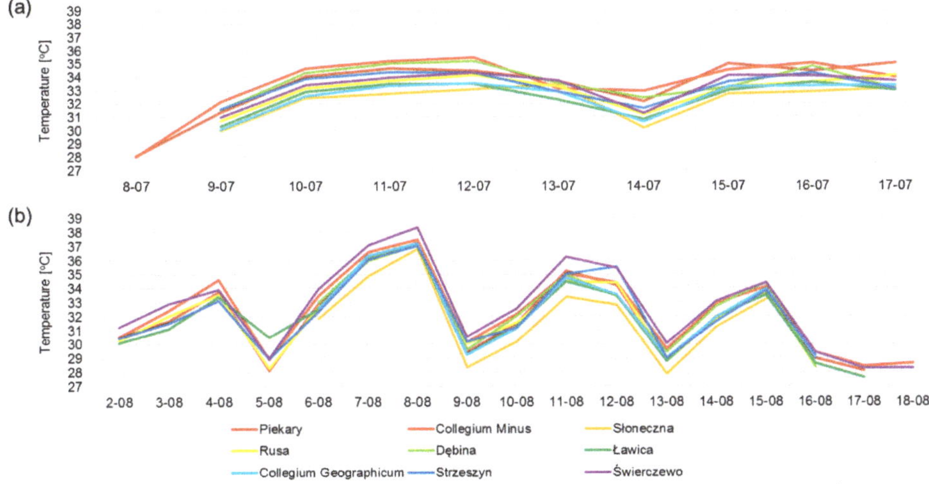

Figure 5. T_{max} during the heat wave of 8–17 July 2010 (**a**) and 2–18 August 2015 (**b**).

3.4.2. Daily Air Temperature on 12 July 2010 and 15 August 2015

On 12 July 2010, the course of daily air temperature was found to be similar at all measuring points. The lowest air temperature was observed at 2–3:30 UTC, while the highest temperature was found at 13–15 UTC (Figure 6a). During the day, the air temperature differences between the individual measuring points were lower than at night. In the hours before noon, the range of air temperature fluctuations was less than 2 °C (minimum 1.7 °C at 10:30 UTC), while in the night and early morning they exceeded 5 °C (maximum 5.2 °C at 1:00 UTC). On the aforementioned day the lowest daily amplitude of the air temperature was found at Słoneczna (10.6 °C), while the highest was at Dębina (16.8 °C).

On 11 August 2015, the course of the daily air temperature was similar in all measuring points. The lowest air temperature was observed at 4–4:30 UTC, while the highest temperature was found at 13–15:30 UTC (Figure 6b). During the day, the air temperature differences between the individual measuring points were significantly lower than at night. The smallest differences (about 1.5 °C) occurred in the morning (6:30 UTC) and the afternoon (16:00 UTC), while the largest were in the evening and night hours with a maximum at 21:00 UTC (4.8 °C). On the day analysed, the lowest daily amplitude of air temperature was observed at Słoneczna (12.5 °C), while the largest were in Dębina and Collegium Geographicum (16.3 °C and 16.2 °C, respectively).

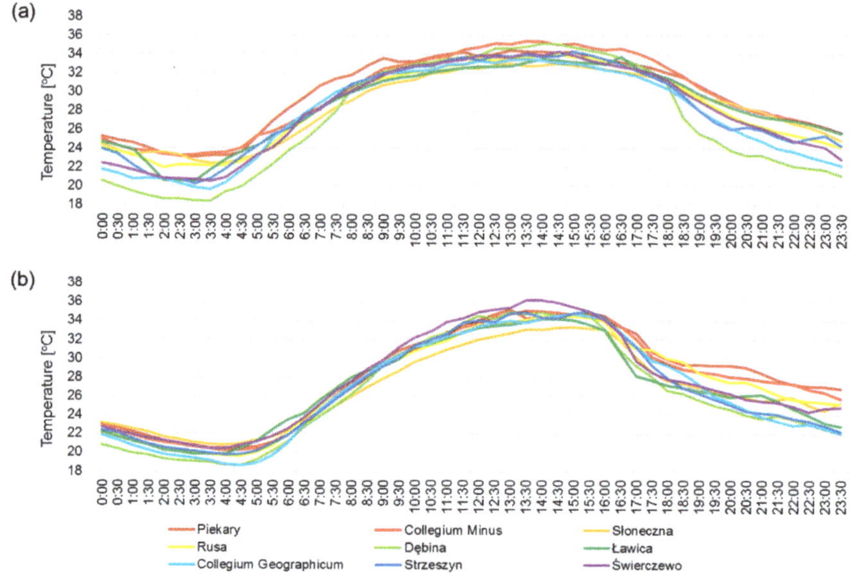

Figure 6. Daily course of air temperature in Poznań on 12 July 2010 (**a**) and 11 August 2015 (**b**).

3.4.3. Air Temperature and Land Use

To obtain the air temperature on 2 m high above land surface (Tasl) conversion of the land surface temperature (LST) using a linear regression model for two, taking into consideration the days used. The regression equations for 12 July 2010 (1) and 11 August 2015 (2) are:

$$\text{Tasl} = 0.20234 \times \text{LST} + 24.22906 \tag{1}$$

$$\text{Tasl} = 0.3278714 \times \text{LST} + 11.68150. \tag{2}$$

The coefficient of determination (R^2) was 0.65 (for probability level $p = 0.008467$) and 0.68 (for probability level $p = 0.01112$), respectively, for the first and second day. It should be emphasised that the regression models used are only an attempt to quantify the dependencies (Tasl and LST) according to the limited possibilities of their verification, due to difficulties in obtaining the subsequent observations (a limited number of satellite images).

On 12 July 2010 at 9:34 UTC there was a significant diversification of the thermal conditions in the surveyed area (Figure 7a). The spatial distribution of air temperature clearly showed the impact of land development and use. The lowest mean air temperature was 29.8 °C over water bodies, and the highest 32.6 °C, within the industrial and commercial, public, military and private unit areas (Figure 8a). Among the warmest areas in the city were those with continuous urban fabric, industrial and commercial, public, military and private unit areas, fast transit roads and associated land, other roads and associated land, while among the coldest areas were water bodies, herbaceous vegetation associations and forests (Figure 8a).

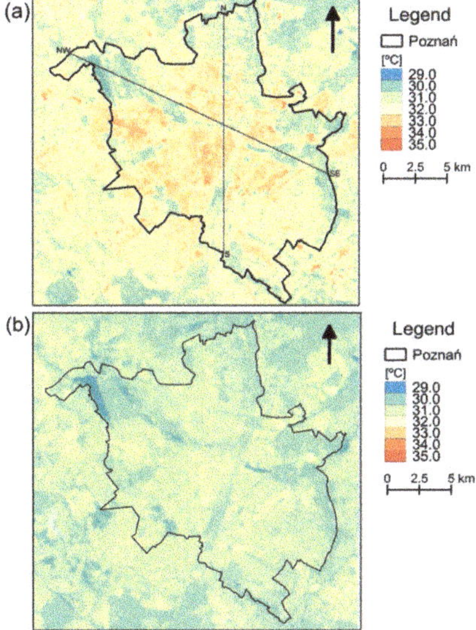

Figure 7. Air temperature in Poznań on 12 July 2010 (**a**) and 11 August 2015 (**b**) at 9:34. Profile lines (**a**) used in Figure 9.

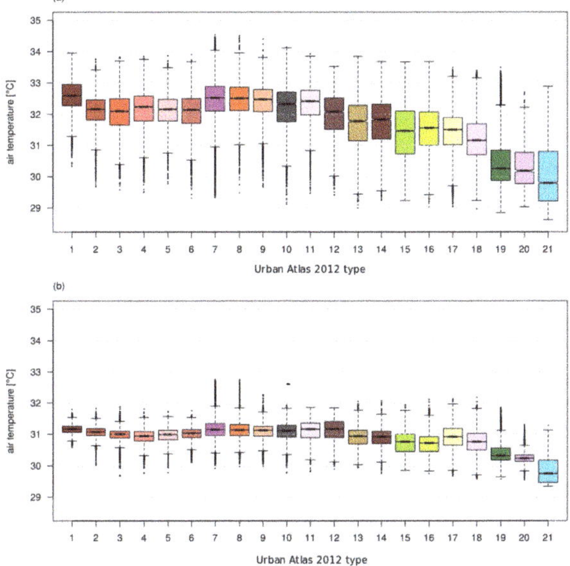

Figure 8. Statistic of air temperature in Poznań ((**a**)—12 July 2010, (**b**)—11 August 2015) on the basis of Landsat images according to Urban Atlas 2012 types (colours and order of types according to legend in Figure 1). In the boxplot, the middle values denote medians; the box extends to the Q1 (first quartile) and Q3 (third quartile), while the whiskers show the range (99.3%). The upper whisker shows Q3 + 1.5 × IQR (the interquartile range) and the lower shows Q1 − 1.5 × IQR. The notches extend to +/−1.58 IQR/sqrt(n) and the dots represent outliers.

On 11 August 2015 at 9:34 UTC, there was less variation in air temperature than on 12 July 2010 (Figure 7b). The impact of land development and use on the thermal conditions was significantly weaker. The lowest mean value of air temperature was 29.7 °C over water bodies and the highest, 31.2 °C, in continuous urban fabric, airports and mineral extraction and dump sites (Figure 8b).

The effect of land use on thermal conditions is evident in the air temperature profile from the northwest of the city to the southeast (about 22 km) (Figure 9). Along the profile, the air temperature varied from 28.9 °C to 33.2 °C. The coldest areas were recorded in the northwestern part of the city, i.e., in the western green wedge area, where Kierskie Lake is situated; then there is a systematic increase in air temperature with a maximum in the city centre (>32.0 °C), where a slightly colder area of the Warta River is also noted.

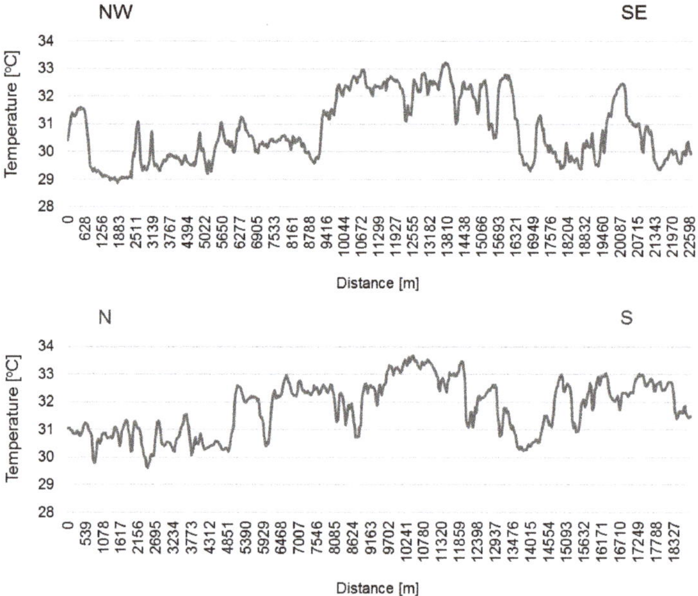

Figure 9. Air temperature in NW–SE and N–S profiles on 12 July 2010 at 9:43 UTC.

A significant decrease in the air temperature is observed in forest areas in the eastern part of Poznań, forming the eastern green wedge. In turn, along the profile made along the north–south line (about 19 km length), the air temperature changed from 29.6 °C to 33.7 °C (Figure 9). The coldest areas were found in the northern part of the city, where farmland dominates. A significant increase in air temperature (>32.0 °C) was found in areas with low-density, compact urban development occurring in the Naramowice, Winogrady, Stare Miasto and Wilda housing estates. A clear drop in air temperature was recorded in the Warta River Valley.

3.4.4. Atmospheric Circulation on 12 July 2010 and 11 August 2015

The analysis showed the different effects of land development on the thermal conditions in the city on selected days; the weather situation in Poland occurring on these days was therefore analysed. On 12 July 2010, Poland was within the weak-gradient high-pressure system, while on 11 August 2015 it was within the reach of the trough of low pressure (Figure 10). Both pressure situations provided an influx of warm air masses from the east and southeast sector. Recorded anomalies indicate the presence of warm air masses, whose temperature was higher by more than 8 °C than average in the summer. The 500 hPa isobaric surface on both days over central Poland settled over 140 m higher than average during the summer season. On 11 August, a warm weather front moved from the south to the north of the country, providing the exchange of air masses over the area. Advection of tropical air masses took place after the weather front (Figure 11). The different thermal conditions in the city during the two selected days were due to the differentiation of atmospheric circulation. On 12 July 2010, the air over the city was additionally warmed up by settling for longer over the urban area with anticyclonic weather. In turn, the passage of the weather front on the second of the days analysed resulted in significantly lower temperature values.

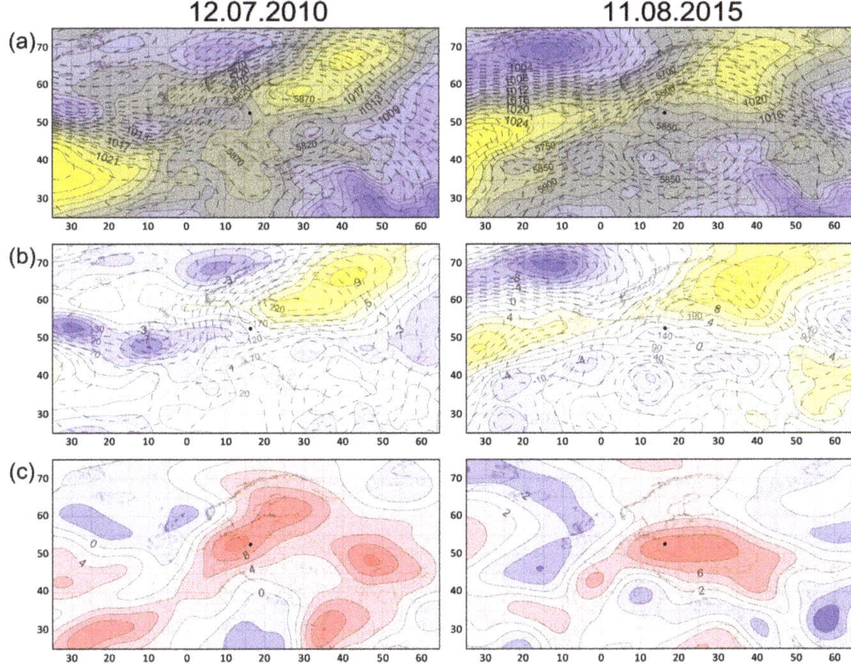

Figure 10. SLP and z500 hPa (**a**), SLP and z500 hPa anomalies (**b**) and T850 anomalies (**c**) on 12 July 2010 (**right column**) and 11 August 2015 (**left column**).

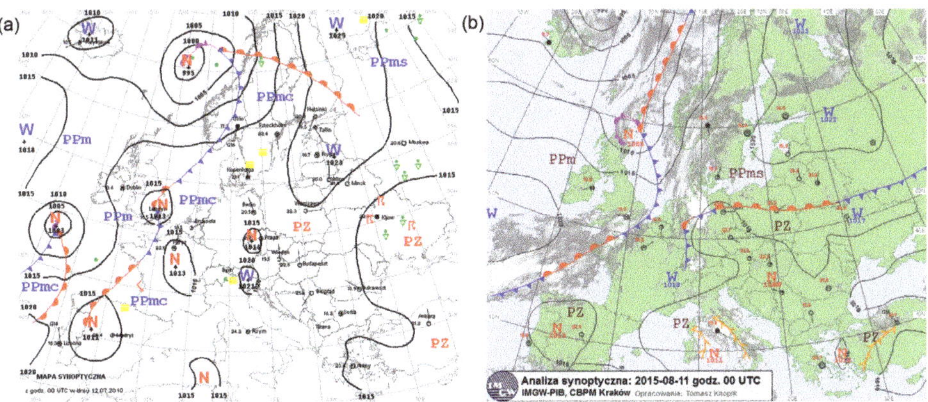

Figure 11. Weather situation on 12 July 2010 (**a**) and 11 August 2015 (**b**). Source: Meteorology and Water Management–National Research Institute.

4. Discussion

The research showed the T_{max} increase in the summer season in Poznań over the last 50 years. The results are consistent with previous studies conducted in Poznań [27,38] and other Polish cities [27,39], as well as in the rest of Europe [38,40,41].

The observed increase in T_{max} translated into an increase in the number of hot days and heat waves. Of the exceptionally hot summer seasons in Poznań, one should especially mention the summers of 1992, 1994, 2006 and 2015. Similar results were also obtained in other regions of Poland [42–44]. As Sulikowska et al. [27] have shown, the summer of 2015 was exceptionally hot in Poland, especially in the southwest of the country.

In the city districts, there was a great diversification of thermal conditions during the summer season and during the analysed heat waves. The highest air temperature was recorded in heavily transformed areas, i.e., in the developed city centre or industrial and commercial and public areas, while natural areas such as forests and areas near water bodies were characterised by the lowest air temperature. On this basis, within the city, one may distinguish areas with favourable biometeorological conditions during heat waves, such as green wedges, where the air temperature was significantly lower than in the centre and the air temperature differences were several degrees. A similar distribution of areas with highest and lowest temperatures was determined by Majkowska et al. [33] by analysing the urban heat island in Poznań, and Walawender [32], who analysed the urban heat island in Kraków based on satellite images.

In the course of daily air temperature in Poznań, the greater difference between the individual measuring points in the early hours, which gradually decreased during the day, was clearly visible. These differences are due to numerous shadowing effects of urban space and slower heating of the artificial surfaces [11,45–47] and faster cooling of natural areas, especially in clear and windless weather [11].

The research conducted using satellite images showed different variations in air temperature in the city area on the days analysed. On the first day, the average air temperature differences in the studied area were 2.8 °C, while on the second day they were smaller and amounted to 1.5 °C. On the second day, there was also less variation in air temperature within the various types of land use. No significant impact of the city's development on thermal conditions on 11 August 2015 was caused by the exchange of air masses.

This situation was a consequence of the passage of a weather front, while on 12 July 2010 the anticyclonic weather caused tropical air to hold for a few days. When stable pressure systems hold for a longer time, this leads to intensification of thermal conditions in highly transformed areas, which results in deterioration of biometeorological conditions and poses risks to human health and life. This highlights the significant role of atmospheric circulation in shaping weather conditions that can be modified by local factors. Similar results were obtained by Półrolniczak et al. [48], who demonstrated that the greatest intensity of the urban heat island is at night, especially during anti-cyclone circulation.

5. Conclusions

The results of the research conducted thus show how diverse thermal conditions occur in the city and how in particular types of land use they may be compared to those recorded at the meteorological station representing the mesoclimatic conditions (Poznań–Ławica). As the article has demonstrated, this diversity may result from the dissimilarity of a given part of the city due to the type of land development, but also it may result from meteorological conditions, i.e., barometric situation and the passage of atmospheric fronts, as well as advection or settling of the same air mass for a long time. The correct recognition of dependencies governing the occurrence of extreme phenomena in the city, such as heat waves, will hence translate into the possibility of more accurate forecasting of this phenomenon, which may consequently have a considerable social dimension.

Acknowledgments: This work was supported by the Polish National Science Centre under grant number UMO-2014/15/N/ST10/00717.

Author Contributions: Marek Półrolniczak prepared satellite images, conducted the analysis of land surface cover and its influence on LST and air temperature in the city and complemented the description of the results. Arkadiusz M. Tomczyk inspired and carried out the research as well as prepared the manuscript. Leszek Kolendowicz assisted with the results and discussion. Leszek Kolendowicz and Marek Półrolniczak have delivered temperature data from Climatology Department measurement net located in the Poznań area.

Conflicts of Interest: The authors declare no conflict of interest.

References

1. IPCC. *Climate Change 2013: The Physical Science Basis. Contribution of Working Group I to the Fifth Assessment Report of the Intergovernmental Panel in Climate Change*; Cambridge University Press: Cambridge, UK, 2013.
2. Lhotka, O.; Kyselý, J. Characterizing joint effects of spatial extent, temperature magnitude and duration of heat waves and cold spells over Central Europe. *Int. J. Climatol.* **2015**, *35*, 1232–1244. [CrossRef]
3. Keggenhoff, I.; Elizbarashvili, M.; King, L. Heat Wave Events over Georgia since 1961: Climatology, Changes and Severity. *Climate* **2015**, *3*, 308–328. [CrossRef]
4. Unkašević, M.; Tošić, I. Seasonal analysis of cold and heat waves in Serbia during the period 1949–2012. *Theor. Appl. Climatol.* **2015**, *120*, 29–40. [CrossRef]
5. Tomczyk, A.M. Hot weather in Potsdam in the years 1896–2015. *Meteorol. Atmos. Phys.* **2017**. [CrossRef]
6. Tomczyk, A.M.; Sulikowska, A. Heat waves in lowland Germany and their circulation-related conditions. *Meteorol. Atmos. Phys.* **2017**. [CrossRef]
7. Koffi, B.; Koffi, E. Heat waves across Europe by the end of the 21st century: Multiregional climate simulations. *Clim. Res.* **2008**, *36*, 153–168. [CrossRef]
8. Kyselý, J. Recent severe heat waves in central Europe: How to view them in a long-term prospect? *Int. J. Climatol.* **2010**, *30*, 89–109. [CrossRef]
9. Piniewski, M.; Mezghani, A.; Szcześniak, M.; Kundzewicz, Z. Regional projections of temperature and precipitation changes: Robustness and uncertainty aspects. *Meteorol. Z.* **2017**, *26*, 223–234. [CrossRef]
10. Luber, G.; McGeehin, M. Climate change and extreme heat events. *Am. J. Prev. Med.* **2008**, *35*, 429–435. [CrossRef] [PubMed]
11. Błażejczyk, K.; Kuchcik, M.; Milewski, P.; Dudek, W.; Kręcisz, B.; Błażejczyk, A.; Szmyd, J.; Degórska, B.; Pałczyński, C. *Urban Heat Island in Warsaw—Climatic and Urban Conditionings*; Wyd. Akademickie SEDNO: Warszawa, Poland, 2014; ISBN 978-83-7963-018-9.
12. Gabriel, K.M.A.; Endlicher, W.R. Urban and rural mortality rates during heat waves in Berlin and Brandenburg, Germany. *Environ. Pollut.* **2011**, *159*, 2044–2050. [CrossRef] [PubMed]
13. Kuchcik, M. Mortality in Warsaw: Is there any connection with Feather and air pollution? *Geogr. Pol.* **2011**, *74*, 29–45.
14. D'lppoliti, D.; Michelozzi, P.; Marino, C.; De'Donato, F.; Menne, B.; Katsouyanni, K.; Kirchmayer, U.; Analitis, A.; Medina-Ramón, M.; Paldy, A.; et al. The impact of heat waves on mortality in 9 European cities: Results from the EuroHEAT project. *Environ. Health* **2010**, *9*. [CrossRef]
15. Vandentorren, S.; Suzan, R.; Medina, S.; Pascal, M.; Maulpoix, A.; Cohen, J.C.; Ledrans, M. Mortality in 13 French cities during the August 2003 heat wave. *Am. J. Public Health* **2004**, *94*, 1518–1520. [CrossRef] [PubMed]
16. Kawashima, S.; Ishida, T.; Minomura, M.; Miwa, T. Relations between surface temperature and air temperature on a local scale during winter nights. *J. Appl. Meteorol. Climatol.* **2000**, *39*, 1570–1579. [CrossRef]
17. Voogt, J.A.; Oke, T.R. Thermal remote sensing of urban climates. *Remote Sens. Environ.* **2003**, *86*, 370–384. [CrossRef]
18. Sun, Y.J.; Wang, J.F.; Zhang, R.H.; Gillies, R.R.; Xue, Y.; Bo, Y.C. Air temperature retrieval from remote sensing data based on thermodynamics. *Theor. Appl. Climatol.* **2005**, *80*, 37–48. [CrossRef]
19. Dobrovolny, P. The surface urban heat island in the city of Brno (Czech Republic) derived from land surface temperatures and selected reasons for its spatial variability. *Theor. Appl. Climatol.* **2013**, *112*, 89–98. [CrossRef]
20. Schwarz, N.; Schlink, U.; Franck, U.; Grossmann, K. Relationship of land surface and air temperature and its implications for quantifying urban heat island indicators—An application for the city of Leipzig (Germany). *Ecol. Indic.* **2012**, *18*, 693–704. [CrossRef]

21. Ptak, M.; Choiński, A.; Piekarczyk, J.; Pryłowski, T. Applying Landsat Satellite Thermal Images in the Analysis of Polish Lake Temperatures. *Pol. J. Environ. Stud.* **2017**, *26*, 2159–2165. [CrossRef]

22. United Nations. World Population Prospects 2017. Available online: https://esa.un.org/unpd/wpp/Publications/Files/WPP2017_KeyFindings.pdf (accessed on 5 November 2017).

23. Główny Urząd Statystyczny. Rocznik Statystyczny Rzeczypospolitej Polskiej 2015. Available online: https://stat.gov.pl/obszary-tematyczne/roczniki-statystyczne/roczniki-statystyczne/rocznik-statystyczny-rzeczypospolitej-polskiej-2015,2,10.html (accessed on 5 November 2017).

24. Woś, A. *Climate of Poland in the Second Half of the 20th Century*; Wyd. Naukowe UAM: Poznań, Poland, 2010; ISBN 978-83-232-2180-7.

25. Abaurrea, J.; Asín, J.; Cebrián, A.; Centelles, A. Modeling and forecasting extreme heat events in the central Ebro valley, a continental-mediterranean area. *Glob. Planet Chang.* **2007**, *57*, 43–58. [CrossRef]

26. Rey, G.; Jougla, E.; Fouillet, A.; Pavillon, G.; Bessemoulin, P.; Frayssinet, P.; Clavel, J.; Hémon, D. The impact of major heat waves on all-cause and cause specific mortality in France 1971–2003. *Int. Arch. Occup. Environ. Health* **2007**, *80*, 615–626. [CrossRef] [PubMed]

27. Sulikowska, A.; Wypych, A.; Woszczek, I. The 2015 summer heatwaves in Poland and their synoptic background. *Badania Fizjograficzne Seria A Geografia Fizyczna* **2016**, *67*, 205–223.

28. Tomczyk, A.M.; Półrolniczak, M.; Bednorz, E. Circulation Conditions' Effect on the Occurrence of Heat Waves in Western and Southwestern Europe. *Atmosphere* **2017**, *8*, 31. [CrossRef]

29. Salmi, T.; Maiittii, A.; Anttila, P.; Ruoho-Airola, T.; Amnel, T. *Detecting Trends of Annual Values of Atmospheric Pollutants by the Mann-Kendall Test and Sen's Slope Estimates—The Excel Template Application MAKESENS*; Publications on Air Quality No. 31; Finnish Meteorological Institute: Helsinki, Finland, 2002; pp. 1–35.

30. Sobrino, J.A.; Raissouni, N. Toward remote sensing methods for land cover dynamic monitoring: Application to Morocco. *Int. J. Remote Sens.* **2000**, *21*, 353–366. [CrossRef]

31. Jiménez-Muñoz, J.C.; Sobrino, J.A. A generalized single-channel method for retrieving land surface temperature from remote sensing data. *J. Geophys. Res.* **2003**, *108*, 4688. [CrossRef]

32. Walawender, J. Application of LANDSAT satellite data and GIS techniques for estimation of thermal conditions in urban area (using an example of Krakow agglomeration). *Prace Geogr.* **2009**, *122*, 81–98.

33. Majkowska, A.; Kolendowicz, L.; Półrolniczak, M.; Hauke, J.; Czernecki, B. The urban heat island in the city of Poznań as derived from Landsat 5 TM. *Theor. Appl. Climatol.* **2017**, *128*, 769–783. [CrossRef]

34. Kalnay, E.; Kanamistu, M.; Kistler, R.; Collins, W.; Deaven, D.; Gandin, L.; Iredell, M.; Saha, S.; White, G.; Woollen, J.; et al. The NMC/NCAR 40-Year Reanalysis Project. *Bull. Am. Meteorol. Soc.* **1996**, *77*, 437–471. [CrossRef]

35. Urban Atlas 2012. Available online: https://www.eea.europa.eu/data-and-maps/data/urban-atlas (accessed on 5 December 2017).

36. Hijmans, R.J. Raster: Geographic Data Analysis and Modeling. R Package Version 2.5-8. 2016. Available online: http://CRAN.R-project.org/package=raster (accessed on 5 December 2017).

37. Bivand, R.; Keitt, T.; Rowlingson, B. Rgdal: Bindings for the Geospatial Data Abstraction Library. R Package Version 1.1-10. 2016. Available online: http://CRAN.R-project.org/package=rgdal (accessed on 5 December 2017).

38. Tomczyk, A.M. Circulation-related conditioning of the occurrence of heatwaves in Poznań. *Prz. Geogr.* **2014**, *86*, 41–52. [CrossRef]

39. Tomczyk, A.M.; Bednorz, E. Heat waves in Central Europe and their circulation conditions. *Int. J. Climatol.* **2016**, *36*, 770–782. [CrossRef]

40. Vidal, J.P.; Martin, E.; Franchisteguy, L.; Baillon, M.; Soubeyroux, J.M. A 50-year high-resolution atmospheric reanalysis over France with the Safran system. *Int. J. Climatol.* **2010**, *30*, 1627–1644. [CrossRef]

41. Del Río, S.; Cano-Ortiz, A.; Herrero, N.; Penas, A. Recent trends in mean maximum and minimum air temperatures over Spain (1961–2006). *Theor. Appl. Climatol.* **2012**, *109*, 605–626. [CrossRef]

42. Wibig, J.; Podstawczyńska, A.; Rzepa, M.; Piotrowski, P. Heatwaves in Poland—Frequency, trends and relationships with atmospheric circulation. *Geogr. Pol.* **2009**, *82*, 33–46. [CrossRef]

43. Porębska, M.; Zdunek, M. Analysis of extreme temperature events in Central Europe related to high pressure blocking situations in 2001–2011. *Meteorol. Z.* **2013**, *22*, 533–540. [CrossRef]

44. Wypych, A.; Sulikowska, A.; Ustrnul, Z.; Czekierda, D. Temporal Variability of Summer Temperature Extremes in Poland. *Atmosphere* **2017**, *8*, 51. [CrossRef]

45. Landsberg, H.E. *The Urban Climate*; Academic Press: New York, NY, USA, 1981.
46. Oke, T.R. The energetic basis of the urban heat island. *Q. J. R. Meteorol. Soc.* **1982**, *108*, 1–24. [CrossRef]
47. Voogt, J.A. Urban heat island. In *Encyclopedia of Global Environmental Change*, 3rd ed.; Munn, T., Ed.; Wiley: Chichester, UK, 2002; pp. 660–666.
48. Półrolniczak, M.; Kolendowicz, L.; Majkowska, A.; Czernecki, B. The influence of atmospheric circulation on the intensity of urban heat island and urban cold island in Poznań, Poland. *Theor. Appl. Climatol.* **2017**, *127*, 611–625. [CrossRef]

 *atmosphere*

Article

Estimating the Influence of Housing Energy Efficiency and Overheating Adaptations on Heat-Related Mortality in the West Midlands, UK

Jonathon Taylor [1,*], Phil Symonds [1], Paul Wilkinson [2], Clare Heaviside [2,3,4], Helen Macintyre [3,4], Michael Davies [1], Anna Mavrogianni [1] and Emma Hutchinson [2]

1   UCL Institute for Environmental Design and Engineering, Central House, 14 Upper Woburn Place, London WC1H 0NN, UK; p.symonds@ucl.ac.uk (P.S.); michael.davies@ucl.ac.uk (M.D.); a.mavrogianni@ucl.ac.uk (A.M.)
2   Department of Social and Environmental Health Research, London School of Hygiene and Tropical Medicine, 15-17 Tavistock Place, London WC1H 9SH, UK; paul.wilkinson@lshtm.ac.uk (P.W.); clare.heaviside@phe.gov.uk (C.H.); emma.hutchinson@lshtm.ac.uk (E.H.)
3   Climate Change Group, Centre for Radiation, Chemical and Environmental Hazards, Public Health England, Chilton OX11 0RQ, UK; helen.macintyre@phe.gov.uk
4   School of Geography, Earth and Environmental Sciences, University of Birmingham, Edgbaston, Birmingham B15 2TT, UK
*   Correspondence: j.g.taylor@ucl.ac.uk; Tel.: +44-(0)20-3108-59174

Received: 20 April 2018; Accepted: 15 May 2018; Published: 16 May 2018

Abstract: Mortality rates rise during hot weather in England, and projected future increases in heatwave frequency and intensity require the development of heat protection measures such as the adaptation of housing to reduce indoor overheating. We apply a combined building physics and health model to dwellings in the West Midlands, UK, using an English Housing Survey (EHS)-derived stock model. Regional temperature exposures, heat-related mortality risk, and space heating energy consumption were estimated for 2030s, 2050s, and 2080s medium emissions climates prior to and following heat mitigating, energy-efficiency, and occupant behaviour adaptations. Risk variation across adaptations, dwellings, and occupant types were assessed. Indoor temperatures were greatest in converted flats, while heat mortality rates were highest in bungalows due to the occupant age profiles. Full energy efficiency retrofit reduced regional domestic space heating energy use by 26% but increased summertime heat mortality 3–4%, while reduced façade absorptance decreased heat mortality 12–15% but increased energy consumption by 4%. External shutters provided the largest reduction in heat mortality (37–43%), while closed windows caused a large increase in risk (29–64%). Ensuring adequate post-retrofit ventilation, targeted installation of shutters, and ensuring operable windows in dwellings with heat-vulnerable occupants may save energy and significantly reduce heat-related mortality.

Keywords: heat; mortality; adaptation; dwellings; indoor temperature

---

1. Introduction

In the UK, as in most settings, the risk of mortality increases during hot weather particularly among vulnerable groups such as the elderly [1]. In England and Wales, the heatwaves of 2003 and 2006 led to an estimated 2091 [2] and 680 [3] excess deaths, respectively. Warming temperatures and an increased frequency of extreme temperatures in the future [4], as well as an aging population, are likely to increase the importance of heat as a public health risk in the UK [5]. High temperatures may also lead to increases in population morbidity due to, for example, heat stress and heat exhaustion, kidney failure, and heart attacks [6].

As people in the UK spend the majority of their time indoors, housing is an important determinant of heat exposure and consequent heat-related mortality. A study of the 2003 heat wave in Paris found that living in top-floor flats and in poorly insulated homes were both associated with increased mortality risk [7]. Moreover, large-scale monitoring studies in the UK have suggested that certain housing types exhibit higher indoor temperatures, with flats generally being warmer than most other dwellings, and detached and solid-walled dwellings cooler [8,9]. Similar conclusions arise from modelling studies [10–12], which also indicate the potential effectiveness of adaptations such as shading, use of external shutters on windows, and using solar reflective coatings [13,14] while dwelling energy-efficiency may increase or decrease internal temperatures [10,11]. The Urban Heat Island (UHI) effect—where urban areas are significantly hotter than surrounding rural areas primarily due to the modification of land surfaces and waste heat—may exacerbate heat exposure during hot weather and increase heat mortality risk [15].

A number of studies have incorporated building overheating markers to predict heat exposure at the population-level. Dwelling characteristics and population demographics have been used to develop a heat risk index for London [16,17], while the same have been used alongside UHI data to identify vulnerable areas in Birmingham, UK [18,19]. Outside of the UK, housing, UHI, and population data have been combined in heat exposure studies in Melbourne, Australia [20], New York City [21], and across the U.S [22].

In the UK, modelled temperature exposures from dwellings and/or the UHI has been combined to estimate heat-related mortality in London, Sheffield, and the West Midlands. Taylor et al. [23] estimated the spatial variation in heat exposure using simulated UHI temperatures, and building physics models of indoor temperatures for individual dwellings in the London housing stock; an age-specific heat-mortality function was then used to estimate heat attributable mortality using underlying census population and age data. Liu et al. [24] also used building physics models of buildings, in combination with high resolution climate projections, to estimate the spatial variation in heat-related mortality risks across the city of Sheffield. Finally, modelled indoor temperatures were used to estimate the changes in population mortality in the West Midlands, UK, prior to and following a number of different energy-efficiency and overheating adaptations to dwellings and the built environment [25].

Using the underlying indoor temperature and health model described in Taylor et al. [25], this paper aims to explore the variation in heat mortality risk across building types in the West Midlands region of the UK, based on dwelling indoor overheating risks and occupant characteristics. The effects of energy efficiency (including wall, floor, or roof insulation, and full retrofit), behavioural (window-opening), and heat adaptations (external shutters and low absorptance surface coatings) on heat exposure, mortality, and energy use are explored. In addition, the reduction in mortality through more realistic implementation of adaptations targeted at dwellings or residents is also examined.

2. Methods

2.1. Building Modelling

The West Midlands region has a population of 5.6 million [26], and contains Birmingham, the second-most populous urban area in the UK. During the 2003 heatwave, the region had an estimated 130 excess deaths due to hot weather [2].

The baseline housing stock and population model is the 2010–2011 English Housing Survey (EHS), which contains regionally-representative housing and occupant data for 1558 dwellings in the West Midlands [27]. Resident age data is available within the database for each dwelling occupant. Data within the EHS is used to inform the geometry, floor area, glazing area, construction type, and insulation levels of each dwelling, while the energy efficiency and airtightness of each dwelling has been estimated using standardised methods [28].

Readers are referred to Symonds et al. (2016) for further information on the indoor temperature model. Briefly, indoor temperatures and energy use for space heating in the West Midlands housing stock were estimated using a series of metamodels derived from the results of simulation studies [29,30] using the building physics model EnergyPlus. Previous studies have compared the underlying EnergyPlus model outputs against a large dataset of monitored indoor temperatures, showing that the model is able to capture the trends in overheating risk between dwelling variants [30]. Here, the model is adapted to model dwellings with and without energy efficiency, occupant behavior, and passive overheating interventions. Air conditioners (A/C) were not modelled due to their rarity in English housing stock—estimated at 3% of dwellings [31]—and because of their high energy demands.

Metamodels were developed for each combination of dwelling geometry (end terrace, mid-terrace, semidetached, detached, bungalow, converted flats, low-rise flats, and high-rise flats), wall type (cavity or solid), and heat adaptation (shutters or no shutters). For each combination of the above, EnergyPlus models were developed with fabric energy efficiency levels, permeability, floor area, glazing area, and local wind exposure randomly sampled from distributions available from representative samples [27,32] of English dwellings. For this study, roof and wall absorptance was also randomly selected in the range of 0.1–0.6, representing the painting of external surfaces with a low absorptance paint; the indoor temperature threshold above which windows are opened were also randomly selected (18 °C–35 °C) to represent extreme ranges in occupant behaviour. Archetypes, used to represent dwelling geometries, can be seen in Appendix A. Models were run using Test Reference Year weather data, representing the "average" climate for 2030 under a medium emission scenario (A1B-50th percentile) [33], assumed to be representative for the region.

From the results of these simulations, we computed the mean maximum daytime living room temperature at different two-day rolling mean maximum outdoor summer temperatures, as well as annual energy use (kWh) for space heating. A metamodel was then generated using artificial neural networks [34] to determine energy use and indoor temperatures from dwelling characteristics across the West Midlands housing stock. The metamodel was applied to obtain indoor temperature and energy use estimates for individual dwellings in the West Midlands under the adaptation scenarios in Table 1, using weather data that describes "average" Birmingham summers in 2030s, 2050s, and 2080s (A1B-medium emissions scenario, 90% probability [4,33]).

Table 1. The adaptations and underlying assumptions modelled. The reduction in fabric U-value is based on the UK Government's Standard Assessment Procedure (SAP) for Energy Rating of Dwellings, with the lowest possible U-value for the fabric component selected based on the fabric type and dwelling age [35]. The change in permeability is estimated based on the work by [a] Hong et al. [36] or [b] UK SAP, following the methods described by Hamilton et al. [37].

	Adaptation	Details
Energy Efficiency	Cavity wall insulation (CWI)	All cavity walls are modelled as insulated, reducing wall U-value, and the infiltration rate in previously uninsulated dwellings reduced by 0.2 air changes per hour (ach) [a]
	Internal solid wall insulation (SWI)	All solid walls are modelled as internally insulated, reducing wall U-value, and the infiltration rate in previously uninsulated dwellings reduced by 0.3 ach [a]
	Floor insulation (FI)	All floors are modelled as insulated, reducing floor U-value, and the infiltration rate in previously uninsulated dwellings reduced by 0.1 ach [b]
	Loft insulation (LI)	All lofts are modelled as insulated, reducing loft U-value, and the infiltration rate in previously uninsulated dwellings reduced by 0.1 ach [a]
	Full Retrofit	A full retrofit (floors, loft, walls, and triple-glazed windows) is modelled, with reductions in U-value as above and the infiltration rate reduced by 0.7 ach to a Building Regulations minimum permeability of 3 $m^3/h/m^2$
Heat Adaptation	Shutters	External shutters are closed daily between 9 a.m. and 6 p.m. during the summer
	Absorptance	The solar absorptance of the building façade is reduced from 0.7 to 0.1

Table 1. *Cont.*

	Adaptation	Details
Occupant Behaviour	Windows Open	Windows are opened when indoor temperatures exceed 18 °C during summer, representing a scenario were windows are continuously open
	Windows Closed	Windows are opened when indoor temperatures exceed 35 °C, representing a scenario were windows are continuously closed

2.2. Mortality Calculations

Household weighting values in the EHS were used to estimate the mean and distribution of occupant temperature exposures across the West Midlands housing stock. As there is no spatial information for the EHS, we assume the modelled households have equal distributions of UHI temperature exposures. From this, a dwelling-specific indoor temperature anomaly relative to the regional population-weighted mean was calculated:

$$T^*_{max,k,d} = T_{max,out,d} + T_{Indoor\ Anomaly,k,d} \tag{1}$$

where $T_{max,out,d}$ is the two-day rolling mean maximum outdoor temperature for day d; $T_{Indoor\ Anomaly,k,d}$ is a positive or negative temperature anomaly representing the deviation in estimated two-day rolling mean maximum indoor temperature for dwelling k from the population-mean rolling maximum indoor temperature on day d for the West Midlands; and $T^*_{max,k,d}$ is the temperature to which occupants are exposed on day d in household k. For adaptation scenarios, anomalies were calculated relative to the mean of the unadapted stock.

Calculations of heat-related mortality were based on applying region-specific temperature-mortality functions for the West Midlands to the EHS occupant age data and corresponding estimated dwelling indoor temperatures. Heat-associated mortality is described in terms of Relative Risk (RR), or the ratio of the probability of mortality occurring in a heat-exposed group to the probability of mortality in an unexposed group. The all-age heat-mortality RR for the West Midlands was derived from Armstrong et al. [38] from which age-specific (0–64, 65–74, 75–84, 85+) temperature-mortality slopes were derived using the age-specific RRs for England and Wales published by Gasparrini et al. [1]. The underlying age-specific all-cause mortality rates by season were obtained from the Office for National Statistics (ONS); here we adjust these to reflect summer rates. Dwelling-specific heat mortality was then calculated as:

$$D_{k,d} = \sum_i \left[occupants_{i,k} \times deathrate_i \times \left(RR_{heat,i}^{(T^*_{max,k,d} - 23\ °C)} - 1 \right) \right] \tag{2}$$

where $occupants_{i,k}$ is the number of individuals of age-group i in dwelling k; $deathrate_i$ is the summertime daily mortality rate per person for age-group i; $RR_{heat,i}$ is the relative risk (RR) of mortality due to temperature for age group i; and 23 °C is the estimated regional heat mortality threshold for the West Midlands [38]. Readers are referred to Taylor et al. [25] for a detailed description of mortality calculations.

3. Results

3.1. Heat Exposure across Dwelling Variants

The average living room temperatures across a range of dwelling variants when $T_{max,out,d}$ exceeds 23 °C during the 2030s summer can be seen in Figure 1. Certain dwelling variants are hotter than others, including mid-terraced dwellings and flats. Multiple distributions are provided for flats to represent the different floor levels of the flats within the buildings. Buildings with higher indoor temperatures were also found to exceed the 23 °C temperature threshold more regularly during hot weather, with the coolest dwellings exceeding it 39 times and the hottest exceeding it 101 times during the 123 days of summer, 2030s (median 69 days). The sample size for each dwelling variant within the EHS is also shown.

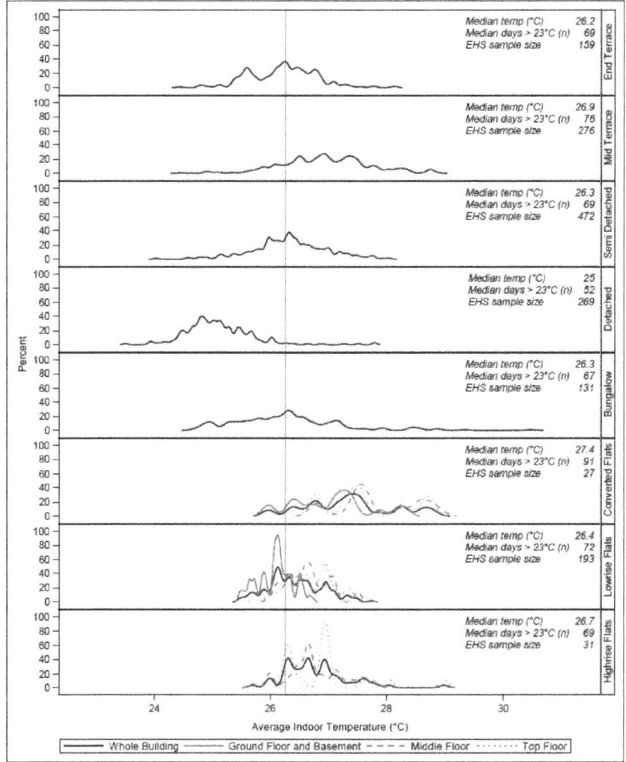

Figure 1. Density plots of the average indoor temperature when $T_{max,out,d}$ exceeds 23 °C during the 2030 summer. The red vertical line shows the median for the stock. There is a lack of data for ground-floor high-rise flats, and so these are not shown. English Housing Survey (EHS).

3.2. Housing Adaptations and Heat Exposure

The distribution of average indoor temperature exposures for a 2030s summer when $T_{max,out,d}$ exceeds 23 °C can be seen in Figure 2, before and after various energy efficiency, heat mitigating, and behavioural adaptations. Individual energy efficiency retrofits generally do not lead to a significant increase in temperature exposure, with the exception of internal solid wall insulation which causes a median temperature increase of 0.1 °C (range: −0.4 to 0.9 °C) in solid-walled dwellings. While individual fabric interventions to not lead to a significant increase in median temperatures, the cumulative effects of different energy efficiency interventions on permeability is reflected in an increase following the full retrofit of all buildings in the stock (median 0.2 °C, range: −1.0–1.7 °C). Full retrofit is predicted to reduce 2030s energy use by 25.5% relative to the current stock, and individual retrofits show comparatively more modest reductions in energy use for space heating. This energy saving is due to reduced ventilative and fabric heat losses only, as changes to heating systems are not modelled. Shutters are able to significantly reduce indoor temperature exposure across the stock (median: −1.4 °C, range: −4.1–0.2 °C), and lead to a small increase in space heating energy consumption as the absence of solar gains means heating is occasionally required to meet setpoint temperatures.

Decreasing the surface absorptance of the external surfaces led to a smaller reduction (median: −0.5 °C, range: −1.5 °C–0.6 °C) as well as an increase in space heating energy use of 4.1% during the 2030 heating season (September–May). Regarding occupant behaviours, keeping windows open when internal temperatures exceed 18 °C had a modest impact on reducing temperatures compared to the threshold of 22 °C modelled in the 'current' stock (median: −0.4 °C, range: −1.1–0.3 °C). The largest risk-factor for heat exposure is keeping windows closed at all times (median 1.4 °C, range: 0.1–3.4 °C).

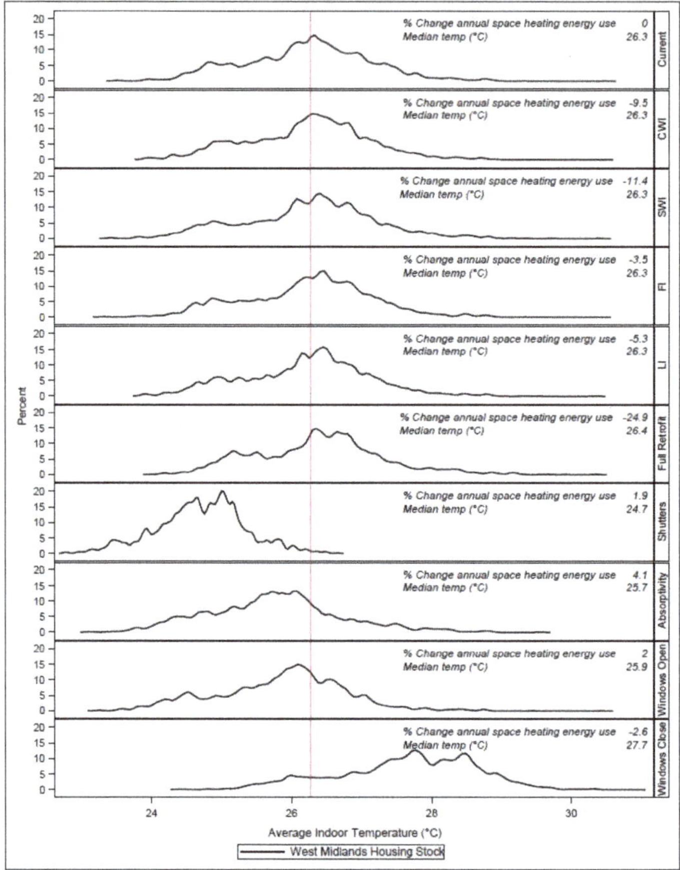

Figure 2. Density plots of the average indoor temperature when $T_{max,out,d}$ exceeds 23 °C during the 2030 summer following energy efficiency, heat, and behavioural adaptations. The red vertical line shows the median for the current (unadapted) stock (26.3 °C).

3.3. Mortality across Dwelling Variants

Analysis of the West Midlands population by age group and dwelling type in the West Midlands from the EHS can be seen in Table 2. A higher proportion of elderly occupants inhabit bungalows and converted flats, while more young individuals live in purpose-built flats and terraced dwellings. The mortality rate per million occupants of each dwelling type at increasing temperatures shows how the relative heat mortality risk varies by dwelling variant (Figure 3A). The rate of increase reflects the housing overheating characteristics and the age profiles of the occupant population. Bungalows show the greatest rate increase in mortality risk with increasing outdoor temperatures, followed by converted and low-rise flats. The different age profiles in dwelling types and their variation in indoor temperatures are reflected in the absolute estimated mortality and risk of mortality across variants. The largest predicted mortality under increasing temperatures were residents of houses rather than flats, primarily semi-detached dwellings, followed by bungalows and detached properties (Figure 3B). This is due to semi-detached and detached properties housing the largest number of individuals in the West Midlands population (34% and 21%, respectively), while the age effects of the occupant population play a significant role in bungalows despite their moderate indoor overheating risk and relative infrequency in the housing stock (5.5%). The mortality in flats was predicted to represent only a small fraction of overall mortality, also due to their infrequency across the West Midlands housing stock.

Table 2. Dwelling type by percent of residents within each age group, West Midlands.

Dwelling Type	Age Group				
	0–64	65–74	75–85	85+	TOTAL
End Terrace	88.3%	7.8%	3.3%	0.6%	541,917
Mid Terrace	91.7%	3.7%	3.8%	0.9%	1,060,700
Semi Detached	85.3%	8.7%	5.6%	0.7%	1,826,175
Detached	83.0%	10.3%	5.8%	0.9%	1,145,396
Bungalow	42.5%	28.1%	22.3%	7.1%	297,168
Converted Flat	78.6%	11.2%	8.0%	2.1%	56,047
Low-rise Flat	83.3%	9.0%	6.1%	1.7%	366,618
High-rise Flat	88.1%	4.8%	7.1%	0.0%	65,942
TOTAL	4,495,026	484,724	325,129	64,301	5,359,963

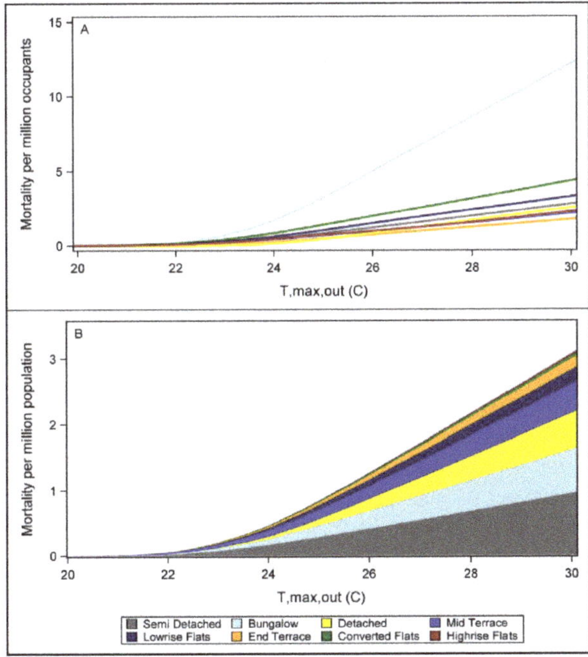

Figure 3. (**A**) The mortality per million occupants per day of each dwelling type at increasing outdoor temperatures; (**B**) The mortality per million population per day in the West Midlands at increasing outdoor temperatures, stacked by dwelling variant.

At the population level, appreciable risk of heat mortality (here defined as 100 per million occupants) under 2030s conditions exists only in 15% of the population, with risks increasing with age band (Figure 4). The wide range of risk within each age group is attributable to indoor temperature exposures, with the differences between the coolest and hottest dwellings causing a fivefold increase in mortality risk amongst the 75–84 age group, and a fourfold increase in those over 85.

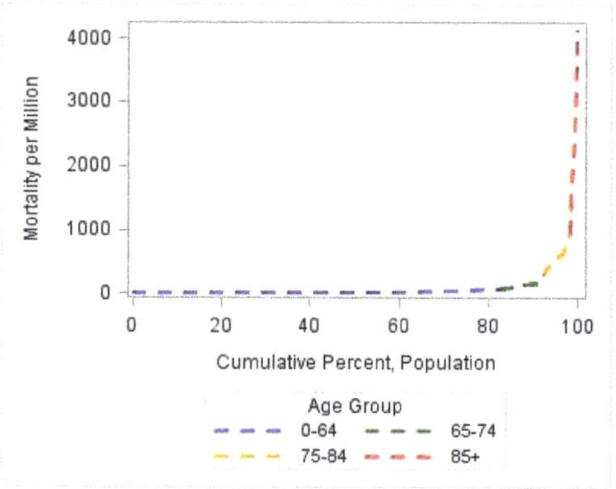

Figure 4. The heat-related mortality risk per million population for summer 2030, by cumulative population percent and age classification.

3.4. Housing Stock Adaptation and Mortality

The estimated mortality of occupants in the current stock under typical 2030s, 2050s, and 2080s climates, and following a range of adaptations, can be seen in Table 3. Individually, energy efficiency adaptations did not cause significant changes in heat mortality relative to the current stock, apart from loft insulation which reduced mortality. Full retrofit led to a small increase in heat mortality risk (2.5–4.4%), driven primarily by the increased indoor temperatures associated with internal solid wall insulation and the cumulative reduction in permeability that restricts ventilation and convective heat dissipation. Any small increase in summertime heat-related mortality from full retrofit is likely to be offset by a much larger reduction in winter mortality due to warmer housing, as well as benefits from the significant energy savings for space heating.

Of the modelled heat-mitigation scenarios, installation of external shutters was the most effective, causing an estimated reduction in heat-related mortality of 43%, 40%, and 37% in weather conditions representative of typical 2030s, 2050s, and 2080s summers, respectively, while reducing absorptance was less effective (15%, 14%, and 12%). Of the occupant behaviours modelled, reducing the window opening threshold to 18 °C had only modest reduction in heat mortality risk (6–10%), while keeping windows closed led to a substantial increase in population heat mortality risk of 29–64%. The significant increase in risk associated with closed windows indicates that occupant behaviour or housing where windows cannot be opened due to inadequate windows, outdoor pollution, crime, or noise-may be the single largest modifier of indoor heat exposure and consequent heat-related mortality risk.

Targeted interventions were assessed to determine how population heat-related mortality might decrease under more realistic levels of adaptation. Installing shutters in properties with residents over the age of 85 (2.8% of the stock) decreased heat-related mortality risk by 5–9% (Scenario 1), while installing them in the 12.1% of dwellings with residents over 75 decreased heat-related mortality risk by 28–33% (Scenario 2). It may not be straightforward to install shutters on certain dwellings (e.g., high-rise flats), or there may be local regulations that prevent changes to the external façade (e.g., listed buildings, assumed here to be all buildings built prior to 1918). Installation of shutters in all buildings, excluding these, is estimated to reduce population heat-related mortality by 32–38% (Scenario 3).

Table 3. The estimated heat-related mortality per million population in the West Midlands prior to and following adaptation.

	Mortality Per Million (Percent Change from Unadapted Stock)					
Adaptation	*2030*		*2050*		*2080*	
Current	*93*		*126*		*194*	
Adaptation-Energy						
CWI	*93*	*(0%)*	*126*	*(0%)*	*194*	*(0%)*
SWI	*93*	*(0%)*	*126*	*(0%)*	*194*	*(0%)*
FI	*93*	*(0%)*	*126*	*(0%)*	*194*	*(0%)*
LI	*92*	*(−0.6%)*	*126*	*(0%)*	*193*	*(−0.6%)*
Full Retrofit	*97*	*(4.4%)*	*131*	*(3.5%)*	*199*	*(2.5%)*
Adaptation–Heat						
Shutters	*53*	*(−42.9%)*	*76*	*(−39.9%)*	*122*	*(−36.9%)*
Absorptivity	*78*	*(−15.3%)*	*108*	*(−14.2%)*	*170*	*(−12.4%)*
Adaptation-Behaviour						
Windows Open	*83*	*(−10.1%)*	*116*	*(−8.1%)*	*182*	*(−5.9%)*
Windows Closed	*151*	*(63.5%)*	*184*	*(45.9%)*	*249*	*(28.7%)*
Targeted Intervention						
Scenario 1	*85*	*(−8.5%)*	*117*	*(−7.3%)*	*184*	*(−4.8%)*
Scenario 2	*62*	*(−32.7%)*	*88*	*(−30.3%)*	*139*	*(−28.1%)*
Scenario 3	*58*	*(−37.5%)*	*82*	*(−34.8%)*	*131*	*(−32.2%)*

3.5. Mortality across Adapted Dwelling Variants

Figure 5 shows the variation in mean heat mortality risk by dwelling, occupant age group, and heat mitigating (shutters) or deleterious behaviour (windows closed). The greatest risk is estimated in those aged over 85 living in mid terraced dwellings. Bungalows, which showed a rapid increase in mortality with increasing temperatures, have a relatively low rate of mortality within each age category. This low rate of mortality is due to the modest temperatures within the modelled bungalows, while the rapid rise in mortality per thousand occupants is due to bungalows being homes to the largest proportion of elderly in the West Midlands. The dwelling types do not show a consistent order of risk within each age classification due to the variation in the fabric and geometry characteristics of the dwellings.

Heat mitigating (shutters) or deleterious behaviour (windows closed) (Figure 5) also has a significant impact on mortality risk, varying by dwelling and occupant vulnerability. In most cases, application of shutters greatly reduces the risk of heat mortality in dwellings with vulnerable occupants. For those aged over 85, shutters led to a median reduction of 40% mortality risk, with a range of 37–78%, while for those between 75 and 85 the median reduction was 59% (range: 37–90%). Shutters were less effective in ground floor flats and dwellings with low glazing areas. Leaving windows closed led to a median increase in risk for occupants over 85 of 164% (range: 130–219%), and a median increase of 164% (range: 119–257%) for those between 75 and 85.

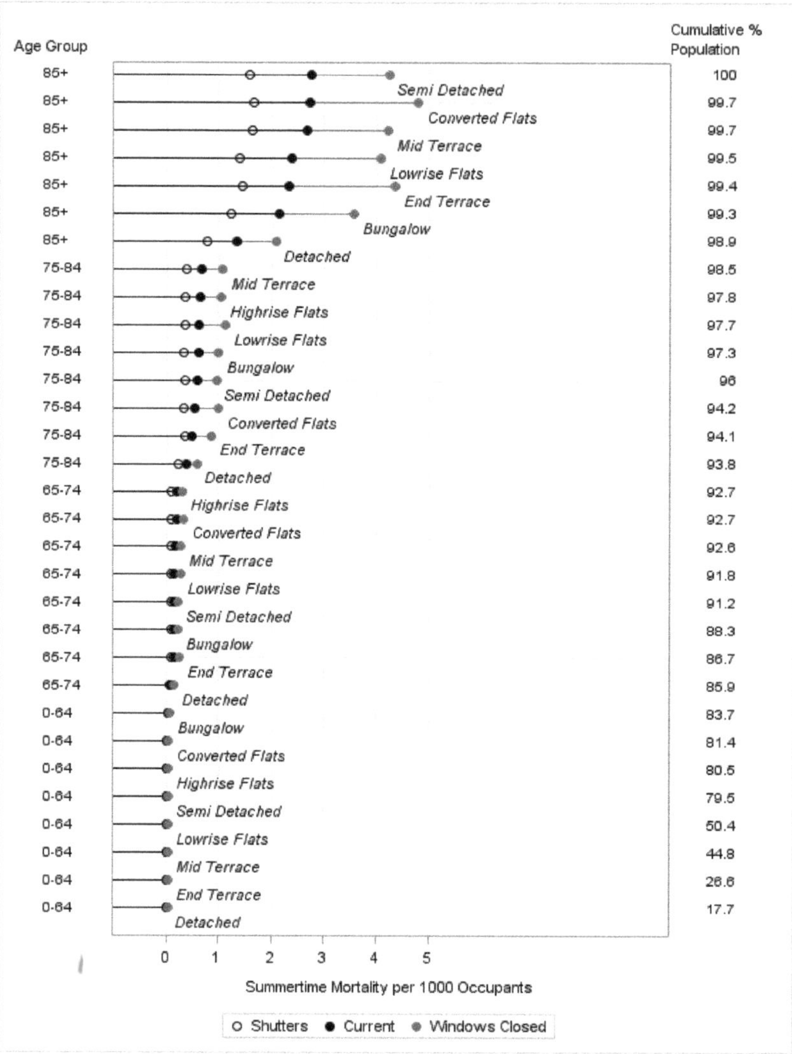

Figure 5. The median mortality per 1000 occupants over the 2030 summer by occupant age classification and dwelling type for the current stock, and with shutters or windows closed. The cumulative frequency of each age class/dwelling variant is shown on the secondary axis.

4. Discussion and Conclusions

With a changing climate and aging population, there is an urgent need to identify ways to mitigate against heat exposure without increasing greenhouse gas emissions in order to reduce population heat mortality. There is, however, little empirical evidence to help identify the best solution with respect to population heat mortality and energy consumption. We have described the application of a heat risk model to the housing stock and population of the West Midlands, UK, and estimated how adaptations to the housing stock may alter the risk of heat-related mortality. Modelled indoor temperatures showed a wide variation across different dwelling variants, indicating that housing type is a significant modifier of heat exposure risk during hot weather. Certain dwelling variants, such as flats, mid terraced houses, and bungalows were found to be at elevated risk of high indoor temperatures. Heat-related adaptations to dwellings showed decreases in indoor temperatures in line with previous modelling studies on overheating in housing [13,14], while changes to indoor temperatures following energy efficiency adaptations also reflect those from previous modelling studies [11,12].

The application of the mortality model to the indoor temperature estimates indicates that building adaptations have the potential to alter the mortality of building occupants during warm and hot weather. The most effective adaptation to reduce heat-related health effects was using external shutters during the daytime, which was able to reduce heat-related mortality by over 43% under the 2030, 40% under 2050, and 37% under 2080 summer weather scenarios. Reducing the absorptance of the external façade led to a more modest reduction in estimated mortality of 12%, 14%, and 15% under the same climate scenarios, but with the unintended consequence of increasing 2030 winter space heating energy consumption by 4%. These results therefore indicate that external shutters are a more effective and efficient means of reducing internal temperature exposure during summer months.

Full retrofits led to a small increase in overheating risk and heat mortality, driven primarily by internal solid walled insulation and reductions in ventilation due to decreased permeability. The impacts of energy efficiency improvements on energy use for space heating were significant, particularly following the whole-building retrofit. It should be noted that, while these adaptations may marginally increase risks during hot weather, they may significantly reduce mortality risks during cold weather [39]. Cold weather is currently associated with a much higher burden of mortality than hot weather in the UK, and while heat-related mortality is predicted to increase in the future due to climate change, cold-related mortality is expected to remain the greater risk [5,40]. Consequently, modest increases in heat-related risks should not discourage the installation of energy-efficient retrofits, but retrofits should ensure adequate ventilation and in certain cases would be best done in conjunction with adaptations to reduce overheating risk.

While individual dwelling variants showed a range of indoor temperatures, occupant age was the largest risk factor for heat mortality. Targeted interventions found that installation of shutters in dwellings with vulnerable elderly occupants could significantly reduce summertime mortality risk, by 5–33% while only requiring adaptation of 3–12% of the housing stock. Similarly, the scenario where windows are closed increased heat-related mortality risk by as much as 260% in certain dwellings amongst the elderly, and so interventions should also ensure that windows are openable and operable by occupants and that support is provided at a community level for heat vulnerable or low mobility individuals. Housing interventions offer an advantage over local built environment adaptations such as urban greening, in that they may be targeted specifically at the homes of the most vulnerable with lower financial costs. Based on the results, future policies may wish to encourage energy-efficient retrofits in parallel with adaptations to prevent overheating, prioritise the installation of external shutters in dwellings with vulnerable, elderly occupants, and to ensure that the vulnerable can adequately ventilate their houses during hot weather.

There were a number of assumptions necessary in the building physics modelling. We assume a complete implementation of adaptations in either the whole or targeted stock. In the case of shutters, it is assumed that they are functional and closed throughout the day. In reality, this is unlikely to be the case: it may not be possible to install operable shutters in all dwellings, and occupant shutter closing behaviour is likely to have a similar broad range as window-opening behaviours. Amongst the most vulnerable, those currently unable to open windows due to mobility issues will also likely be unable to operate shutters. We have not modelled active heat adaptations such as Air Conditioning (A/C), as they require significant energy expenditure and should be discouraged, and because we assume perfect installation and operation across the housing stock would reduce heat mortality to very low levels. Energy saving calculations from retrofits are presented as an indicator of the maximum potential energy savings, and do not account for occupant 'take-back', where occupants opt for increased thermal comfort rather than the energy savings provided by such adaptations.

The application of the mortality model also has limitations which should be acknowledged. We assume heat exposure occurs in the home. While it is likely much of the population will be out during the day, mortality is dominated by deaths among the more vulnerable groups, who are more likely to spend the day at home. We also assume that the Armstrong heat-mortality risk function—derived using outdoor temperatures—applies to exposures in the indoor environment, a necessary assumption due to a lack of direct evidence on indoor temperatures and mortality. We do not include local variations in outdoor temperature from Urban Heat Island effects in the model due to the lack of spatial information in the EHS, however previous studies have estimated that the UHI leads to an increase of 21–50% in heat-related mortality during hot weather in the West Midlands [15,25]. We therefore assumed that all dwelling variants and occupant age groups have equivalent exposures to elevated UHI temperatures, which may not be the case [19]. Modelled dwelling adaptations, such as white roofs, may themselves affect the UHI. Some dwelling variants, particularly high-rise and converted flats, have small sample sizes (Figure 1), while the sample of these dwelling variants with occupants over 85 is smaller still. This means that mortality estimates for elderly occupants is subject to a large amount of uncertainty due to limited data on building characteristics.

The results highlight the importance of shading and adequate ventilation in housing as temperatures increase, and that targeted adaptation of vulnerable dwellings can reduce summertime heat mortality risk without needing to adapt a large proportion of the existing stock. Adaptations to buildings should be performed in conjunction with other public health measures, such as providing public cool spaces, heatwave advice, and UHI mitigation, while active adaptations such as A/C should be discouraged as this may increase energy consumption. While this study has focused on the West Midlands, UK, the results can provide insight into potential heat exposure, mitigation, and mortality risk in other temperate regions with housing stocks dominated by naturally-ventilated, older dwellings. Areas with large or increasing elderly populations may be at greater risk of heat mortality effects during hot weather. Studies have shown the existence of heat-mortality relationships worldwide [41], and while the threshold and age-specific slope of this relationship may vary internationally, there are opportunities to passively modify housing in order to reduce heat exposure and subsequent heat mortality. While we have estimated mortality—and reductions in heat mortality—under future climate scenarios, we have not accounted for population aging, adaptation to heat, nor any transformation of the housing stock due to demolition and construction. Future research could refine the model to enable predictions of future mortality under a range of climate, population, and adaptation scenarios. In addition, the model will be applied nationally, and using spatially-varying housing data and local air temperatures which include the urban heat island effect.

Author Contributions: J.T. performed the modelling, results analysis, and lead the writing of the paper. P.S., J.T., and A.M. developed the underlying building physics model, while P.W. developed the means of linking indoor temperature exposures to health outcomes. Funding was obtained by P.W. and M.D. All authors contributed to the writing the manuscript.

Acknowledgments: The research was funded by the National Institute for Health Research Health Protection Research Unit (NIHR HPRU) in Environmental Change and Health at the London School of Hygiene and Tropical Medicine in partnership with Public Health England (PHE), and in collaboration with the University of Exeter, University College London, and the Met Office. The views expressed are those of the author(s) and not necessarily those of the NHS, the NIHR, the Department of Health or Public Health England.

Conflicts of Interest: The authors declare no conflict of interest. The founding sponsors had no role in the design of the study; in the collection, analyses, or interpretation of data; in the writing of the manuscript, and in the decision to publish the results.

Appendix A

Figure A1. *Cont.*

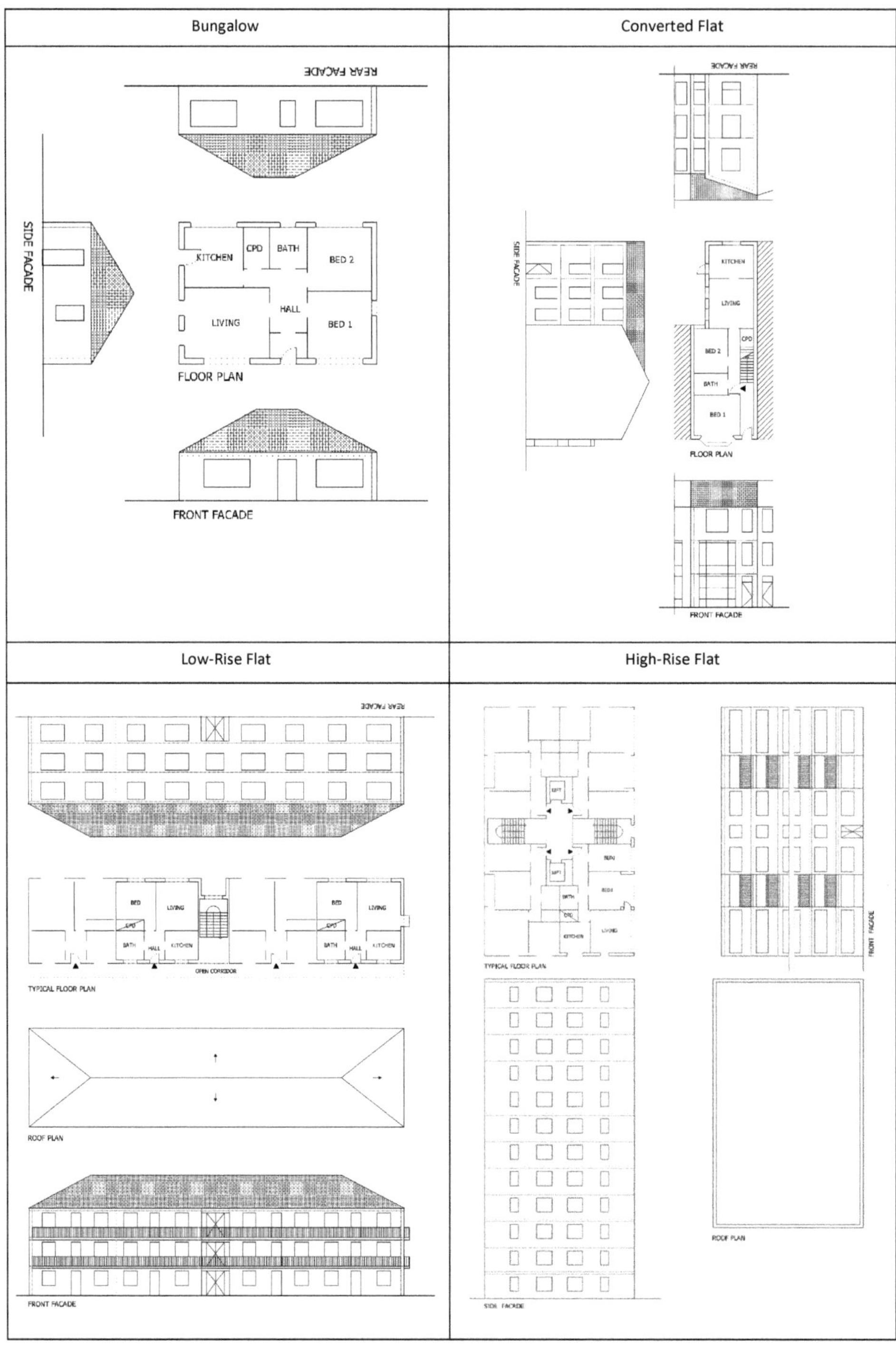

Figure A1. Building Geometries.

References

1. Gasparrini, A.; Armstrong, B.; Kovats, S.; Wilkinson, P. The effect of high temperatures on cause-specific mortality in England and Wales. *Occup. Environ. Med.* **2012**, *69*, 56–61. [CrossRef] [PubMed]
2. Johnson, H.; Kovats, R.S.; McGregor, G.; Stedman, J.; Gibbs, M.; Walton, H.; Cook, L.; Black, E. The impact of the 2003 heat wave on mortality and hospital admissions in England. *Health Stat. Q.* **2005**, *25*, 6–11. [CrossRef]
3. *PHE Heatwave Plan for England 2015*; Public Health England: London, UK, 2015.
4. Murphy, J.; Sexton, D.; Jenkins, G.; Boorman, P.; Booth, B.; Brown, K.; Clark, R.; Collins, M.; Harris, G.; Kendon, E. *UKCP09 Climate Change Projections*; Met Office, Hadley Centre: Exeter, UK, 2009. Available online: http://ukclimateprojections.metoffice.gov.uk/media.jsp?mediaid=87894&filetype=pdf (accessed on 1 October 2017).
5. Hajat, S.; Vardoulakis, S.; Heaviside, C.; Eggen, B. Climate change effects on human health: Projections of temperature-related mortality for the UK during the 2020s, 2050s and 2080s. *J. Epidemiol. Community Health* **2014**, *68*, 641–648. [CrossRef] [PubMed]
6. Zuo, J.; Pullen, S.; Palmer, J.; Bennetts, H.; Chileshe, N.; Ma, T. Impacts of heat waves and corresponding measures: A review. *J. Clean. Prod.* **2015**, *92*, 1–2. [CrossRef]
7. Vandentorren, S.; Bretin, P.; Zeghnoun, A.; Mandereau-Bruno, L.; Croisier, A.; Cochet, C.; Ribéron, J.; Siberan, I.; Declercq, B.; Ledrans, M. August 2003 heat wave in France: Risk factors for death of elderly people living at home. *Eur. J. Public Health* **2006**, *16*, 583–591. [CrossRef] [PubMed]
8. Beizaee, A.; Lomas, K.J.; Firth, S.K. National survey of summertime temperatures and overheating risk in English homes. *Build. Environ.* **2013**, *65*, 1–17. [CrossRef]
9. Lomas, K.; Kane, T. Summertime temperatures in 282 UK homes: Thermal comfort and overheating risk. In Proceedings of the 7th Windsor Conference, the Changing Context of Comfort in an Unpredictable World, Windsor, UK, 12–15 April 2012.
10. Mavrogianni, A.; Wilkinson, P.; Davies, M.; Biddulph, P.; Oikonomou, E. Building characteristics as determinants of propensity to high indoor summer temperatures in London dwellings. *Build. Environ.* **2012**, *55*, 117–130. [CrossRef]
11. Taylor, J.; Mavrogianni, A.; Davies, M.; Das, P.; Shrubsole, C.; Biddulph, P.; Oikonomou, E. Understanding and mitigating overheating and indoor $PM_{2.5}$ risks using coupled temperature and indoor air quality models. *Build. Serv. Eng. Res. Technol.* **2015**, *36*, 275–289. [CrossRef]
12. Taylor, J.; Davies, M.; Mavrogianni, A.; Shrubsole, C.; Hamilton, I.; Das, P.; Jones, B.; Oikonomou, E.; Biddulph, P. Mapping indoor overheating and air pollution risk modification across Great Britain: A modelling study. *Build. Environ.* 2016. [CrossRef]
13. Gupta, R.; Gregg, M. Preventing the overheating of English suburban homes in a warming climate. *Build. Res. Inf.* **2013**, *41*, 281–300. [CrossRef]
14. Porritt, S.M.; Cropper, P.C.; Shao, L.; Goodier, C.I. Ranking of interventions to reduce dwelling overheating during heat waves. *Energy Build.* **2012**, *55*, 16–27. [CrossRef]
15. Heaviside, C.; Cai, X.-M.; Vardoulakis, S. Attribution of mortality to the Urban Heat Island during heatwaves in the West Midlands, UK. *Environ. Health* **2016**, *15*, S27. [CrossRef] [PubMed]
16. Wolf, T.; McGregor, G. The development of a heat wave vulnerability index for London, United Kingdom. *Weather Clim. Extremes* **2013**, *1*, 59–68. [CrossRef]
17. Jenkins, K.; Hall, J.; Glenis, V.; Kilsby, C.; McCarthy, M.; Goodess, C.; Smith, D.; Malleson, N.; Birkin, M. Probabilistic spatial risk assessment of heat impacts and adaptations for London. *Clim. Chang.* **2014**, *124*, 105–117. [CrossRef]
18. Tomlinson, C.J.; Chapman, L.; Thornes, J.E.; Baker, C.J. Including the urban heat island in spatial heat health risk assessment strategies: A case study for Birmingham, UK. *Int. J. Health Geogr.* **2011**, *10*, 42. [CrossRef] [PubMed]
19. Macintyre, H.L.; Heaviside, C.; Taylor, J.; Picetti, R.; Symonds, P.; Cai, X.-M.; Vardoulakis, S. Assessing urban population vulnerability and environmental risks across an urban area during heatwaves—Implications for health protection. *Sci. Total Environ.* **2018**, 610–611. [CrossRef] [PubMed]

20. Loughnan, M.; Nicholls, N.; Tapper, N. Mapping heat health risks in urban areas. *Int. J. Popul. Res.* **2012**, *2012*, 518687. [CrossRef]

21. Klein Rosenthal, J.; Kinney, P.L.; Metzger, K.B. Intra-urban vulnerability to heat-related mortality in New York City, 1997–2006. *Health Place* **2014**, *30*, 45–60. [CrossRef] [PubMed]

22. Reid, C.E.; O'Neill, M.S.; Gronlund, C.J.; Brines, S.J.; Brown, D.G.; Diez-Roux, A.V.; Schwartz, J. Mapping community determinants of heat vulnerability. *Environ. Health Perspect.* **2009**, *117*, 1730–1736. [CrossRef] [PubMed]

23. Taylor, J.; Wilkinson, P.; Davies, M.; Armstrong, B.; Chalabi, Z.; Mavrogianni, A.; Symonds, P.; Oikonomou, E.; Bohnenstengel, S.I. Mapping the effects of Urban Heat Island, housing, and age on excess heat-related mortality in London. *Urban Clim.* **2015**, *14*, 517–528. [CrossRef]

24. Liu, C.; Kershaw, T.; Fosas, D.; Ramallo Gonzalez, A.P.; Natarajan, S.; Coley, D.A. High resolution mapping of overheating and mortality risk. *Build. Environ.* **2017**, *122*, 1–14. [CrossRef]

25. Taylor, J.; Wilkinson, P.; Picetti, R.; Symonds, P.; Heaviside, C.; Macintyre, H.; Davies, M.; Mavrogianni, A.; Hutchinson, E. Comparison of built environment adaptations to heat exposure and mortality during hot weather, West Midlands region, UK. *Environ. Int.* **2017**. [CrossRef] [PubMed]

26. ONS. *2011 Census: Aggregate Data (England and Wales)*; Office for National Statistics: London, UK, 2011; Available online: https://census.ukdataservice.ac.uk/get-data/aggregatedata (accessed on 9 May 2018).

27. DCLG. *English Housing Survey 2010–2011*; Department for Communities and Local Government: London, UK, 2011.

28. Hughes, M.; Armitage, P.; Palmer, J.; Stone, A. *Converting English Housing Survey Data for Use in Energy Models*; Cambridge Architectural Research Ltd.: Cambridge, UK; University College London: London, UK, 2012.

29. Symonds, P.; Taylor, J.; Chalabi, Z.; Mavrogianni, A.; Davies, M.; Hamilton, I.; Vardoulakis, S.; Heaviside, C.; MacIntyre, H. Development of an adaptable England-wide indoor overheating and air pollution model. *J. Build. Perform. Simul.* **2016**, *9*, 606–619. [CrossRef]

30. Symonds, P.; Taylor, J.; Mavrogianni, A.; Davies, M.; Shrubsole, C.; Hamilton, I.; Chalabi, Z. Overheating in English dwellings: Comparing modelled and monitored large-scale datasets. *Build. Res. Inf.* **2016**, *45*, 195–208. [CrossRef]

31. BRE. *Energy Follow Up Survey 2011: Report 9: Domestic Appliances, Cooking, and Cooling Equipment*; Building Research Establishment: Watford, UK, 2013.

32. Stephen, R. *Airtightness in UK Dwellings*; Building Research Establishment: Watford, UK, 2000.

33. Eames, M.; Kershaw, T.; Coley, D. On the creation of future probabilistic design weather years from UKCP09. *Build. Serv. Eng. Res. Technol.* **2010**, *32*, 127–142. [CrossRef]

34. Schaul, T.; Bayer, J.; Wierstra, D.; Sun, Y.; Felder, M.; Sehnke, F.; Rückstieß, T.; Schmidhuber, J. PyBrain. *J. Mach. Learn. Res.* **2010**, *11*, 743–746.

35. BRE. *The Government's Standard Assessment Procedure for Energy Rating of Dwellings*; Building Research Establishment: Watford, UK, 2009.

36. Hong, S.; Ridley, I.; Oreszczyn, T.; Warm Front Study Group. The impact of energy efficient refurbishment on the airtightness in English dwellings. In Proceedings of the 25th AVIC Conference, Prague, Czech Republic, 23–29 August 2004; pp. 7–12.

37. Hamilton, I.; Milner, J.; Chalabi, Z.; Das, P.; Jones, B.; Shrubsole, C.; Davies, M.; Wilkinson, P. Health effects of home energy efficiency interventions in England: A modelling study. *BMJ Open* **2015**, *5*, e007298. [CrossRef] [PubMed]

38. Armstrong, B.G.; Chalabi, Z.; Fenn, B.; Hajat, S.; Kovats, S.; Milojevic, A.; Wilkinson, P. Association of mortality with high temperatures in a temperate climate: England and Wales. *J. Epidemiol. Community Health* **2011**, *65*, 340–345. [CrossRef] [PubMed]

39. Vardoulakis, S.; Dimitroulopoulou, C.; Thornes, J.; Lai, K.-M.; Taylor, J.; Myers, I.; Heaviside, C.; Mavrogianni, A.; Shrubsole, C.; Chalabi, Z.; et al. Impact of climate change on the domestic indoor environment and associated health risks in the UK. *Environ. Int.* **2015**, *85*, 299–313. [CrossRef] [PubMed]

40. Vardoulakis, S.; Heaviside, C. *Health Effects of Climate Change in the UK 2012: Current Evidence, Recommendations and Research Gaps*; Health Protection Agency: London, UK, 2012; pp. 1–242.
41. Gasparrini, A.; Guo, Y.; Hashizume, M.; Lavigne, E.; Zanobetti, A.; Schwartz, J.; Tobias, A.; Tong, S.; Rocklöv, J.; Forsberg, B.; et al. Mortality risk attributable to high and low ambient temperature: A multicountry observational study. *Lancet* **2015**, *386*, 369–375. [CrossRef]

 *atmosphere*

Article

Ambulance Service Resource Planning for Extreme Temperatures: Analysis of Ambulance 999 Calls during Episodes of Extreme Temperature in London, UK

Giorgos Papadakis [1,*], **Zaid Chalabi** [1] **and John E. Thornes** [2,3]

[1] Department of Public Health, Environments and Society, Faculty of Public Health and Policy, London School of Hygiene and Tropical Medicine, 15-17 Tavistock Place, WC1H 9SH London, UK; zaid.chalabi@lshtm.ac.uk
[2] School of Geography, Earth and Environmental Sciences, University of Birmingham, B15 2TT Birmingham, UK; j.e.thornes@bham.ac.uk
[3] Chemicals and Environmental Effects Department, Centre for Radiation, Chemical and Environmental Hazards, Public Health England, Chilton, OX11 0RQ Oxon, UK
* Correspondence: papadakis_giorgos@yahoo.com

Received: 18 March 2018; Accepted: 8 May 2018; Published: 11 May 2018

Abstract: The association between episodes of extreme temperature and ambulance 999 calls has not yet been properly quantified. In this study we propose a statistical physics-based method to estimate the true mean number of ambulance 999 calls during episodes of extreme temperatures. Simple arithmetic mean overestimates the true number of calls during such episodes. Specifically, we apply the physics-based framework of nonextensive statistical mechanics (NESM) for estimating the probability distribution of extreme events to model the positive daily variation of ambulance calls. In addition, we combine NESM with the partitioned multiobjective method (PMRM) to determine the true mean of the positive daily difference of calls during periods of extreme temperature. We show that the use of the standard mean overestimates the true mean number of ambulance calls during episodes of extreme temperature. It is important to correctly estimate the mean value of ambulance 999 calls during such episodes in order for the ambulance service to efficiently manage their resources.

Keywords: ambulance 999 calls; extreme weather; resource planning; London; UK

---

1. Introduction

The impact of extreme weather on the number of ambulance 999 calls has been reported in many studies but is rarely quantified [1–8]. These studies show a significant increase in the number of calls during periods of extreme weather (e.g., heat waves, cold waves).

Alessandrini et al. [1] applied time series analysis to examine the associations between emergency ambulance dispatches and biometeorological discomfort conditions in Emilia-Romagna, Italy. Their study showed a strong relationship between ambulance dispatches and temperature. Dolney and Sheridan [2] studied the relationship between extreme heat and ambulance data response calls for the city of Toronto, Ontario, Canada. They reported that over a four-year period (from 1999 to 2002), the average number of ambulance calls increased by 10 percent over normal levels on those days considered oppressively hot. In further studies, Mahmood et al. [3] analyzed the impact of air temperature on London ambulance call-out incidents and response times and Nitschke et al. [4] analyzed the impact of two extreme heat episodes on morbidity and mortality in Adelaide, South Australia. Schaffer et al. [5] examined emergency department visits, ambulance calls, and mortality associated with the 2011 heat wave in Sydney, Australia. They concluded that the heat wave resulted in an increase in the number of emergency department visits and ambulance calls, particularly in older persons, as well as an increase in all-cause mortality. Ambulance call-outs and response times in Birmingham and the impact of extreme weather and climate change were studied by Thornes et al. [6], who also considered the impact of cold episodes. Turner et al.'s [7] time-series analysis of the association between hot and cold temperatures and ambulance attendances in Brisbane, Australia, suggested that ambulance attendance records can be used in the development of local weather/health early warning systems. Wong and Lai [8] examined the effect of strong weather on the daily demand for ambulance services in Hong Kong suggesting the potential value of developing of a short-term forecast system of daily ambulance demand using weather variables.

The aim of this study is to examine records of recent extreme temperature periods and to estimate and interpret the expected (mean) value of the positive daily variation of ambulance calls in London. More specifically, it is to estimate how many more 999 calls are likely to be received in London during extreme temperature weather events. We argue and show that use of the standard mean of ambulance 999 calls for examining episodes of extreme temperature overestimates the true level and that a combination of statistical physics and risk analysis-based methods provides a better estimate of the mean number of 999 calls.

It should be also noted that not all 999 ambulance calls are acted upon; between the years 2000 and 2014, London Ambulance Service (LAS) only responded to two-thirds of the 999 calls on average. The remaining third include cases where it is clear that the caller does not require an ambulance and/or can be advised to consult 111 or General Practice (GP) services etc.

When calculating the standard mean of a random variable, it is assumed that the variable is "well-behaved". As an example, consider a random variable which exhibits the random values x_1 to x_n. If the random variable is behaving "normally", then the mean value is given by $(x_1 + \ldots + x_n)/n$. In other words, we are assigning equal probabilities $(1/n)$ for each occurrence. However, when the random variable in question exhibits extreme or complex behavior, we can no longer attach equal probability weights (i.e., $1/n$) to the occurrences of the random variable when calculating the mean. We need to introduce instead some bias to the probability weights to account for extreme behavior of the variable. The standard method for calculating means assumes that the random variable during periods when it exhibits extremes has an equal probability of occurrence [9]. In reality, as the value of the random variable increases, the probability of occurrence decreases. The theory of nonextensive statistical mechanics (NESM) is concerned with understanding and analyzing this complexity using Tsallis probabilistic context [10,11]. In addition, a series of publications demonstrate the effectiveness of NESM for the study of extreme phenomena. Our approach is similar to that of Basili [12] who showed that using NESM gives better forecasts of influenza pandemic outbreaks.

2. Method

The description of the method divided into three sections. The first two sections describe the two theories underpinning the analysis, and the third section outlines the application of the methods in six steps.

2.1. Nonextensive Statistical Mechanics

The theory of nonextensive statistical mechanics (NESM), originally introduced by Tsallis [10], is a method for modelling complex systems exhibiting extreme behavior. These include natural processes such as earthquakes, floods, extreme weather events etc., and non-natural phenomena such as extreme fluctuations in the financial market. Tsallis distribution (q-exponential distribution) characterizes such systems and derives its principles from the concept of nonadditive entropy, which is a generalization of the classical Boltzmann–Gibbs (GB) entropy [10,11]. GB entropy characterizes systems which exhibit normal (Brownian) nonextreme fluctuations. In this study, we apply NESM to characterize the ambulance 999 calls. In this case, the Tsallis complementary cumulative distribution function (CCDF) is expressed as:

$$P(D > x) = \left[1 - (1 - q)\frac{x}{x_0}\right]^{\frac{1}{1-q}},\tag{1}$$

where $P(D > x)$ is the probability that the positive daily difference of calls is greater than x; q and x_0 are parameters of the distribution. The key parameter is q which measures deviations from classical entropy i.e., it characterizes extreme and behavior. The closer it is to unity the more normally distributed is x and the further it deviates upward from unity the more extreme x is [13–15]. In other words, the estimated value of q reflects the absence or existence of extreme values of ambulance 999 calls (low or high values). If q deviates significantly from unity, the time-series of ambulance 999 calls exhibits extreme values whereas if q tends to one the 999 calls portray normal (i.e., BG-type) fluctuations. Note that Tsallis distribution (i.e., the q-exponential distribution) recovers the exponential distribution in the limit $q \to 1$. Note also that in this study the calculation of the CCDF is based on the concept of escort probability which is a transformation of the ordinary physical probability when dealing with complex systems (for more information see [11,16,17]). Overall, the Tsallis framework is capable of calculating the degree of correlations in a dynamic system and is also capable of describing transitions of a complex system from a normal random phase (Poissonian processes) to a phase where the system self-organizes towards an extreme phase.

2.2. The Partitioned Multi-Objective Risk Method

The partitioned multiobjective risk method (PMRM) [18] is a risk analysis method to handle extreme and potentially catastrophic risks. The PMRM calculates conditional expected values for different ranges of the variable of interest from the normal to the extreme range. The ranges are specified according to some criteria which are explained below.

2.3. Application of Methods

We outline the methods in six steps:

1. We focus on four extreme temperature episodes in London (see below) and we calculate the positive daily difference (i.e., difference between successive days) of the total number of 999 emergency incidents (calls) for each episode.
2. We divide each of the four extreme temperature episodes into three periods that correspond to the period before (Period A), during (Period B) and straight after (Period C) the extreme weather episode.
3. We estimate two nonclassical statistical-mechanics-based measures of the daily difference in 999 calls: Tsallis exceedance probability distribution function and Tsallis mean value.
4. We partition the estimated Tsallis probability distribution function for each period into three ranges by using the partitioned multiobjective risk method (PMRM). These ranges correspond to low positive daily difference of calls/high exceedance probability (Range 1), medium positive daily difference of calls/medium exceedance probability (Range 2) and high positive daily difference of calls/low exceedance probability (Range 3). The whole range (Range 4) is defined as the union of the three ranges.
5. We show that (i) the calculated Tsallis mean value in Range 3 of Period B is a measure of the increase of the mean value of the positive daily difference of calls during an extreme weather period; (ii) the calculated Tsallis mean value in Range 4 (unconditional mean value of Period B) is an accurate measure of the mean value of the positive daily difference of calls during extreme weather periods. In other words, we propose that Tsallis mean value estimates the true mean of the underlying distribution which describes extreme events.
6. We estimate the exceedance probability for various thresholds of the positive daily difference of calls.

Figure 1 captures the aforementioned steps.

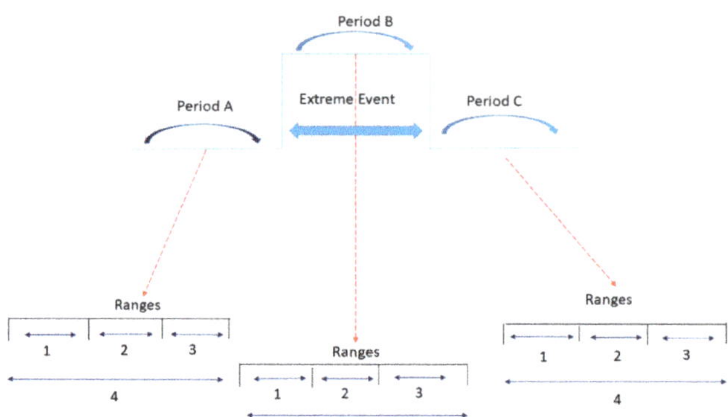

Figure 1. Schematic representation of the methodological steps followed in this study.

3. Data

We examined 999 emergency ambulance calls in London from 1 April 2000 to 31 December 2014. The daily average number of ambulance calls during this period is 3904. Instead of analyzing the whole sequence of 999 emergency ambulance calls (see Figure 2), we chose to focus on four well-reported extreme temperature episodes in the UK as described in Thornes [19,20]. These specific episodes have health relevance. The Heat wave Plan for England [21] defines a heat wave in London expressed as two consecutive days at 32 °C with the night time temperature in between not dropping below 18 °C. On the other hand, the Cold Weather Plan for England [22] gives warnings when the mean temperature for the day is 2 °C or below, which was the case for 16 days out of 31 in December 2010 (Cold spell in December 2010, see Table 1). The choice to study well-documented extreme temperature episodes enables us to scrutinize the probabilistic nature of such episodes. In order to examine the fluctuations of the number of ambulance calls before, during and after the extreme temperature episodes, we selected broader time periods centered at the peak of each episode (for a detailed description see the Analysis section). These periods and the corresponding extreme temperature episodes are given in Table 1.

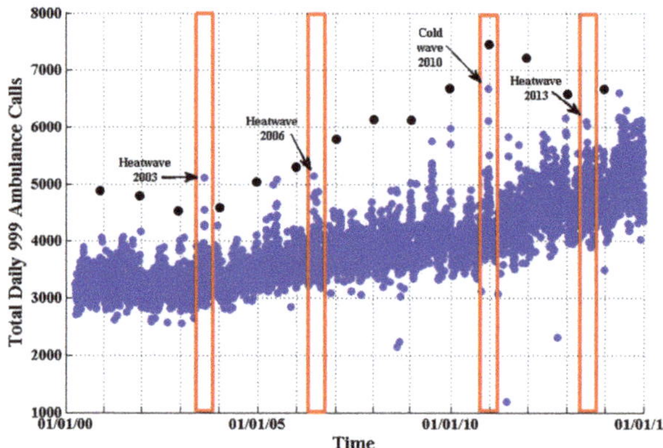

Figure 2. The daily total number of 999 emergency ambulance incidents (calls—blue dots) in London from 1 April 2000 to 31 December 2014. The red rectangles delimit the four extreme temperature episodes which we study in this paper. Seasonal peaks which are the spikes on the 1 January each year are marked with black dots (see text above).

Table 1. Selected periods of study and the associated number of days for four extreme temperature episodes.

Extreme Temperature Episode	Selected Period	Number of Days
Cold spell in December 2010	1 October 2010–28 February 2011	151
Heat wave in August 2003	25 May 2003–24 October 2003	153
Heat wave in July 2006	15 April 2006–16 September 2006	155
Heat wave in July 2013	19 April 2013–21 September 2013	156

Figure 2 shows the daily total number of 999 emergency ambulance incidents (blue dots) in London from 1 April 2000 to 31 December 2014. The red rectangles delimit the four extreme weather episodes which we study in this paper. Significant peaks in the number of calls are observed in December 2010 [6], in August 2003, in July 2006 and in July 2013. It is worth noting that the daily average number of calls has dramatically increased from 2000 to 2014. For example, the daily average number of calls during 2000 was 3235 whereas for 2014 was 5055. Note that this increasing trend has been removed from our analysis by windowing the extreme weather episodes of interest (see next section). Figure 2 suggests that rapid daily variations in urban temperature are strongly related to the 999 ambulance calls particularly during extreme temperature periods. These temperature increases or decreases over short time periods are fully reflected in Figure 3, which shows the positive daily difference of the total number of 999 emergency calls in London from 1 April 2000 to 31 December 2014. The positive daily difference is equal to the difference between the number of calls on one day and that of the previous day if the difference is positive, otherwise it is zero. We used the positive daily difference instead of the absolute number of calls because the time-series is nonstationary and because we are only interested in extreme positive daily deviations. There are also seasonal peaks that correspond to surges in calls on the 1 January each year during New Year's Eve celebrations [19] (see Figure 2) and which seem to disappear from 2013 onwards. We substituted these peaks by the average number of calls. The rationale for windowing the extreme weather episodes (red rectangles) is explained in detail in the following section.

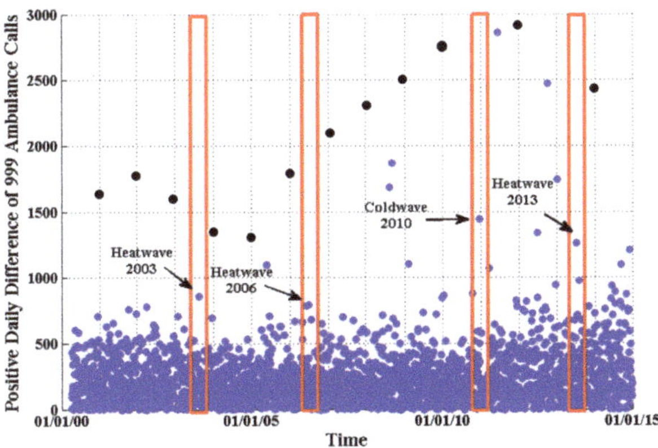

Figure 3. The positive daily difference of the total number of 999 emergency ambulance calls (blue dots) in London from 1 April 2000 to 31 December 2014. The red rectangles delimit the four extreme temperature episodes which we study in this paper. Seasonal peaks which are the spikes on the 1 January each year are marked with black dots (see text above).

4. Analysis

This section presents the method we use to estimate the Tsallis mean value of the positive daily variations of calls during an extreme weather period. We describe the method giving as an example the 2010 cold wave period. Figure 4 shows the daily total number of 999 emergency ambulance calls in London from 1 July 2010 to 31 May 2011. Note that the minimum number of calls for this period is chosen to be 3500. We need to set a minimum threshold (baseline) for the number of calls for each extreme weather episode to avoid analyzing very low numbers of calls that are not caused by temperature variations. They could be due to other causes such as an ambulance operations problem. Overall, despite the fact that our study gives special attention to the occurrence of high numbers of calls, we try to avoid analyzing extremely low numbers of calls unrelated to temperature changes. In other words, we do not calculate positive daily differences for calls below this baseline. For example, we can clearly see in Figure 1 that a plausible baseline for the number of ambulance calls during the 2010 cold wave period is 3500. However, there are two days within this episode which had 3000 calls. If we keep those two days in the dataset the positive daily difference of calls would be erroneously high and unrelated to temperature. In addition, to focus only on peaks that are caused solely by extreme temperature, we substituted the peak observed on the 1 January 2011 (7455 calls) by the average number of calls (4201).

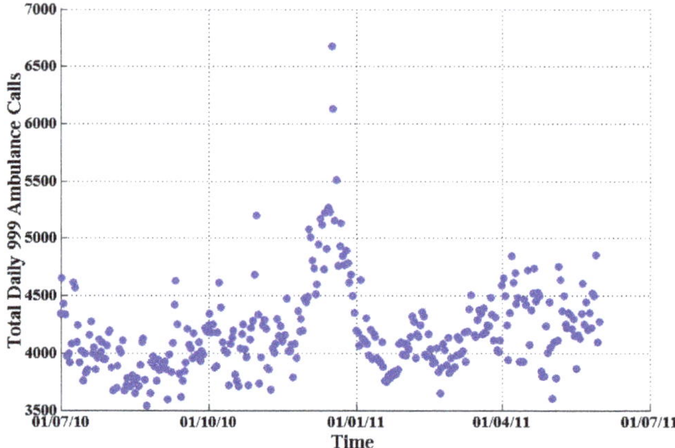

Figure 4. The daily total number of 999 emergency ambulance calls (blue dots) in London from 1 July 2010 to 31 May 2011.

For each extreme weather episode, we specify three periods. We define the extreme weather period as Period B. This is a period of one month in duration centered at the peak of the positive daily difference of calls (Figure 5). Other durations were chosen for sensitivity analysis purposes (see below). Although usually extreme temperature episodes last for a few days (few events) we choose a period of one month in order to have a sufficient number of events (31 events) for calculating the associated probability distribution. Using the actual number of days that concern the extreme temperature episode would lead us to a poor estimation of the probability distribution due to the small number of events selected. We define further two associated periods, one before (Period A) and one straight after (Period C) the extreme weather period (Period B). Periods A and C are two-month periods where we expect "normal" numbers of calls (i.e., corresponding to "normal weather conditions") to be recorded.

As we have explained, we use the positive daily difference of ambulance 999 calls (x) as the measure for analysis and we estimate the parameter q and the Tsallis CCDF for each one of the extreme temperature periods. We estimate the Tsallis probability distribution of the time series and the associated Tsallis expected value (referred as generalized q-expectation value in NESM's theoretical framework, see Tsallis 2009) for each of the datasets. Tsallis probability distribution and the associated expected value are calculated for each period of the four selected extreme weather episodes (i.e., before (Period A), during (Period B) and right after (Period C) the episode). We focus on the Tsallis expected value of the extreme temperature in Period B because it measures the increase of the mean value of the positive daily difference of calls relative to periods A and C. Moreover, as we show in the following sections, by analyzing Period B we obtain an accurate measure of the true mean of the positive daily difference of calls.

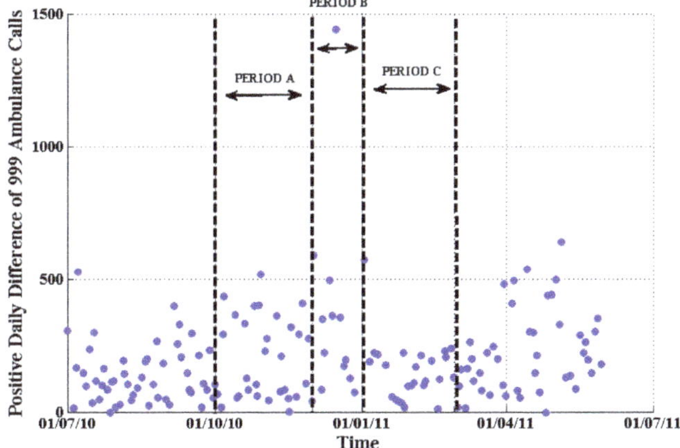

Figure 5. The positive daily difference of calls of 999 emergency ambulance incidents (blue dots) in London from 1 July 2010 to 31 May 2011. The dataset is divided into three periods. Period A corresponds to the period before the extreme temperature, Period B is the extreme temperature period and Period C is after the extreme temperature.

Parameter q is calculated by fitting Tsallis CCDF to the dataset using the Levenberg–Marquardt (LM) algorithm [23] (Figure 6). The LM algorithm is an iterative numerical procedure that is suited to solving nonlinear least squares problems [24–26]. Figure 7 shows the exceedance probability distribution function corresponding to each period for the 2010 cold wave. When the time series exhibits extremes (i.e., during the extreme weather period) the q value acquires its highest value and the distribution becomes "fatter" at the tail end, which means that the probability of getting high values of calls (x) is not small.

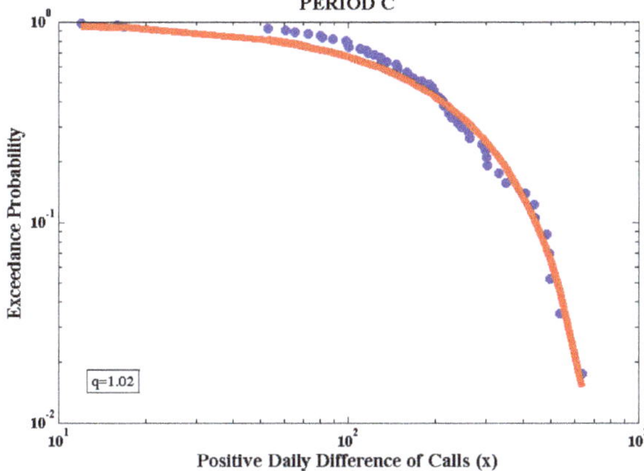

Figure 6. Log–log plot of the exceedance probability distribution function. The dataset (blue dots) and the Tsallis fitting curve (Equation (1), red line) for period C of the 2010 cold wave. The q value is calculated equal to 1.02.

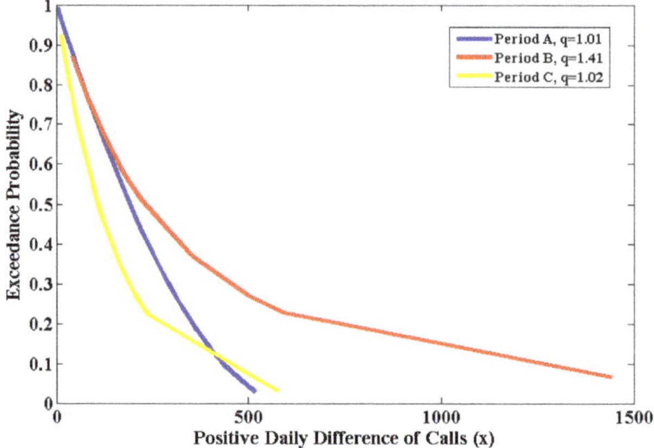

Figure 7. The exceedance probability distribution function corresponding to each period. It should be noted that the axis in this plot is not in logarithmic scale as in Figure 6.

Based on the PMRM method, we partition the exceedance probability distribution function of the positive daily difference of ambulance calls-outs (D) into a number of ranges; three ranges are selected here: low D/high exceedance probability (Range 1), medium D/medium exceedance probability (Range 2) and high D/low exceedance probability (Range 3) (Figure 8). In terms of extreme events, Range 3 is the most important, and the PMRM provides a robust method for its interpretation and analysis. Range 4 includes all the three ranges 1 to 3 (Figure 8).

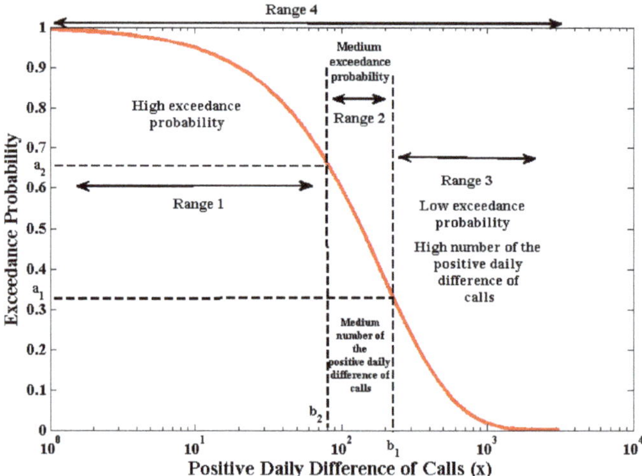

Figure 8. The partition of the positive daily difference of calls into three ranges of the exceedance probability (red line), i.e., high, medium and low exceedance probability, for the whole dataset, in London from 1 July 2010 to 31 May 2011. The x-axis is in logarithmic scale.

Figure 8 shows the partition of the exceedance probability function of the positive daily difference of calls into three ranges of the exceedance probability, i.e., high, medium and low exceedance probability, for the whole dataset, in London from 1 July 2010 to 31 May 2011. Without loss of generality, the partitioning probabilities a_1 and a_2 are selected to be 0.35 and 0.68 respectively (y-axis).

The associated positive daily difference of calls (x-axis) is $b_1 = 220$ and $b_2 = 80$. This selection of the partition probabilities defines Range 2 (medium exceedance probability) as the linear part of the curve of the exceedance probability (red line) shown in Figure 8. The same partitioning probabilities are used throughout this paper to calculate the unconditional and conditional expected values of the positive daily difference of the ambulance 999 calls in the four ranges.

In the analysis, the conditional and unconditional Tsallis expected values are calculated for each of the three periods (A, B, and C) and four ranges by using NESM and the partitioned multi-objective risk method. The results are presented in the following section.

5. Results

5.1. Cold Wave: December 2010

Table 2 shows the estimated q value and Tsallis expected value for the four ranges of each period.

Table 2. The estimated q value and Tsallis expected value for the four ranges of each period.

Period	q	Range 1 (Tsallis Mean Value)	Range 2 (Tsallis Mean Value)	Range 3 (Tsallis Mean Value)	Range 4 (Tsallis Mean Value)
A					
1 October 2010–30 November 2010	1.01	46	28	42	116
B					
1 December 2010–31 December 2010	1.41	37	72	98	207
C					
1 January 2011–28 February 2011	1.02	14	43	45	102

Sensitivity Analysis

To test the sensitivity of the results to the selected duration of periods A, B and C, we performed a sensitivity analysis by changing the duration of Period B. Table 3 shows the estimated q value and Tsallis expected value for the three ranges of each period when Period B is set to last 2 months (instead of one month) centered at the peak of the positive daily difference of calls. Comparing results between Tables 2 and 3 we observe that the estimated values are not affected by the change of the duration of Period B.

Table 3. Sensitivity analysis. The estimated q value and Tsallis expected value for the four ranges of each period when Period B is set to last 2 months.

Period	q	Range 1 (Tsallis Mean Value)	Range 2 (Tsallis Mean Value)	Range 3 (Tsallis Mean Value)	Range 4 (Tsallis Mean Value)
A					
15 September 2010–14 November 2010	1.01	50	26	34	110
B					
15 November 2010–15 January 2011	1.43	25	82	86	193
C					
16 January 2011–15 March 2011	1.02	14	40	41	95

5.2. Heat Wave: August 2003

Figure 9 shows the daily total number of 999 emergency ambulance calls in London from 1 March 2003 to 31 December 2003. The minimum number of calls for this period is chosen to be equal to 2800.

Figure 10 shows the positive daily difference of calls of 999 emergency ambulance calls (blue dots) in London from 1 March 2003 to 31 December 2003.

Table 4 shows the estimated q value and Tsallis expected value for the three ranges of each period.

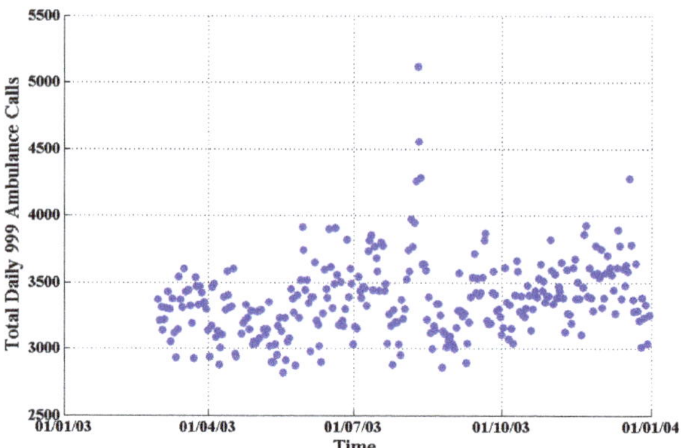

Figure 9. The daily total number of 999 emergency ambulance calls (blue dots) in London from 1 March 2003 to 31 December 2003.

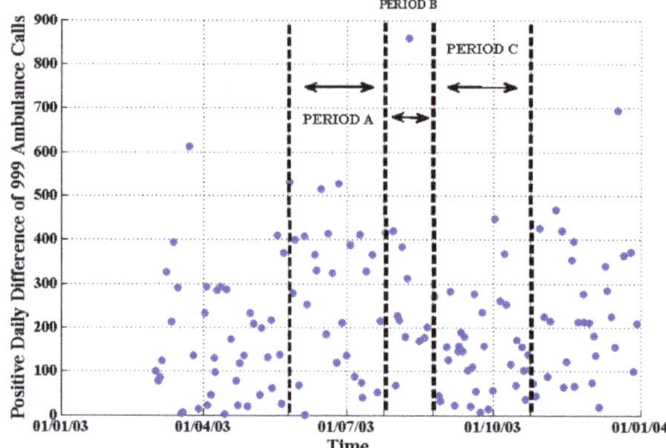

Figure 10. The positive daily difference of calls of 999 emergency ambulance incidents (blue dots) in London from 1 March 2003 to 31 December 2003.

Table 4. The estimated q value and Tsallis expected value for each range of each period.

Period	q	Range 1 (Tsallis Mean Value)	Range 2 (Tsallis Mean Value)	Range 3 (Tsallis Mean Value)	Range 4 (Tsallis Mean Value)
A					
25 May 2003–24 July 2003	1.01	25	60	45	130
B					
25 July 2003–24 August 2003	1.10	47	80	100	227
C					
25 August 2003–24 October 2003	1.01	17	49	28	94

5.3. Heat wave: July 2006

Figure 11 shows the daily total number of 999 emergency ambulance calls in London from 1 February 2006 to 30 November 2006. The minimum number of calls for this period is chosen to be equal to 3200.

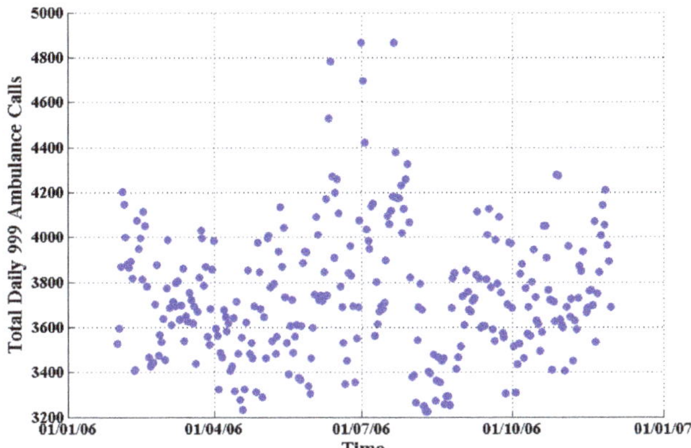

Figure 11. The daily total number of 999 emergency ambulance calls (blue dots) in London from 1 February 2006 to 30 November 2006.

Figure 12 shows the positive daily difference of calls of 999 emergency ambulance calls (blue dots) in London from 1 February 2006 to 30 November 2006.

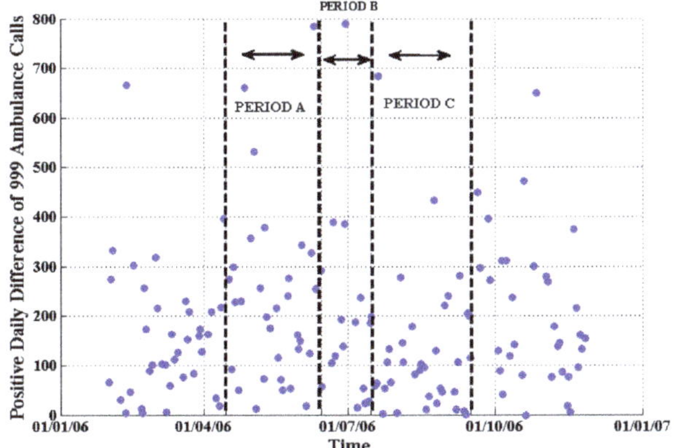

Figure 12. The positive daily difference of calls of 999 emergency ambulance calls (blue dots) in London from 1 February 2006 to 30 November 2006.

Table 5 shows the estimated q value and Tsallis expected value for the four ranges of each period.

Table 5. The estimated q value and Tsallis expected value for the four ranges of each period.

Period	q	Range 1 (Tsallis Mean Value)	Range 2 (Tsallis Mean Value)	Range 3 (Tsallis Mean Value)	Range 4 (Tsallis Mean Value)
A					
15 April 2006–14 June 2006	1.01	18	39	33	90
B					
15 June 2006–15 July 2006	1.38	27	60	61	148
C					
16 July 2006–16 September 2006	1.04	10	34	33	77

5.4. Heat wave: July 2013

Figure 13 shows the daily total number of 999 emergency ambulance calls in London from 1 May 2013 to 30 September 2013. The minimum number of calls for this period is chosen to be equal to 4100.

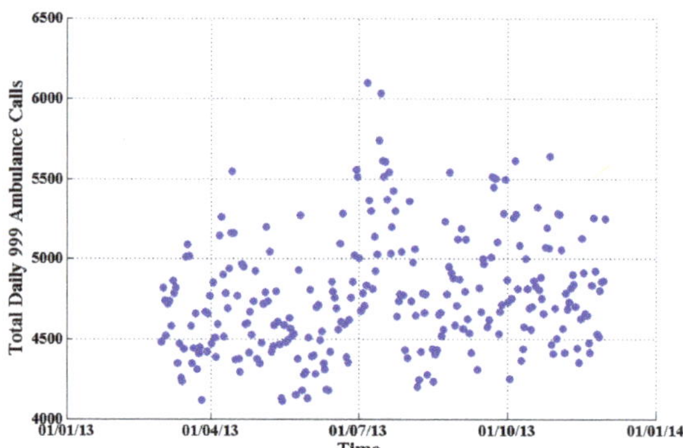

Figure 13. The daily total number of 999 emergency ambulance calls (blue dots) in London from 1 March 2013 to 30 November 2013.

Figure 14 shows the positive daily difference of calls of 999 emergency ambulance calls (blue dots) in London from 1 March 2013 to 30 November 2013.

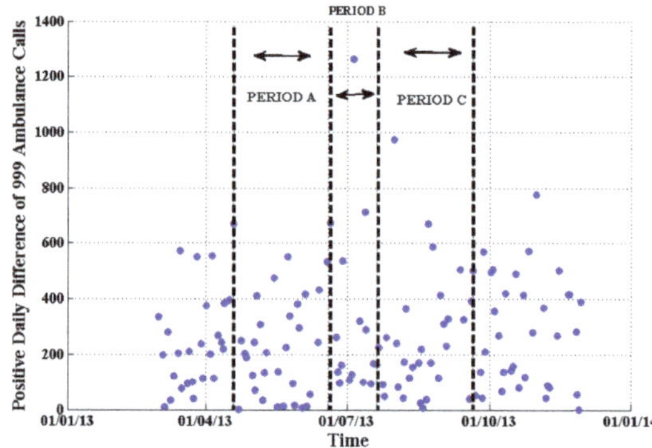

Figure 14. The positive daily difference of calls of 999 emergency ambulance calls (blue dots) in London from 1 March 2013 to 30 November 2013.

Table 6 shows the estimated q value and Tsallis expected value for the four ranges of each period.

Table 6. The estimated q value and Tsallis expected value for the four ranges of each period.

Period	q	Range 1 (Tsallis Mean Value)	Range 2 (Tsallis Mean Value)	Range 3 (Tsallis Mean Value)	Range 4 (Tsallis Mean Value)
A					
19 March 2013–19 June 2013	1.01	21	54	50	125
B					
20 June 2013–20 July 2013	1.24	63	62	77	202
C					
21 July 2013–26 September 2013	1.03	22	60	52	134

6. Discussion and Conclusions

Results show that during extreme temperature periods (Period B) the q parameter deviates from unity and varies from 1.10–1.41. Moreover, Tables 2–6 show that the estimated Tsallis expected value in Range 3 of Period B equals approximately the difference between the estimated Tsallis expected values in Range 4 of Periods B and A. This finding signifies that the calculated expected value in Range 3 of Period B is a measure of the increase of the expected value of the positive daily difference of calls during an extreme temperature period in respect to periods A and C. In addition, the estimated Tsallis expected value in Range 4 (unconditional expected mean of Period B) is a true measure of the expected positive daily difference of calls during extreme weather periods. Table 7 shows that for the four case studies, the mean daily difference of ambulance 999 calls in Range 4 and Range 3 varies between 148–227 and 61–100, respectively. In other words, the positive daily difference of calls (Period B, Range 4) varies between 148 and 227 and the increase of the positive daily difference of calls during an extreme temperature period (Period B, Range 3) varies between 61 and 100. Another significant finding presented in Table 7 is that the use of the standard mean overestimates the true positive daily difference of calls in cases of extreme weather. The standard mean is simply the sum of the positive daily difference of calls divided by their number. This finding shows the importance of using Tsallis distribution to derive the expected values of calls during extreme temperature periods. Assuming that for London each ambulance call-out costs about £250, it is straightforward to appreciate that an inaccurate estimate of the mean of calls when planning ahead for resources can lead the ambulance authorities to overestimate the required resources. In addition, we can approximately determine how many staff and ambulances need to be available in reserve in a near future event of an extreme temperature episode.

Table 7 also shows that with the exception of results for 2006 heat wave, the Tsallis expected value (Range 4) was close to 200. Moreover, Tsallis expected value associated with extreme temperature conditions (Range 3), i.e., the increase of the expected increase in the positive daily difference of calls during an extreme weather period, was estimated to be close to 90.

Table 8 shows the exceedance probability for various thresholds of the positive daily difference of calls. With the exception of the 2006 heat wave similar exceedance probabilities are estimated for the rest of the extreme temperature episodes.

Table 7. The standard and Tsallis values, and the additional Tsallis value due to extreme weather (Range) 3 for the four case studies (extreme weather periods).

Period B	Standard Mean	Tsallis Mean (Mean Value of Calls for Range 4)	Additional Tsallis Mean at Extreme Temperature (Mean Values of Calls for Range 3)
Cold wave: December 2010	348	207	98
Heat wave: August 2003	303	227	100
Heat wave: July 2006	195	148	61
Heat wave: July 2013	338	202	77

Table 8. The exceedance probability for various thresholds of the positive daily difference of calls.

Period B	Exceedance Probability for			
	Positive Daily Difference of Calls $\geq$ 100	Positive Daily Difference of Calls $\geq$ 200	Positive Daily Difference of Calls $\geq$ 400	Positive Daily Difference of Calls $\geq$ 700
Cold wave: December 2010	0.72	0.54	0.34	0.2
Heat wave: August 2003	0.76	0.56	0.31	0.15
Heat wave: July 2006	0.59	0.37	0.16	0.09
Heat wave: July 2013	0.74	0.56	0.34	0.17

How Would the Method Be Used in Practice for Planning Resources? An Illustrative Example

Assume that we are planning ahead now for a future extreme temperature event and we wish to estimate the expected variation in the positive daily difference of ambulance calls in London. We have at our disposal a dataset consisting of the 999 calls on the cold wave in December 2010, the heat wave in August 2003, the heat wave in July 2006 and the heat wave in July 2013. Following the methodology described in this paper, we end up with Tables 7 and 8. By studying these tables, we ascertain that the true positive daily difference of calls during the next extreme temperature episode will vary between 148 and 227 (Table 7 Period B, Range 4). Moreover, we estimate that the increase in the positive daily difference of calls during the next extreme temperature period will vary between 61 and 100 (Period B, Range 3). We can also estimate the probability of occurrence of different levels of the positive daily difference of calls. If instead of the Tsallis mean we used the standard mean, we would have overestimated the true mean value of the positive daily difference of calls for both ranges. Tsallis distribution secures that we assign a legitimate probabilistic context to all calls without neglecting the probability of extreme events (in our case positive daily difference of calls $\geq$700, see Table 8).

Stakeholders can use our findings for resource planning. For example, the ambulance service could estimate the likely consequences of a future extreme temperature weather event as follows. Assuming that the average positive daily difference of 999 calls during 'normal temperature periods' in London is 125 (Table 6, Period A, Range 4), then the ambulance service would expect an increase between of 61 and 100 calls (Table7, Range 3) on day 1 (and for each day) of the extreme temperature episode. If the extreme temperature period lasts for 7 days, then the total increase of the average positive daily difference will vary between 427 and 700 calls. The expected total number of the average positive daily difference of 999 calls during each day of this period will vary between 148 and 227 (Table 7 Period B, Range 4).

By using the estimates obtained in Table 7, we present in Table 9 the expected total number of the positive daily difference of 999 calls during a hypothetical seven-day extreme temperature period using the standard and Tsallis mean for comparison. In addition, we present the expected number of the positive daily difference of additional 999 calls due to the seven-day extreme temperature period relative to a seven-day "normal temperature" period. To obtain Table 9, we simply multiplied the values obtained in Table 7 by seven (number of days of the extreme temperature period). Because the values in Table 7 cover a range over the four modelled extreme events, we present the values in Table 9 as ranges too.

Table 9. The expected number of the positive daily difference of calls during a hypothetical seven-day extreme temperature period using the standard and Tsallis means for comparison (first and second rows). The third row gives the expected number of the positive daily difference of additional calls during the seven-day extreme temperature period relative to a normal seven-day period.

Seven-Day Period of Extreme Temperature	
Standard Mean	1365–2436 is the spread of the total number of the positive daily difference of calls during the seven-day period of extreme temperature.
Tsallis Mean (Range 4)	1036–1589 is the spread of the total number of the positive daily difference of calls during the seven-day period of extreme temperature.
Additional Tsallis mean at extreme temperature (Range 3)	427–700 is the spread of the positive daily difference of additional calls due to the seven-day extreme temperature compared to a normal temperature period.

The spread of values in the cells in Table 9 correspond to those of Table 7 across the four extreme events. It is obvious that the standard mean (first row) overestimates the expected number of 999 calls during this extreme seven-day period compared to the Tsallis mean (second row). Furthermore, in this approach we are able to calculate the true expected additional number of 999 calls compared to a normal period (third row). This shows that the standard method of calculating means during extreme temperature events overestimates the true number.

The conclusions of this study would be strengthened by examining more records of 999 calls during episodes of extreme temperature in other cities in the UK and worldwide. Moreover, future work should include analyzing the temporal variation of the Tsallis q parameter by using event-based moving windows. Such analysis should elucidate further the behavior of 999 calls and the existence of possible patterns or distinct correlations before, during and after an extreme temperature period.

Our findings are not meant to be used for day-to-day operational planning. Table 8 shows that the probability of occurrence of an extreme positive daily difference of calls is not negligible and can be used as a guide for the authorities to plan ahead for the occurrence of the extreme temperature period. Using the current standard mean methods, the ambulance call-out rate during periods of extreme weather is being overestimated. This increases costs and utilized scarce resources. By using the Tsallis mean, a better estimation can be derived that will improve the planning of ambulance resources.

Author Contributions: G.P. and Z.C. designed this study and prepared the first draft of the manuscript. J.E.T. provided the dataset and helped in its interpretation. G.P. performed the data analysis. All authors contributed to the subsequent drafts of the manuscript and approved the final draft.

Acknowledgments: Research funded by the National Institute for Health Research Health Protection Research Unit (NIHR HPRU) in Environmental Change and Health at the London School of Hygiene and Tropical Medicine in partnership with Public Health England (PHE), and in collaboration with the University of Exeter, University College London, and the Met Office. The views expressed are those of the author(s) and not necessarily those of the NHS, the NIHR, the Department of Health or Public Health England. We would like to thank Leanne Smith, Dave Clarke, and Robin Hutchings of London Ambulance Service for their very helpful contributions and enthusiastic support.

Conflicts of Interest: The authors declare no conflict of interest.

References

1. Alessandrini, E.; Sajani, S.Z.; Scotto, F.; Miglio, R.; Marchesi, S.; Lauriola, P. Emergency ambulance dispatches and apparent temperature: A time series analysis in Emilia-Romagna, Italy. *Environ. Res.* **2011**, *111*, 1192–1200. [CrossRef] [PubMed]
2. Dolney, T.J.; Sheridan, S.C. The relationship between extreme heat and ambulance data response calls for the city of Toronto, Ontario, Canada. *Environ. Res.* **2006**, *101*, 94–103. [CrossRef] [PubMed]
3. Mahmood, M.A.; Thornes, J.E.; Pope, F.D.; Fisher, P.A.; Vardoulakis, S. Impact of air temperature on London ambulance call-out incidents and response times. *Climate* **2017**, *5*, 61. [CrossRef]
4. Nitschke, M.; Tucker, G.R.; Hansen, A.L.; Williams, S.; Zhang, Y.; Peng, B. Impact of two recent extreme heat episodes on morbidity and mortality in Adelaide, South Australia: A case-series analysis. *Environ. Health* **2011**, *10*, 42. [CrossRef] [PubMed]
5. Schaffer, A.; Muscatello, D.; Broome, R.; Corbett, S.; Smith, W. Emergency department visits, ambulance calls, and mortality associated with an exceptional heat wave in Sydney, Australia, 2011: A time-series analysis. *Environ. Health* **2012**, *11*, 3. [CrossRef] [PubMed]
6. Thornes, J.E.; Fisher, P.A.; Raymwnt-Bishop, T.; Smith, C. Ambulance calls-outs and response times in Birmingham and the impact of extreme weather and climate change. *Emerg. Med. J.* **2014**, *31*, 220–228. [CrossRef] [PubMed]
7. Turner, L.R.; Connell, D.; Tong, S. Exposure to hot and cold temperatures and ambulance attendances in Brisbane, Australia: A time-series study. *BMJ Open* **2012**, *2*, e001074. [CrossRef] [PubMed]
8. Wong, H.T.; Lai, P.C. Weather inference and daily demand for emergency ambulance services. *Emerg. Med. J.* **2012**, *29*, 60–64. [CrossRef] [PubMed]
9. Gao, J.; Liu, F.; Zhang, J.; Hu, J.; Cao, Y. Information entropy as a basic building block of complexity theory. *Entropy* **2013**, *15*, 3396–3418. [CrossRef]
10. Tsallis, C. Possible generalization of Boltzmann-Gibbs Statistics. *J. Stat. Phys.* **1988**, *52*, 479–487. [CrossRef]
11. Tsallis, C. *Introduction to Nonextensive Statistical Mechanics: Approaching a Complex World*; Springer: New York, NY, USA, 2009; ISBN 978-0-387-85358-1.
12. Basili, M.; Ferrini, S.; Montomoli, E. Swine influenza and vaccines: An alternative approach for decision making. *Eur. J. Public Health* **2013**, *23*, 669–673. [CrossRef] [PubMed]
13. Papadakis, G.; Vallianatos, F.; Sammonds, P. Evidence of nonextensive statistical physics behavior of the Hellenic subduction zone seismicity. *Tectonophysics* **2013**, *608*, 1037–1048. [CrossRef]
14. Papadakis, G.; Vallianatos, F.; Sammonds, P. A nonextensive statistical physics analysis of the 1995 Kobe, Japan earthquake. *Pure Appl. Geophys.* **2015**, *172*, 1923–1931. [CrossRef]
15. Papadakis, G.; Vallianatos, F.; Sammonds, P. Non-extensive statistical physics applied to heat flow and the earthquake frequency-magnitude distribution in Greece. *Phys. A* **2016**, *456*, 135–144. [CrossRef]
16. Abe, S. Geometry of escort distributions. *Phys. Rev. E* **2003**, *68*. [CrossRef] [PubMed]
17. Beck, C.; Schlogl, F. *Thermodynamics of Chaotic Systems: An Introduction*; Cambridge University Press: New York, NY, USA, 1993; ISBN 0-521-43367-3.
18. Haimes, Y.Y. *Risk Modelling, Assessment and Management*, 4th ed.; John Wiley & Sons: Hoboken, NJ, USA, 2016; ISBN 978-1-119-01798-1.
19. *The Impact of Extreme Weather and Climate Change on Ambulance Incidents and Response Times in London*; Pilot Report for the London Ambulance Service; Public Health England: Leeds, UK, 2014; pp. 1–30.
20. Thornes, J.E. Predicting demand through climate data. *Ambulancetoday* **2013**, *10*, 39–41.
21. Heat Wave Plan for England. Available online: https://www.gov.uk/government/publications/heatwave-plan-for-england (accessed on 21 April 2018).
22. Cold Weather Plan for England. Available online: https://www.gov.uk/government/publications/cold-weather-plan-cwp-for-england (accessed on 21 April 2018).
23. Seber, G.A.F.; Wild, C.J. *Nonlinear Regression*, 2nd ed.; John Wiley & Sons: Hoboken, NJ, USA, 2003; ISBN 978-0-471-47135-6.
24. Levenberg, K. A method for the solution of certain non-linear problems in least squares. *Quart. Appl. Math.* **1944**, *2*, 164–168. [CrossRef]

25. Marquardt, D.W. An algorithm for least-squares estimation of nonlinear parameters. *SIAM J. Appl. Math.* **1963**, *11*, 431–441. [CrossRef]
26. More, J.J. The Levenberg-Marquardt algorithm: Implementation and theory. *Lect. Notes Math.* **1978**, *630*, 105–116. [CrossRef]

 *atmosphere*

Article

The Summers 2003 and 2015 in South-West Germany: Heat Waves and Heat-Related Mortality in the Context of Climate Change

Stefan Muthers *, Gudrun Laschewski and Andreas Matzarakis

Deutscher Wetterdienst, Research Center Human Biometeorologie, 79108 Freiburg, Germany; gudrun.laschewski@dwd.de (G.L.); andreas.matzarakis@dwd.de (A.M.)
* Correspondence: stefan.muthers@dwd.de

Academic Editor: Clare Heaviside
Received: 15 August 2017; Accepted: 7 November 2017; Published: 15 November 2017

Abstract: After 2003, another hot summer took place in Western and Central Europe in 2015. In this study, we compare the characteristics of the two major heat waves of these two summers and their effect on the heat related mortality. The analysis is performed with focus on South-West Germany (Baden–Württemberg). With an additional mean summer mortality of +7.9% (2003) and +5.8% (2015) both years mark the top-two records of the summer mortality in the period 1968–2015. In each summer, one major heat wave contributed strongly to the excess summer mortality: In August 2003, daily mortality reached anomalies of +70% and in July 2015 maximum deviations of +56% were observed. The August 2003 heat wave was very long-lasting and characterized by exceptional high maximum and minimum temperatures. In July 2015, temperatures were slightly lower than in 2003, however, the high air humidity during the day and night, lead to comparable heat loads. Furthermore, the heat wave occurred earlier during the summer, when the population was less acclimated to heat stress. Using regional climate models we project an increasing probability for future 2003- and 2015-like heat waves already in the near future (2021–2050), with a 2015-like event occurring about every second summer. In the far future (2070–2099) pronounced increases with more than two 2015-like heat waves per summer are possible.

Keywords: heat-waves; heat-related mortality; 2003; 2015; climate change; Germany

---

1. Introduction

High temperatures are an established risk factor for human health [1]. Heat waves, usually understood as periods of persistent hot conditions, have been found to be associated with sharp increases in mortality all over the globe [2]. Similar effects are found using morbidity data [3]. Understanding how heat waves affect human health is an important key for the preparation and adaptation to the projected increase in heat wave durations and frequencies [4].

The risk of suffering from adverse health effects due to heat stress is related to different physiological or socioeconomic factors. In particular, older, frail people are highly vulnerable to heat stress [5] and patients with chronic deseases, e.g., cardiovascular or respiratory diseases [6]. The physiological mechanisms of the adverse health effects are well known and documented, e.g., dehydration and reduced blood viscocity, which increases the risk for thrombosis [7]. The general stress for the cardiovascular system associated with the work required to maintain thermoregulation induces another risk factor [8]. Moreover, socioeconomic ascpects like living alone [9] or living in dense urban areas [10] can increase the risk of dying during a heat wave.

In Europe, the record breaking summer of 2003 has received particular attention due to the pronounced health impacts. In particular, the August 2003 heat wave lead to large increases in the mortality rates from

Spain [11] over France [12], Italy [13], and Germany [14,15] to Austria [16]. Overall, the heat waves of the summer 2003 caused more than 80,000 additional deaths in 12 European countries [17].

As a consequence of the summer 2003, several European countries installed heat health warning systems (HHWS) to inform the health system and the public of possible threads due to upcoming heat waves [18]. In Germany, for instance, a HHWS is operational since 2005 and raises heat warnings based on the human-biometeorological index Perceived Temperature and a building simulation model [19].

In 2015, another very warm summer took place in Europe. Across Europe several temperature records were set from London over Paris and Berlin to Dobřichovice in the Czech Republic [20,21]. For Germany, the nationwide highest temperature was observed in Kitzingen (central Germany) with 40.3 °C on 5 July. and again on 7 August. While the 2003 heat waves were centered over Western and Central Europe, the heat waves of the summer 2015 were more pronounced over Central and Eastern Europe (Figure 1) [20,22].

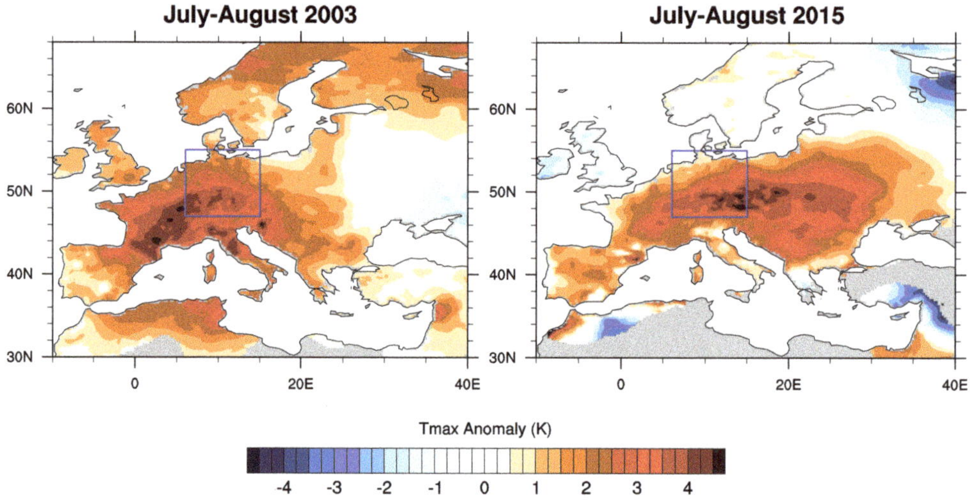

Figure 1. July–August 2003 (**left**) and July–August 2015 (**right**) daily maximum temperature (Tmax) anomalies with respect to 1981–2010 based on the E-OBS dataset (version 16.0) [23]. Within the bounding box of Germany, which is shown by the blue box, a mean July–August daily Tmax anomaly of 2.8 °C is found for both summers.

For Germany, daily maximum temperature July–August anomalies of 2.8 °C with respect to 1981–2010 are found for both summers, when using the E-OBS data set and averaging over an area in Central Europe covering Germany (Figure 1).

In general, an increasing frequency of record breaking heat waves has been observed in the recent decades: from the 2003 heat wave in Western Europe, to the 2010 heat wave in Eastern Europe and Russia, to the heat waves of the summer 2015 [24–26]. For several of these events, an anthropogenic fingerprint could be found: Stott et al. [27], for instance, estimated that the human influence on the climate system has doubled the probability for temperatures extremes as found in 2003. Similar findings apply for the summer heat waves of 2015 [28]. With the ongoing anthropogenic climate change, a further increase in the number of heat waves in the upcoming decades is very likely [4]. Given the pronounced negative health effects of heat waves, improving the adaptation of the population to heat waves is therefore crucial.

The 2003 and 2015 heat waves have been ranked as the second and sixth most severe European events with respect to their intensity and spatial extend [26]. Germany was affected by both heat waves. Here, we present a comparison of the two summers 2003 and 2015 focusing on similarities and differences with respect to the meteorological conditions and the health impact with a focus on South-West Germany. In the context of climate change, we furthermore assess the likelihood of similar heat waves in the near and far future, using regional climate model (RCM) simulations.

2. Data and Methods

2.1. Health Data

Daily mortality data for the federal state Baden–Württemberg (South-West Germany, population 10.9 Mio in 2015) for the period 1968 to 2015 were provided by the Statistisches Landesamt Baden–Württemberg. This absolute all-cause mortality was transformed to mortality rates (deaths per 100.000 inhabitants) to reduce possible biases due to changes in the population size. Therefore, daily population data were linearly interpolated from the annual values. To estimate the additional mortality associated with extreme weather conditions, we calculated the baseline mortality following an approach of Koppe and Jendritzky [29]. Therefore, a 365-day Gaussian smoothing was fitted to the mortality rates in a first step. The smoothing removed any long-term trends from the data, but is only weakly influenced by single influenza episodes or heat waves. The beginning/end of the time series were padded with artificial data, resembling the average annual cycle of the first/last five years, to allow for the computation of the Gaussian filter for the full data period. Since the Gaussian smoothing with a one-year window resulted in an underestimation of the annual cycle, the smoothed time series was adjusted with a correction factor in a second step. The correction factor was chosen in the way, which minimizes the differences between the original values and the smoothed curve. In the following, the baseline mortality is represented by the adjusted 365-day smoothed mortality rates and mortality anomalies were calculated by deviations of the absolute mortality rates from this baseline.

Although this analysis focuses only on the years 2003 and 2015, the full 48 yr period was used to estimate the singularity of the years 2003 and 2015. For comparison to the reference period of the meteorological data, the same period 1971–2000 was used to estimate the standard deviation (σ) of the mortality data. Any long-term trends in the mortality record were removed by the complex approach to estimate the baseline mortality, and do not affect the calculation of σ over the 30 yr period.

2.2. Meteorological Data

Since mortality data was available for South-West Germany, the analysis of the meteorological conditions also focused on this region. Hourly observational measurements of 2 m air temperature and humidity for the weather stations Freiburg, Stuttgart, and Mannheim were extracted from the database of the Deutscher Wetterdienst. For each observation time, the average over the three stations was calculated to represent the average for the federal state Baden–Württemberg. Note, this average can not resemble an area weighted average over the entire federal state, due to the large topographical differences within the federal state, reaching from the Upper-Rhine valley (about 200 m. a.s.l.) to the highest altitudes of the Black Forest and the Swabian Alb (above 1000 m a.s.l). By focusing on the three larger cities—with two of them being situated in the south and the north of the warm and densely populated Rhine valley—this approach allows to estimate the meteorological conditions perceived by the majority of the population of Baden–Württemberg. Estimates for the mean climate, the standard deviation, and percentiles were calculated over the reference period 1971–2000.

In addition to the direct meteorological observations we calculated the index HUMIDEX, which combines the thermal load due to air temperatures and water vapour pressures. HUMIDEX was calculated using hourly air temperature and dew point observations [30]. In a first step the HUMIDEX was computed for each station. In a second step the Baden–Württemberg average was derived as described above. The daily minimum and maximum values of the parameters considered were calculated based on the hourly average values of the Baden–Württemberg average.

2.3. Climate Model Data

16 RCM experiments for Europe from the World Climate Research Program Coordinated Regional Downscaling Experiment (EURO-CORDEX) initiative [31] were used to estimate future changes in the number of extreme heat waves per summer (Table 1). All simulations were performed with a high resolution of 12.5 km for the past (historical) and the future under the Representative Concentration Pathways (RCPs) RCP4.5 and RCP8.5. For two of the 16 RCMs, the RCP4.5 simulation was not available. The boundary conditions for the RCMs were taken from different global models (compare Table 1).

Table 1. Overview of the CORDEX EUR-11 simulations used in this study [31].

RCM	Driving GCM	Modelling Center	Scenarios
CLM	CanESM2		historical, RCP8.5
CLM	CNRM-CM5		historical, RCP4.5, RCP8.5
CLM	EC-EARTH	CLM Community with contributions	historical, RCP4.5, RCP8.5
CLM	MIROC5	by BTU, DWD, ETHZ, UCD, WEGC	historical, RCP8.5
CLM	HadGEM2-ES		historical, RCP4.5, RCP8.5
CLM	MPI-ESM-LR		historical, RCP4.5, RCP8.5
CNRM	CNRM-RM5	Météo France	historical, RCP4.5, RCP8.5
HIRHAM5	EC-EARTH	Danish Meteorological Institute, Copenhagen, Denmark	historical, RCP4.5, RCP8.5
WRF	IPSL-CM5A-MR	Laboratoire des Sciences du Climat et de l'Environnement, IPSL, CEA/CNRS/UVSQ and Contributors INERIS Institut National de l'Environnement Industriel et des Risques, Verneuil en Halatte, France/Institut Pierre Simon Laplace, CNRS, France	historical, RCP4.5, RCP8.5
RACMO22E	EC-EARTH	Royal Netherlands Meteorological Institute,	historical, RCP4.5, RCP8.5
RACMO22E	HadGEM2-ES	Ministry of Infrastructure and the Environment	historical, RCP4.5, RCP8.5
RCA4	CNRM-CM5		historical, RCP4.5, RCP8.5
RCA4	EC-EARTH		historical, RCP4.5, RCP8.5
RCA4	CM5A-MR	Rossby Centre, Swedish Meteorological and	historical, RCP4.5, RCP8.5
RCA4	MPI-ESM-LR	Hydrological Institute, Norrkoping Sweden	historical, RCP4.5, RCP8.5
RCA4	HadGEM2-ES		historical, RCP4.5, RCP8.5

For each simulation, the data at the grid points of Freiburg, Stuttgart, and Mannheim was extracted and the HUMIDEX was calculated based on the modeled daily mean 2 m temperature and dew point temperature. The Baden–Württemberg average was calculated by averaging over the three stations.

The analysis of the RCM data was performed for three different time frames: 1971–2000 was used as reference period to identify possible biases between model simulations and observations. The future projections were evaluated for the near (2021–2050) and far (2070–2099) future. The uneven period for the far future is related to the fact that not all model simulations were available until the year 2100.

The model spread is quantified by the interquantile range (IQR, resembling the box area of a boxplot), which is defined by the range which contains exactly 50% of the values of a distribution. Model results and range are given as multi-model mean, separately for each scenario and time period.

3. Results

3.1. Comparison of the 2003 and 2015 Heat Waves

Focusing first on the meteorological summer season (Figure 2), we found an JJA daily maximum temperature anomaly of 5.6 °C and 3.2 °C for 2003 and 2015, respectively, corresponding to 4.7 and 2.7 standard deviations of the interannual variability of the 30 yr period 1971–2000. With these anomalies, the two summers correspond to the two warmest summers since 1968, with respect to the average daily maximum temperature.

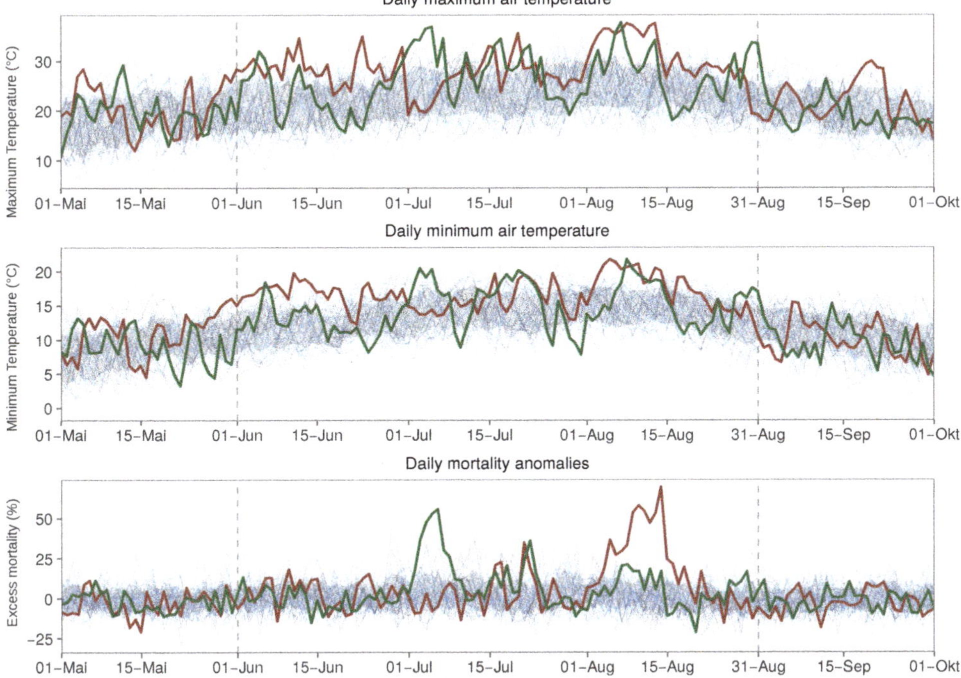

Figure 2. Daily maximum temperature (**top**), daily minimum temperature (**middle**), and daily excess mortality (**bottom**) from May to September in Baden-Württemberg. Beginning and end of the climatological summer season is highlighted by the vertical gray dashed lines. The summer 2003 is displayed in red; green lines denote the conditions of the summer 2015. Other years of the period 1968–2015 are shown by the thin gray lines in the background. The gray shading denotes the standard deviation of the corresponding parameter, estimated over the reference period 1971–2000.

In the JJA mortality data similar anomalies occurred. Averaged over the summer season a daily excess mortality of 7.9% and 5.8% was found for 2003 and 2015, respectively (3.9 and 2.6 standard deviations), making the two summers the summers with the highest summer mortality since 1968. With respect to the average population size of the years 2003 and 2015 these anomalies correspond to about 1770 additional deaths during the summer 2003 and about 1380 additional deaths in 2015.

To identify possible anomalies in the preceding seasons or potential effects of mortality displacement in the seasons following the summer, we briefly assessed the mortality anomalies for all seasons of the years 2003 and 2015. In the winter seasons (December to February) prior to the summers 2003 and 2015, a flu epidemic took place, however, only in 2015 a pronounced increase in the mortality anomalies of the winter season was found (compare Table 2). During the spring season (March to May) of 2003 and 2015, mortality was close to normal. Similarly, the mortality anomalies during autumn (September to November) and winter are within the normal range of variability.

Table 2. Seasonal mortality anomalies (%) with respect to the baseline for the years 2003 and 2015.
[†]: For 2015 only December was available, therefore, no DJF anomaly could be calculated.

Season	2003	2015
December–February (previous)	0.9	7.2
March–May	1.1	0.9
June–July	7.9	5.8
September–November	1.3	0.3
December–February (following)	1.1	–[†]

Furthermore, no pronounced periods of negative anomalies took place within the summer seasons (Figure 2). In 2015 a few days with mortality anomalies below the long term range of variability are visible around August 20, and in 2003, a tendency towards below average mortality values can be found towards the end of August. Overall, however, negative anomalies are very rare, suggesting no distinct mortality displacement.

Heat related mortality, however, is particularly sensitive to multi-day periods with enhanced heat stress, i.e., heat waves (e.g., [32]). In Figure 2 one major event with enhanced mortality values and high temperatures can be identified per summer: In 2003 a pronounced increase of the mortality rates took place in August, with strongly positive values for up two weeks. In 2015 a period of persistent positive mortality anomalies occurred in early July, lasting for about 9 days. Furthermore, a few days with positive mortality values show up in the second half of July for both summers and a weakly positive event in the first half of August 2015.

The major deviations of the mortality values, however, were found in early July 2015 and early August 2003. The development of the minimum and maximum air temperature, dew point temperature, and HUMIDEX together with the mortality anomalies are shown in Figure 3. The dates are shifted to a common relative time axis with day zero (2 August 2003 and 1 July 2015) being the first day with a daily maximum temperature exceeding the 95th percentile. Since the focus of this analysis is on the heat-related mortality during theses events, we display the time series of the variables, until the mortality anomalies reach again the level of background variables (anomalies $< \sigma$). Daily air temperature maxima (Figure 3a) and minima (Figure 3d) also exceeded the 90th, 95th, and 99th percentile for several consecutive days, e.g., 11 days in a row are above the 99th percentile of the daily maximum temperature in 2003 and 7 above the 99th percentile of the daily minimum temperature. In 2015, the air temperature values were not as extreme as in 2003, however, 5 and 4 consecutive days above the 95th percentile for the daily maximum and minimum temperature, respectively, are found as well.

Besides the high temperature, also the water vapour content of the atmosphere is of major importance for the perception of heat stress, mainly due to its impacts on the heat loss of the body by evaporation. The daily minimum and maximum dew point temperature (Figure 3b,e) reveal some differences between the two heat waves. In 2003, daily maximum dew points temperatures were almost within the normal range and reached the 90th percentile threshold only during the last two days of the heat wave. During the night, the daily minimum temperature was even below the climatological average. In June 2015, however, very high daily maximum and minimum dew points were measured, related to the transport of warm and moist air from the Mediterranean region into the Rhine valley. The 95th percentile was reached for 5 days in a row for the daily maximum dew points and for 3 (not consecutive) days for the daily minimum values.

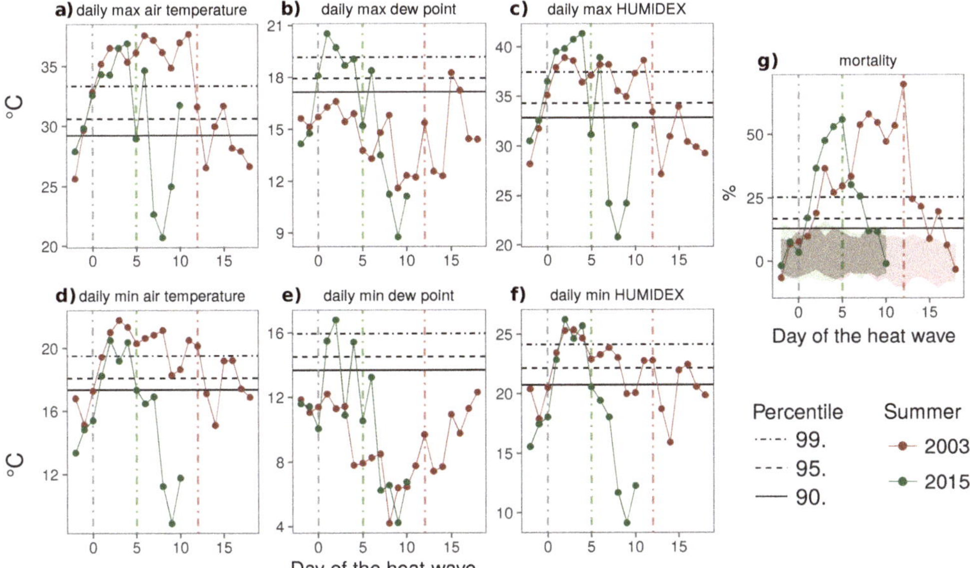

Figure 3. Time series of (**a–f**) a number of heat related meteorological parameters and (**g**) daily mortality anomalies (%) for the major heat wave of (red) 2003 and (green) 2015. For each parameter the daily values are displayed together with the 90th, 95th, and 99th percentile threshold calculated over the summer season (JJA) for the period 1971–2000 shown by the horizontal lines. The gray vertical dashed line indicates the beginning of the heat waves; colored vertical dashed lines mark the maximum of the mortality anomalies during the heat wave. (**a**) Daily maximum air temperature (°C); (**b**) daily maximum dew point (°C); (**c**) daily maximum HUMIDEX (°C); (**d**) daily minimum air temperature (°C); (**e**) daily minimum dew point (°C); and (**f**) daily minimum HUMIDEX (°C); (**g**) daily mortality anomalies (%). Shading denotes the background variability in the mortality values estimated for the period 1971–2000, for the time of the year of the 2003 (red) and the 2015 (green) heat wave.

The combined effect of the air temperature and the humidity is described by the HUMIDEX (Figure 3c,f). Accordingly, the higher temperature in 2003 and the higher humidity in 2015 compensate and led to high HUMIDEX values for both events. From the beginning of the heat wave up to one day before the maximum in the mortality data, all daily maximum HUMIDEX values exceeded the 95th percentile threshold for both heat waves (5 and 12 consecutive days for 2015 and 2003, respectively). Furthermore, the daily minimum HUMIDEX passed the 95th percentile for most of the heat wave days. In terms of absolute values, the highest HUMIDEX values were found in 2015, both for the daily maximum and minimum.

When focusing on the mortality time series, the close relationship between the heat load and the mortality becomes clear. In 2003, the mortality anomalies grew constantly until the maximum of +70% was reached after 11 days (14 August; Figure 3g). After this date, mortality began to decline for four days and almost reached the levels of background variability after 15 days at 19 August. In 2015, mortality increased for five days to a maximum anomaly of +56% (6 July). From this date on, mortality decreased for three days until 9 July. The reduction of the mortality anomalies after the maximum was in both cases very fast.

The extraordinary nature of the health impacts of these two heat waves is highlighted by the fact, that both events cross the 99th percentile of the daily summer (JJA) mortality values for 6 (2015) and 10 (2003) consecutive days. Per definition, less than one day per summer is expected to exceed this threshold in an average summer. With respect to the linear interpolated daily population data, about 1390 additional deaths were counted during the August 2003 heat wave. In the July 2015 heat, about 700 cases above the base line were registered. Both heat waves, therefore, contributed to a great extent to the excess deaths of the corresponding summer season.

Figure 3 furthermore suggest a slight lag between the meteorological conditions and the signal in the mortality data. The mortality anomalies lagged behind the daily maximum HUMIDEX exceeding the 95th percentile by 1 or 2 days, for 2015 and 2003, respectively. Consequently, the maximum in the mortality data occured for both heat waves on days where the meteorological parameters indicate a clear reduction of the heat stress (dashed vertical lines in Figure 3c). These patterns suggest a sustained health effect of the heat wave for at least one day.

3.2. The 2003 and 2015 Heat Waves in the Context of Climate Change

In the context of climate change, the number of heat waves is very likely to increase in the next decades [4]. In the following, we quantify the probability for heat waves comparable to the ones of 2003 and 2015 in the near and far future using RCM simulations.

RCM data is usually not available in a temporal resolution higher than daily means. Therefore, the analysis shown in Figure 3 was repeated using the daily mean HUMIDEX (Figure 4a). This approach led to very comparable results. The heat wave of 2003 was now characterized by 8 consecutive days above the 95th percentile, while for 2015, 5 consecutive days were found. To generalize these findings, we assessed the change in heat wave frequencies for all events where the daily mean HUMIDEX exceeds the 95th percentile for 3 to 10 consecutive days. Therefore, we calculated the 95th temperature percentile for each model separately to reduce the influence of model biases on the analysis [33]. The 95th temperature percentile was calculated using the reference period 1971–2000.

First, the RCM's capability to simulate these heat waves was assessed (Figure 4b). In the reference period the RCMs tend to overestimate the average number of heat waves per year for all durations. Heat waves with a duration of three days, for instance, occured on average every second summer in the observations (0.5 events/yr), while the multi model ensemble median suggest a frequency of about 0.6 events/yr. While for most durations at least a few models simulate frequencies comparable to the observations, the medium-length durations of 5 and 6 days are clearly overestimated by the models. This overestimation of the heat wave persistence in the CORDEX simulations has been discussed earlier [33] and may be attributed to a misrepresentation of land-atmosphere feedbacks (e.g, soil moisture, surface energy fluxes).

The projected changes are calculated separately for each model with respect to the model mean value of the reference period, to reduce some effect of the overestimated persistence. In the near future (2021–2050) and under RCP4.5, a clear increase in the frequency is found for all durations. Short term events (3 days) already increase by 0.9 events/yr (multi-model median). Furthermore, the very strong 10 day events increase by about one event per decade (0.1 events/yr) in comparison to less than one event in 30 yrs in the reference period (multi-model median 0.03 events/yr). No pronounced differences are found between RCP4.5 and RCP8.5 for the near future. Increases are in general slightly larger in RCP8.5, but much smaller than the differences between the models. On this time scale, differences in the emission scenarios have no systematic effect on the heat wave frequencies.

In the far future (2070–2099), however, the differences between the scenarios emerge. For RCP4.5 a further increase is found for all classes. However, at least for the shorter events, a weakening of the trend is found, i.e., the changes 1971–2000 to 2021–2050 are larger than the changes from 2021–2050 to 2070–2099. The three day heat waves, for instance, increase by 1.5 events/yr relative to the reference period (0.6 relative to 2021–2050). In case of the extreme long-lasting heat waves, changes are still small, but nevertheless pronounced. 10 day events may increase by 0.3 events/yr, suggesting one additional extreme heat wave every third summer.

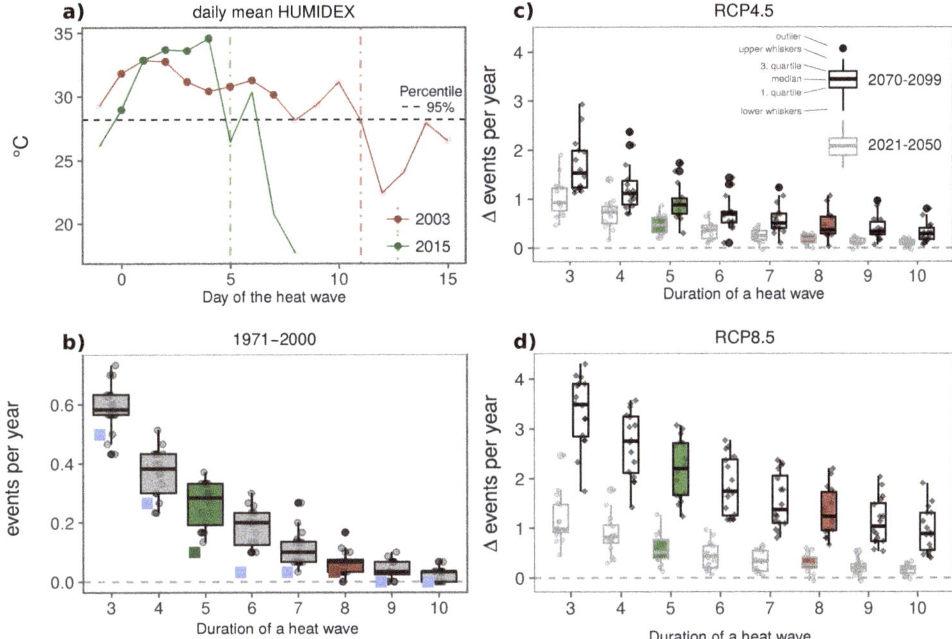

Figure 4. (**a**) Similar to Figure 3 but for the daily mean HUMIDEX; (**b**) Boxplot statistics for number of heat waves per year in the RCM simulations. Heat waves are defined by 3 to 10 consecutive days with a daily mean HUMIDEX above the 95th percentile. Dots resemble the results from the individual simulations; colored squares denote the number of events per year derived from observations. The events of 2003 (8 consecutive days) and 2015 (5 consecutive days) are highlighted by the red and green color; (**c**) Boxplot statistics for the change of heat waves frequencies with a duration between 3 and 10 days in the near (2021–2050) and far (2070–2099) future for the RCP4.5 scenario relative to the model mean value for 1971–2000. Dots resemble the individual RCM simulations; (**d**) Same as (**c**), but for RCP8.5.

Still, in comparison to the RCP8.5 the changes found for RCP4.5 are moderate. For all duration the intensification is always larger than the changes from 1971–2000 to 2021–2050. 3 day heat waves increase by 3.5 events/yr, the 2015-like events with 5 day duration increase by 2.2 events/yr (two additional 2015-like heat waves in every summer), and a 2003-like event is also likely to take place once a summer (multi-model median increase by 1.2 events/yr). The strongest 10 day heat wave, finally, increases by 0.9 events/yr until the end of the 21th century.

4. Discussion and Conclusions

12 years after the extra ordinary heat wave of 2003, another serious heat wave took place in Central Europe. Similar to 2003, the 2015 event lead to a very exceptional increase of the mortality. While in 2003 persistent extreme high air temperature lead to high heat loads, the 2015 event was moreover characterized by high dew point temperatures, causing very sultry conditions.

Similar to 2003, the heat wave of 2015 was not limited to South-West Germany. For large parts of Western Europe exceptional high temperatures and humidity values were observed in 2015 and an increase of the mortality is likely to be found also in other countries. In Switzerland, a comparable mortality increase of +6.9 and +5.4% was reported for summer 2003 and 2015, respectively [34].

For both heat waves our results show, that the maximum in the mortality values occurred one day after the maximum of the wave, when the meteorological parameters indicated a clear reduction of the heat stress. An explanation for this effect can be found in the up to two-day lag, between outdoor heat stress and the indoor conditions [35]. In particular, since many people, especially elderly or sick people which are most vulnerable to heat stress, typically spend a large part of their time inside buildings. Heat warning systems should therefore consider the indoor thermal conditions as well and heat-intervention strategies should focus also to the days following a heat wave.

In this study we applied the HUMIDEX, a combined index which considers the thermal effects of air temperature and humidity. Heat stress perceived by human beings, however, is not only governed by these two parameters. Therefore, the applied HUMIDEX may not cover all possible heat stress situations. This caveat may be avoided by the application of complex human-biometeorological indices, e.g., the Perceived Temperature [36]. In this study, when using RCM simulations to project future changes in heat waves, a heat stress indicator based on temperature and humidity only is a more reliable index, since more sophisticated indices rely on additional input parameters (wind speed, short-wave and long-wave radiation fluxes or cloud cover), which are associated with larger biases in RCMs (e.g., [37]) and are often not available in an appropriate temporal resolution.

Two main differences between the two heat waves should be mentioned. Firstly, the duration of the events. The health impact of heat waves increases with the duration (e.g., [29,38,39]). In this context, the 2003 heat wave, with a duration of two weeks, was very exceptional. Secondly, the timing of a heat wave is important. Heat waves occurring early during the season cause stronger health impacts than heat wave towards the end of the season [32,40,41], due to short-term acclimatisation effects. This is an additional factor explaining the health effects of the 2015 heat wave, which took place in early July, about one months earlier than the 2003 heat wave.

Furthermore the sensitivity of the population to heat stress is not stationary in time. In our study, for instance, where the mortality data was normalized by the population size, a change in the age structure may have changed the sensitivity of the population to heat waves. Between 2003 and 2015, the ratio of elderly (65 years and more), which are more vulnerable to heat stress, has increases from about 17 to 20%. Consequently, the same heat wave is expected to lead to a higher health impact in 2015 in comparison to 2003, if no adaptation has taken place.

One adaptation measure, which was implemented after the 2003 event, is the German HHWS. From 30 June to 7 July 2015 warnings of strong and extreme heat stress were issued by the German Meteorological Service (DWD) for large parts of Germany. A quantification of the influence of these warnings on the heat related mortality in 2015, however, is currently not possible, given the societal changes and the differences between the heat waves with respect to the duration, the time of the year, and the meteorological conditions. More work is needed to quantify this aspect.

For the projected change in heat related mortality, long-term adaptation needs to be considered [42,43] and several different methods exists, to assess long-term adaptation to heat stress for the future [44]. Estimates of long-term adaption for long-lasting heat waves, however, are associated with large uncertainties, given the fact that heat waves with a duration of more than 5 days occured only once or less during the reference period. Therefore, we did not transfer the projected increase in heat wave frequencies into an increase in the heat related mortality. Some long-term adaptation might reduce the future health impact, in particular for the shorter heat waves. The projected increase for all heat waves duration—and in particular the pronounced increases in very long lasting events—and for all scenarios in the near and far future, however demands for an intensification of climate change mitigation and adaptation efforts.

Acknowledgments: We gratefully thank the Statistisches Landesamt Baden-Württemberg for providing the mortality records. Furthermore, we acknowledge the E-OBS dataset from the EU-FP6 project ENSEMBLES (http://ensembles-eu.metoffice.com) and the data providers in the ECA&D project (http://www.ecad.eu).

Atmosphere **2017**, *8*, 224

Author Contributions: Stefan Muthers conceived and designed the study. Stefan Muthers and Gudrun Laschewski gathered data and performed the analysis. Stefan Muthers, Gudrun Laschewski, and Andreas Matzarakis wrote the paper.

Conflicts of Interest: The authors declare no conflict of interest.

References

1. Koppe, C.; Kovats, S.; Jendritzky, G.; Menne, B.; Baumüller, J.; Bitan, A.; Jiménez, J.D.; Ebi, K.L.; Havenith, G.; Santiago, C.L.; et al. *Heat-Waves: Risks and Responses*; Technical Report 2; World Health Organization: Geneva, Switzerland, 2004.
2. Mora, C.; Dousset, B.; Caldwell, I.R.; Powell, F.E.; Geronimo, R.C.; Bielecki, C.R.; Counsell, C.W.W.; Dietrich, B.S.; Johnston, E.T.; Louis, L.V.; et al. Global risk of deadly heat. *Nat. Clim. Chang.* **2017**, *7*, 501–506.
3. Ye, X.; Wolff, R.; Yu, W.; Vaneckova, P.; Pan, X.; Tong, S. Ambient temperature and morbidity: A review of epidemiological evidence. *Environ. Health Perspect.* **2012**, *120*, 19–28.
4. IPCC. Summary for Policymakers. In *Climate Change 2013: The Physical Science Basis. Contribution of Working Group I to the Fifth Assessment Report of the Intergovernmental Panel on Climate Change*; Stocker, T., Qin, D., Plattner, G.K., Tignor, M., Allen, S., Boschung, J., Nauels, A., Xia, Y., Bex, V., Midgley, P., Eds.; Cambridge University Press: Cambridge, UK; New York, NY, USA, 2013; pp. 3–29.
5. Flynn, A.; McGreevy, C.; Mulkerrin, E.C. Why do older patients die in a heatwave. *Int. J. Med.* **2005**, *98*, 227–229.
6. Kovats, S.; Shakoor, H. Heat Stress and Public Health: A Critical Review. *Annu. Rev. Public Health* **2008**, *29*, 41–55.
7. Keatinge, W.R.; Coleshaw, S.R.K.; Easton, J.C.; Cotter, F.; Mattock, M.B.; Chelliah, R. Increased platelet and red cell counts, blood viscosity, and plasma cholesterol levels during heat stress, and mortality from coronary and cerebral thrombosis. *Am. J. Med.* **1986**, *81*, 795–800.
8. Havenith, G. Temperature Regulation, Heat Balance and Climatic Stress. In *Extreme Weather Events and Public Health Responses*; Kirch, W., Menne, B., Eds.; Springer: Berlin/Heidelberg, Germany, 2005; pp. 69–80.
9. Kovats, R.S.; Ebi, K.L. Heatwaves und public health in Europe. *Eur. J. Public Health* **2006**, *16*, 592–599.
10. Gabriel, K.M.A.; Endlicher, W.R. Urban and rural mortality rates during heat waves in Berlin and Brandenburg, Germany. *Environ. Pollut.* **2011**, *159*, 2044–2050.
11. Díaz, J.; García-Herrera, R.; Trigo, R.M.; Linares, C.; Valente, M.A.; De Miguel, J.M.; Hernández, E. The impact of the summer 2003 heat wave in Iberia: How should we measure it? *Int. J. Biometeorol.* **2006**, *50*, 159–166.
12. INVS. Impact sanitaire de la vague de chaleur d'août 2003 en France. In *Rapport Annuel, Institut de Veille Sanitaire*; INVS: Saint-Maurice, France, 2003. Available online: http://www.invs.sante.fr/publications/2003/bilan_chaleur_1103/ (accessed on 14 November 2017).
13. Michelozzi, P.; de'Donato, F.; Accetta, G.; Forastiere, F.; D'Ovidio, M.; Perucci, C.; Kalkstein, L. Impact of Heat Waves on Mortality–Rome, Italy, June–August 2003. *JAMA J. Am. Med. Assoc.* **2004**, *291*, 2537–2538.
14. Heudorf, U.; Meyer, C. Gesundheitliche Auswirkungen extremer Hitze-am Beispiel der Hitzewelle und der Mortalität in Frankfurt am Main im August 2003. *Gesundheitswesen* **2005**, *67*, 369–374.
15. Hoffmann, B.; Hertel, S.; Boes, T.; Weiland, D.; Jöckel, K.H. Increased Cause-Specific Mortality Associated with 2003 Heat Wave in Essen, Germany. *J. Toxicol. Environ. Health A* **2008**, *71*, 759–765.
16. Muthers, S.; Matzarakis, A.; Koch, E. Summer climate and mortality in Vienna—A human-biometeorological approach of heat-related mortality during the heat waves in 2003. *Wien. Klin. Wochenschr.* **2010**, *122*, 525–531.
17. Robine, J.M.; Cheung, S.L.; Le Roy, S.; Van Oyen, H.; Herrmann, F.R. *Report on Excess Mortality in Europe During Summer 2003*; EU Community Action Programme for Public Health; European Commission: Brussels, Belgium, 2007.
18. WMO. *Heatwaves and Health: Guidance on Warning-System Development*; WMO: Geneva, Switzerland, 2015; p. 114.
19. Matzarakis, A. The Heat Health Warning System of DWD—Concept and Lessons learned. In *Perspectives in Atmospheric Sciences*; Karacostas, T., Bais, A., Nastos, P.T., Eds.; Springer International Publishing: Basel, Switzerland, 2017; pp. 191–196.
20. Dong, B.; Sutton, R.; Shaffrey, L.; Wilcox, L. The 2015 European Heat Wave. *Bull. Am. Meteorol. Soc.* **2016**, *97*, S57–S62.

21. Ionita, M.; Tallaksen, L.M.; Kingston, D.; Stagge, J.H.; Laaha, G.; Van Lanen, H.A.; Scholz, P.; Chelcea, S.M.; Haslinger, K. The European 2015 drought from a hydrological perspective. *Hydrol. Earth Syst. Sci.* **2017**, *21*, 1397–1419.

22. Hoy, A.; Hänsel, S.; Skalak, P.; Ustrnul, Z.; Bochníček, O. The extreme European summer of 2015 in a long-term perspective. *Int. J. Climatol.* **2017**, *37*, 943–962.

23. Haylock, M.R.; Hofstra, N.; Klein Tank, A.M.G.; Klok, E.J.; Jones, P.D.; New, M. A European daily high-resolution gridded data set of surface temperature and precipitation for 1950–2006. *J. Geophys. Res. Atmos.* **2008**, *113*. doi:10.1029/2008JD010201.

24. Christidis, N.; Jones, G.S.; Stott, P.A. Dramatically increasing chance of extremely hot summers since the 2003 European heatwave. *Nat. Clim. Chang.* **2014**, *5*, 46–50.

25. Perkins, S.E. A review on the scientific understanding of heatwaves - Their measurement, driving mechanisms, and changes at the global scale. *Atmos. Res.* **2015**, *164*, 242–267.

26. Russo, S.; Sillmann, J.; Fischer, E.M. Top ten European heatwaves since 1950 and their occurrence in the coming decades. *Environ. Res. Lett.* **2015**, *10*, 124003.

27. Stott, P.A.; Stone, D.A.; Allen, M.R. Human contribution to the European heatwave of 2003. *Nature* **2004**, *432*, 610–613.

28. Sippel, S.; Otto, F.E.L.; Flach, M.; van Oldenborgh, G.J. The Role of Anthropogenic Warming in the 2015 Central European Heat Waves. *Bull. Am. Meteorol. Soc.* **2016**, *97*, S51–S56.

29. Koppe, C.; Jendritzky, G. Inclusion of short-term adaptation to thermal stresses in a heat load warning procedure. *Meteorol. Z.* **2005**, *14*, 271–278.

30. Masterton, J.; Richardson, F. *Humidex: A Method of Quantifying Human Discomfort Due to Excessive Heat and Humidity*; Ministere de l'Environnement: Gatineau, QC, Canada, 1979; p. 49.

31. Jacob, D.; Petersen, J.; Eggert, B.; Alias, A.; Christensen, O.B.; Bouwer, L.M.; Braun, A.; Colette, A.; Déqué, M.; Georgievski, G.; et al. EURO-CORDEX: New high-resolution climate change projections for European impact research. *Reg. Environ. Chang.* **2014**, *14*, 563–578.

32. Brooke Anderson, G.; Bell, M.L. Heat waves in the United States: Mortality risk during heat waves and effect modification by heat wave characteristics in 43 U.S. communities. *Environ. Health Perspect.* **2011**, *119*, 210–218.

33. Vautard, R.; Gobiet, A.; Jacob, D.; Belda, M.; Colette, A.; De, M.; Goergen, K.; Gu, I.; Karacostas, T.; Katragkou, E.; et al. The simulation of European heat waves from an ensemble of regional climate models within the EURO-CORDEX project. *Clim. Dyn.* **2013**, *42*, 2555–2575.

34. Vicedo-Cabrera, A.; Ragettli, M.; Schindler, C.; Röösli, M. Excess mortality during the warm summer of 2015 in Switzerland. *Swiss Med. Wkly.* **2016**, *146*, 1–12.

35. Sharag-Eldin, A.; Atchison, K.; Heller, C.; Gharehgozlou, S.; Molana, H.H.; Lucak, W. An Autopsy of an Environmental Tragedy: 1995 Chicago Heat Wave. In Proceedings of the EAAE ARCC 10th International Conference, Lisbon, Portugal, 15–18 June 2016.

36. Staiger, H.; Laschewski, G.; Grätz, A. The perceived temperature—A versatile index for the assessment of the human thermal environment. Part A: Scientific basics. *Int. J. Biometeorol.* **2012**, *56*, 165–176.

37. Schoetter, R.; Hoffmann, P.; Rechid, D.; Schlünzen, K.H. Evaluation and bias correction of regional climate model results using model evaluation measures. *J. Appl. Meteorol. Climatol.* **2012**, *51*, 1670–1684.

38. Díaz, J.; García, R.; de Castro Velázquez, F.; Hernández, E.; López, C.; Otero, A. Effects of extremely hot days on people older than 65 years in Seville (Spain) from 1986 to 1997. *Int. J. Biometeorol.* **2002**, *46*, 145–149.

39. Matzarakis, A.; Muthers, S.; Koch, E. Human biometeorological evaluation of heat-related mortality in Vienna. *Theor. Appl. Climatol.* **2011**, *105*, 1–10.

40. Hajat, S.; Kovats, R.S.; Atkinson, R.W.; Haines, A. Impact of hot temperatures on death in London: A time series approach. *J. Epidemiol. Commun. Health* **2002**, *56*, 367–372.

41. Kyselý, J.; Huth, R. Heat-related mortality in the Czech Republic examined through synoptic and 'traditional' approaches. *Clim. Res.* **2004**, *25*, 265–274.

42. Muthers, S.; Matzarakis, A.; Koch, E. Climate Change and Mortality in Vienna—A Human Biometeorological Analysis Based on Regional Climate Modeling. *Int. J. Environ. Res. Public Health* **2010**, *7*, 2965–2977.

43. Zacharias, S.; Koppe, C.; Mücke, H.G. Climate Change Effects on Heat Waves and Future Heat Wave-Associated IHD Mortality in Germany. *Climate* **2014**, *3*, 100–117.
44. Gosling, S.N.; Hondula, D.M.; Bunker, A.; Iberretta, D.; Liu, J.; Zhang, X.; Sauerborn, R. Adaptation to climate change: A comparative analysis of modelling methods for heat-related mortality. *Environ. Health Perspect.* **2017**, *125*, 087008-1–087008-14.

Review

Temporal Trends in Heat-Related Mortality: Implications for Future Projections

Patrick L. Kinney

Department of Environmental Health, Boston University School of Public Health, 715 Albany Street, Talbot 4W, Boston, MA 02118, USA; pkinney@bu.edu; Tel.: +1-617-358-2469

Received: 30 May 2018; Accepted: 21 September 2018; Published: 18 October 2018

Abstract: High temperatures have large impacts on premature mortality risks across the world, and there is concern that warming temperatures associated with climate change, and in particular larger-than-expected increases in the proportion of days with extremely high temperatures, may lead to increasing mortality risks. Comparisons of heat-related mortality exposure-response functions across different cities show that the effects of heat on mortality risk vary by latitude, with more pronounced heat effects in more northerly climates. Evidence has also emerged in recent years of trends over time in heat-related mortality, suggesting that in many locations, the risk per unit increase in temperature has been declining. Here, I review the emerging literature on these trends, and draw conclusions for studies that seek to project future impacts of heat on mortality. I also make reference to the more general heat-mortality literature, including studies comparing effects across locations. I conclude that climate change projection studies will need to take into account trends over time (and possibly space) in the exposure response function for heat-related mortality. Several potential methods are discussed.

Keywords: heat-related mortality; climate change; trends over time

---

1. Introduction

This paper summarizes and discusses key findings on trends over time in the effects of temperature on human mortality. While not intended to be a comprehensive review of the temperature–mortality relationship literature, I start by briefly summarizing main findings from broad literature. I focus on mortality, although there is also some evidence emerging from the morbidity literature on trends in impacts [1].

The effect of temperature on mortality is the most extensively studied topic within the broad domain of climate and health research, with a reference that covers a broad range of methodologies [2]. A limited number of studies have quantified deaths listed as heat-related on death certificates, most often accumulating individual case reports from medical records [3]. However, heat-related deaths identified by medical records tend to capture only cases with clear heat involvement, which represent only a subset of all heat-related deaths [2]. A recent study found that less than 10% of excess heat-related deaths were labeled as such on death certificates from 1997 to 2013 in New York City [4]. Another type of epidemiology study quantifies excess mortality that occurs in a city or region during a clearly identified heat wave event, as compared to during non heat-wave periods in the same locale. This was for example the approach used by researchers to quantify the impacts of the 1995 Chicago heat wave [5] and the 2003 Paris heat wave [6]. However, heat-related deaths also occur during periods when not obvious heat waves happen, such as when occasional temperature spikes that occur in most of summers. To more comprehensively quantify the overall burden of heat on mortality, a third epidemiologic approach uses regression analysis of multi-year daily time series to quantify exposure–response relationships linking temperature and mortality [7,8]. These latter studies usually include deaths due to all causes, or all causes minus "external" causes such as homicide and suicide, and report substantial impacts in association with high temperatures. Studies that analyze the entire distribution of temperatures also report that cold temperatures are associated with increased mortality risk; however, the extent to which these deaths are caused by low temperature, as opposed to seasonal respiratory infections that co-vary with temperature, has been questioned [9,10]. The specific impact of winter temperature on mortality is a key area of uncertainty in projecting the health effects of climate change. However, the present study only focuses on heat-related deaths.

Multi-city and age-stratified analyses have examined vulnerability factors that can explain differences in heat-mortality effects between cities. Increased heat risk is associated with old or young age, living alone, preexisting chronic diseases, poverty, and low prevalence of air conditioning (A/C) [7,8,11]. Thus, to estimate the quantitative impact of future temperatures on mortality, we need to understand not only how climate may change, but also how these vulnerability factors may evolve in the future.

2. Spatial/Climatic Differences in Temperature Impacts

One very consistent finding from the time series literature is that the shape of the exposure–response function (ERF) differs by latitude (i.e., prevailing climate) [7,8,12–14]. A classic figure from an early study is reproduced in Figure 1, showing the ERFs from 11 U.S. cities [7]. Southern cities show small or non-existent heat effects, but substantial cold effects. Conversely, northern cities show less pronounced cold effects but larger heat effects. Additionally, the lowest point on the curve (termed "minimum mortality temperature" (MMT)) tends to shift to higher temperatures in southern, warmer cities. It is important to note that Figure 1 displays the raw relationship between temperature and mortality, not controlled for seasons. As a result, the "cold" effect is likely substantially overestimated.

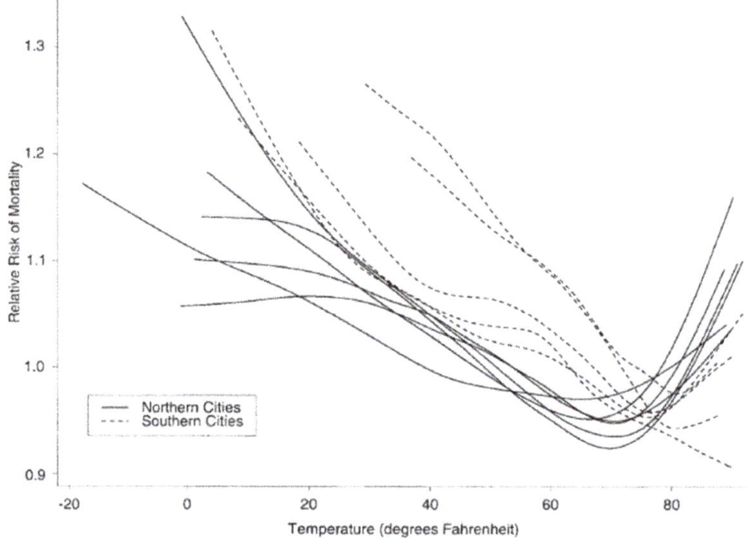

FIGURE 1. Temperature-mortality relative risk functions for 11 US cities, 1973–1994. Northern cities: Boston, Massachusetts; Chicago, Illinois; New York, New York; Philadelphia, Pennsylvania; Baltimore, Maryland; and Washington, DC. Southern cities: Charlotte, North Carolina; Atlanta, Georgia; Jacksonville, Florida; Tampa, Florida; and Miami, Florida. °C = 5/9 × (°F − 32).

Figure 1. From Curriero et al., 2002 [7] showing temperature-mortality risk functions for 11 US cities.

These findings support the concept that populations adapt to climate conditions typical in their cities. This means the populations exhibit health responses mainly at temperatures that are extreme within the local context. Hondula and colleagues defined four classes of climate adaptation from the following aspects: physiological (referred to as acclimatization), behavioral (e.g., avoidance; use of A/C), infrastructural (e.g., white roofs and green infrastructures), and technological (e.g., heat warning systems, and more efficient A/C) [15]. It seems likely that all of these factors play a role, with the relative importance of each varying with settings, populations, and health outcomes of interest. Among the studies that provide empirical evidence of differential temperature effects by location, Anderson and Bell's analysis of 107 U.S. cities from 1987 to 2000 [8] is noteworthy in analyzing factors that modify temperature effects by location. Prevalence of A/C is one significant predictor of the differences across cities in heat effects. Barreca and colleagues [16] also reported higher heat–mortality effects in cooler climates, based on a nationwide, state-level analysis. Further insights into spatial differences in exposure response as a function of local climate were provided by Lee and colleagues who analyzed data from 148 U.S. cities from 1973 to 2006 [17]. Cities were grouped into 8 clusters based on weather patterns. As shown below in Figure 2, heat and cold effects differed across clusters as a function of temperature, with more pronounced cold effects—steeper slopes—in warmer clusters, and lower thresholds for heat effects, but similar slopes, in cooler clusters. It would be tempting to use these findings to develop empirical adaptation functions by relating parameters of the cluster-specific ERFs to cluster-specific climate variables such as seasonal mean temperature. Further evidence supporting the concept that populations adapt to local temperatures has been shown in an international study across over 300 cities [13,14]. Guo and colleagues found that MMTs vary with the mean temperature across countries in a surprisingly consistent way. Still, the authors noted that the exposure–response relationship between climate indicators and temperature-related mortality is not a simple one, and cautioned against using these relationships in a quantitative way to project future impacts.

At a finer spatial scale, one innovative study in France reported an analysis of heat-related mortality within 30×30 km grids across the entire country [18]. This is the only example in the literature to date where health and environmental data have been analyzed within a regular grid over a region, rather than within administrative areas. Within each grid, non-linear exposure-response functions were fit, and the MMTs computed. There was a strong correlation (0.90) between MMTs and mean summer temperatures (MSTs) across grid squares. This suggests that another way to project adaptation might be to model within-country associations between MMTs and MSTs in the current climate, and then adjust future MMTs based on changing future MSTs.

The literature on geographical differences in temperature–mortality ERFs shows that effects vary substantially depending on local climate, and imply that populations eventually adapt to local conditions. They say nothing about the time course over which adaptation occurs. Still, it is tempting to hypothesize based on these findings that future populations would also adapt to changing climatic conditions, at least once a new steady state climate is achieved [19]. A key question is "what does the pace of adaptation look like while climate is on a changing trajectory from historical conditions to a future steady state?".

One way to address this question is by looking at trends over time in temperature–mortality ERFs in a given location as climate changes. However, detecting a climate change-induced adaptation signal from these trends is problematic for several reasons. First, climate has warmed by only about 1 °C over the past century, and health datasets often span only a fraction of this period; thus, the climate-driven trend in adaption would be expected to be small within the observed record. Secondly, there may be trends in other factors that, while not directly related to climate change, can have a profound impact on heat-health effects. These include trends in urbanization, income, housing, the built environment, indoor/outdoor activity patterns, access to healthcare, chronic disease prevalence, and others. One such trend has been the rapid increase in A/C prevalence in the past 3–4 decades in the U.S. In the following section, I examine the literature on temporal trends in temperature–mortality ERFs.

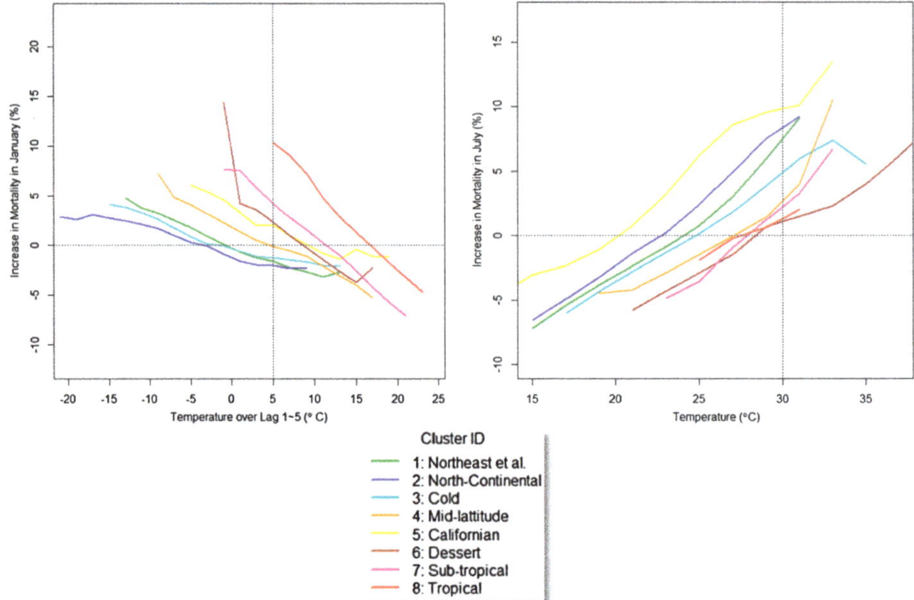

Figure 2. From Lee et al., 2014 [17]. Temperature-mortality risk functions by US region. At (**left**) is effect of temperature on mortality in January. At (**right**) is effect of temperature on mortality in July. Between 8–36 cities are included in each region, with effects summarized using meta regression.

3. Studies of Trends over Time in Heat Impacts

In addition to the literature documenting geographical/climatological differences in temperature–mortality ERFs, recently, there has been growing literature on temporal changes in these effects. Recent review papers document a generally decreasing trend over time in heat-related mortality ERFs, particularly in North America, with less consistent findings in Europe, and Asia [20–22]. Evidence shows that the quantitative effect of heat on mortality has been decreasing in most, but not all, cities where studies have been carried out. The reasons for this decreasing risk have not been clearly identified, but may include: enhanced heat-health awareness and prevention measures, general improvements in population health, and technological changes such as increases in residential A/C prevalence. There are virtually no mortality trend studies for low-income countries, nor for persons who are exposed outdoors because of work, homelessness or recreational activities. Where examined, there has been little evidence that cold effects on mortality have decreased over time [21].

One of the first studies to examine trends over time in heat-related mortality was carried out by Davis and colleagues [23], which documented declining heat effects in 28 U.S. cities over the years 1964–1998. More recently, Bobb and colleagues [24] analyzed data from 105 U.S. cities over 1987–2005 and reported more than a 60% drop in the mortality effect per 5.5 °C (10 °F) rise in same-day temperature. The authors hypothesized that A/C may play a role in this trend, but they were unable to show that rates of decline in health effects by city were related to differences in rates of A/C adoption. Their inability to detect statistically significant A/C effects may have been due in part to the limitations of available A/C data over time, as well as the lack of A/C data that are specific to population groups most affected by high temperatures, such as the poor and elderly. The authors speculated that declines over time in cardiovascular mortality rates may contribute to reduced heat vulnerability.

In contrast to the findings in [24], Barreca et al. [25] reported that A/C prevalence largely explains temporal and spatial differences in heat effects on mortality in U.S. over the 20th century when analyzing at the levels of state and month. This study, and related econometric work by Deschenes and colleagues [26,27] stands apart in methodology by using monthly or annual data aggregated at the state level, rather than the daily time series, city-level approach used in the epidemiologic literature. One advantage of the econometric approach is that, by analyzing data in monthly or annual segments, it likely avoids biases in effect estimates due to short-term harvesting. It could also be argued that annual statistics are more relevant to future climate impact projections, which are usually aggregated to annual or decadal statistics. An interesting review of the advantages and limitations of the econometric approach is provided in [27].

While focusing only on New York City, the work of Petkova and colleagues [28] is noteworthy because it reported heat effects over multiple decades of the 20th century, from 1900 to 2006. They reported a marked decrease in the ERF for heat-related mortality between the first five decades of the 20th century and the most recent four decades. In recent decades, the downward trend appeared to slow somewhat, suggesting a leveling off of the ERF (See Figure 3).

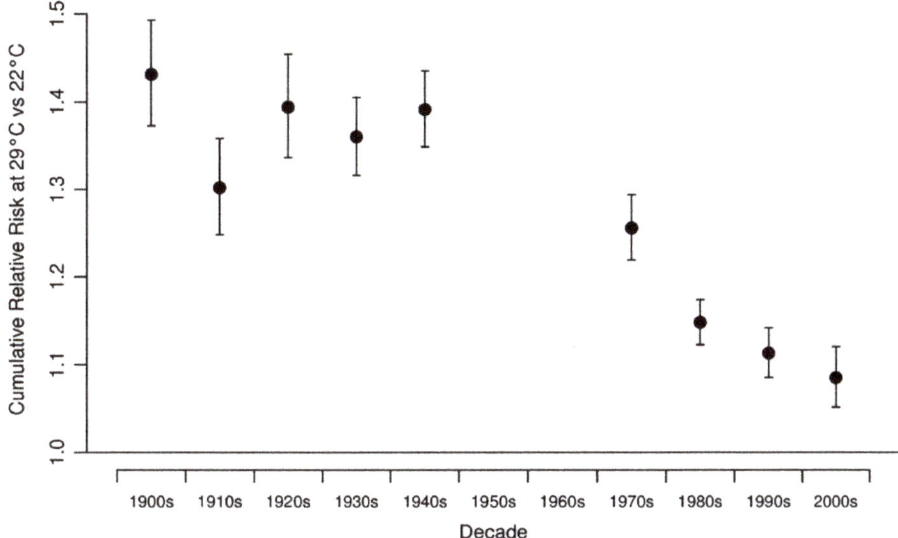

Figure 3. Decadal heat relative risks and 95% confidence intervals in NYC. Note that there is a data gap in mid-century. From Petkova et al., 2014 [28].

Nordio and colleagues [29] reported declines in heat-related mortality but no change in cold related mortality, from 1962 to 2006 in 211 U.S. cities. This is the largest U.S. daily time series study in both spatial and temporal coverages. Cities were divided into 8 climatological clusters, and analyzed in six 7-year segments. Key results are summarized in Figure 4.

Astrom and colleagues [30] reported a decrease in heat effects in Stockholm over the 20th century, and there was some evidence of a leveling off in recent decades. No change in cold effects was observed. In a follow-up study, the same team reported a steady rise in the MMT over the century [31]. As noted above, the MMT may prove to be a useful metric to model changes over time and space in temperature–mortality ERFs. In France, Todd and Valleron also reported a rise in MMTs over time (1968–2009) as well as a strong dependence of the MMT on the MST [18]. (MMT and MST were correlated at 0.9 in all periods cross sectional correlation analysis.) The rise in the MMT with time corresponded to 0.44 °C per degree temporal rise in the MST. The cross-sectional increase in MMT was 0.76 °C per degree rise in MST. The ratio of these quantities (0.58) may offer one measure of the time lag in adaptation due to warming temperatures, in a country where A/C is probably less of an issue in general.

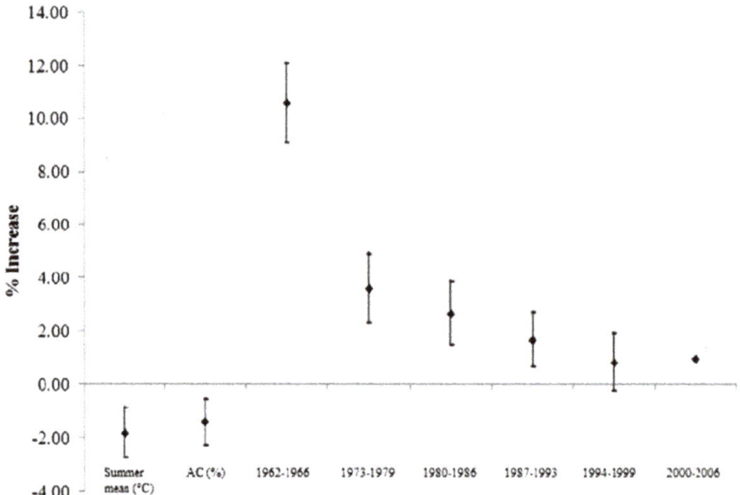

Figure 4. Heat-related mortality effect at 80° vs. 60° F, as a function of summer mean temperature, AC% and time period. From Nordio et al., 2015 [29].

Carson and colleagues analyzed weekly data from London over the 20th century, and found little evidence for heat-related mortality declines, but this may be an artifact of their analysis method, which analyzed data in weekly units, rather than daily data as usually used elsewhere [32]. In contrast, cold-related mortality declined substantially over the 20th century in four discrete time windows. An earlier study of trends in North Carolina, Southern Finland, and Southeastern England reported declines in heat-related mortality in all three locations, including the latter two where A/C was not prevalent [33]. This suggests economic and general health improvements may play a role in the observed declining heat effects. In an internationally combined analysis of data from 272 cities across seven countries from 1985 to 2009 [34], significant declines in heat-related mortality were seen in the U.S., Japan and Spain; however, there was little evidence for declines seen in the UK and the other countries, though the latter analyses were severely limited by statistical power. Declines in heat-related mortality have also been observed in South Korea, Japan, Taiwan and Vienna, Austria [35–39], with little or no change in cold-related mortality effects investigated. One outlier is a study in Shanghai, China that found no decline in heat-related mortality but saw some evidence for decline of cold-related mortality [40]. An examination of 20-year trends in heat-related mortality in nine European cities found some declines and some increases, though interpretation of these findings is limited by the short duration of study [41].

4. Interpretion of Temporal Trends in Temperature Effects

What factors are responsible for the declining ERF for heat-related mortality? While attribution to specific factors remains elusive, studies have speculated that increasing wealth, enhanced heat-health awareness and prevention measures, general improvements in population health, and housing improvements all could play a role [22,23,32,33,42,43]. One leading explanation, at least in the U.S., is the increasing prevalence of A/C usage in recent decades. There is some empirical support for a role of A/C as a modifier of effects, but a great deal of evidence is cross sectional. For example, A/C prevalence can explain some of the city-to-city differences in effect estimates [7,8]. Longitudinally, the evidence remains incomplete, with some studies reporting a strong role for A/C [25] and others not [24]. It seems likely that the power to test for effect modification over time by A/C has been limited by the coarse temporal scale of A/C survey data (e.g., decadal), and also by the problem of not having A/C usage data that are specific to the vulnerable population subset (the ill, elderly and poor). While potentially effective as an adaptive measure, A/C has several important limitations, including capital and energy cost, carbon- and pollution-generating energy demand, and potential for failure during power outages [19].

A related question is the extent, to which heat adaptation trends are being driven by climate warming itself. After analyzing this question, Christidis and colleagues [43] suggested that trends in heat- and cold-related mortality have more to do with economic and cultural trends than with adjustments to the changing climate. In addition, there has been considerable attention paid to reducing heat-health risks by a range of public actors in the U.S. and Europe in the past two decades. This view is supported by the fact that heat effects have declined quite rapidly over the past several decades during a time when climate has warmed only slightly. This is not to say that climate warming will not affect future adaptation trends, but rather that most of the trends observed to date are likely driven by non-climate factors such as those noted above.

Hondula and colleagues provided a thoughtful review of the role of climate in spatial and temporal trends in adaptation [15]. They reviewed ways in which adaptation has been modeled to date in climate and health projection studies. An important caveat to keep in mind is that the ERF, while declining, is only one component of future risk. Rising temperatures and ageing populations could lead to increasing risks in the future [39,44].

5. Projecting Future Temperature Effects

In a great deal of the past literature projecting future mortality impacts of temperature in a warming climate, no adaptation was assumed [45]. Given the mounting evidence for declines over time in heat effects, ignoring "adaptation" trends likely yields overestimates of future heat impacts on mortality. One simple and intuitive approach to incorporating adaptation is to apply ERFs derived from currently hot cities (e.g., Atlanta, GA, USA) to represent the future ERF in currently cool cities (e.g., New York City, NY, USA) that are projected to have hotter temperatures in the future [46]. This has been termed the "analogue city" method. Some have noted that this approach could yield biases if analogue cities differ from the index city in relevant social, economic, or demographic features that affect risk [47]; however, these factors could be theoretically taken into account in a meta regression context. Another limitation of the analogue city method is that it assumes that the ERF from the analogue city is fixed in time, and not itself changing due to trends in other factors. A related method uses ERFs derived from the hottest "analogue summers" in a given location to estimate future risk [48]. However, this method would only capture short-term acclimatization or inter-annual behavioral adaptations. A recent study modeled adaptation based on the mortality risk on "heat wave days" falling above the 99th percentile of temperature [49]. For future projections, a "no adaptation" scenario used the threshold temperatures observed in the historical baseline period to define heat wave days and associated mortality risk in the 2061–2080 period. An "on pace adaptation" scenario used the 99th percentile temperatures for the future time period to define risk. An intermediate, "lagged adaptation" scenario used 99th percentile temperatures for an intermediate time period (2023–2042) to define heat wave days and risks in the 2061–2080 period. This latter approach incorporates the reasonable assumption that it will take some time for adaptation to occur in a rapidly warming climate.

A few projection studies have made adjustments to the heat slope or MMT of the ERF to represent future conditions [50–52]. While in most cases, these adjustments have been made arbitrarily, a more empirical approach was recently reported by Petkova and colleagues, where the ERF in NYC was projected into future, unobserved decades by fitting and extrapolating a non-linear function to the historical trend in effects [53]. Mills and colleagues are the only authors who incorporated both heat and cold adaptation. This is important because even though time trend studies generally do not show measureable changes in cold-related mortality impacts, cross sectional studies show marked differences in the cold ERF depending on the local climate.

How do future mortality projections change when the adaptation assumptions are incorporated? Knowlton et al. provided a useful illustration in [46]. There, heat-related mortality impacts in the 2050s, as compared with those in the 1990s, were modeled with and without an analogue city adjustment to the ERF. Without adaptation, the observed ERF from a time series analysis in NYC was used in both the baseline and future impact assessment. To model adaptation, the ERFs from two analogue cities—Washington DC and Atlanta GA—were averaged. Both cities had current MST within 1 °F of that projected for NYC in the 2050s. Future impacts were reduced by between 28% and 34% depending on the scenario. Other studies using a range of methods have reported a reduction by between 20% and 80% in future impacts under various adaptation assumptions [50,52,53]. A recent comprehensive analysis in 14 European cities applied six different adaptation assumptions for future projections, and concluded that uncertainties related to adaptation were generally larger than those related to climate models and emission scenarios [54].

While available evidence from high-income countries shows that heat effects have been trending downward in recent decades, data gaps and demographic trends add considerable uncertainty to future projections. We have no trend studies in low-income countries, where the epidemiological transition towards increasing chronic disease prevalence, as well as rapid urbanization, may place more people at risk. Additionally, technology-based heat-adaptation measures, such as A/C, may be largely unavailable in low-income settings. We also lack studies of agricultural and construction workers, the homeless, youth athletes and others exposed while engaged in intense physical exertion outdoors [20]. Of particular note is the emerging worldwide epidemic of chronic kidney disease among agricultural workers, which is thought to be partly related to high temperature exposures [55]. In addition, ageing is likely to worsen heat-health risks in the future. Populations are ageing rapidly worldwide, particularly in wealthy countries, and this could lead to increased heat-related mortality risk in the future [56].

6. Summary and Implications

The above review allows us to draw several broad conclusions. Across space, temperature–mortality ERFs for both heat and cold effects differ substantially in ways that appear to depend strongly on prevailing temperatures. In relatively cool climates, the MMT is shifted to the left on the temperature axis, with a shallow cold slope and steep hot slope. In relatively warm climates, the MMT is shifted to the right (higher temperatures), with a steep cold slope and a shallow hot slope. Over time, there is strong evidence that MMTs are rising and that hot slopes are declining, with the particular finding being somewhat dependent on the analytical methods used in individual studies, which are not standardized. There is relatively little evidence for changes in cold slopes over time, in contrast to the cross sectional evidence noted above. Projections of future heat-related mortality that do not take adaptation into account very likely overestimate future heat impacts.

What information can we draw from the current literature to guide future mortality projection studies? There are several possible approaches. As a simple way to incorporate uncertainty regarding adaptation trends, future projections could incorporate sensitivity analyses that apply adjustments to the hot slope ranging from −20% to −80%, a range that is supported by the literature. However, such adjustments would remain somewhat arbitrary, and also would not explicitly take elapsed time into account, which ought to matter in projecting risks of the future.

Alternatively, one could apply a simple time-dependence adjustment by drawing quantitative information on trends in heat slopes based on longitudinal studies such as in [28,29]. The average decadal decline in the heat slope in Petkova's analysis of the past four decades in NYC was about 31% [28]. Nordio's analysis over five decades nationally suggested a decadal decline of about 45% on average per decade [29]. Thus, a range of between 30% and 45% decline in the heat ERF per decade could be applied to projections over the next several decades. This approach has the appeal that it is tied to empirical evidence for trends over time. Additional data to support this approach are available from the supplemental materials provided by Nordio et al., where ERFs by cluster and time period are given, along with corresponding climate data. A recent study applied this approach to project future mortality across the U.S. in a changing climate [57].

Another adaptation model that would be supported by the literature is to adjust the MMT upwards as a function of MST, either keeping the hot slope constant or allowing it to be reduced with rising temperatures. Todd and Valleron [18] provided quantitative backing for this approach in France, although A/C is not yet prevalent in this country. [Their study suggests that the temporal change in MMT vs. MST is about 60% of the magnitude of the spatial change in MMT vs. MST, which hints at the pace of adaptation in a changing climate; in other words, we observe about a 60% adjustment to new climate conditions compared to the observed cross sectional differences.] These findings warrant replication using other national datasets. Heat wave-based mortality models are also amenable to simple adaptation adjustments, as in [49].

More generally, international datasets that include mortality and temperature data observed over multiple decades and locations could be further analyzed to better quantify spatial and temporal patterns in heat-related ERF parameters [57], perhaps using simple parameterizations that include a hot slope and a threshold as in [17]. Cold effects could be similarly modeled. The parameters of city-specific fits could then be analyzed in second stage models that relate them to both time per se, and to spatial and temporal differences in mean temperatures.

Finally, it is important to emphasize that this review has focused on trends in heat-related mortality analyzed at the city scale using administrative data in high-income countries, because that is where the literature has focused until now. While these data are of high relevance to public health planning in the context of a changing climate, they miss important aspects of the problem, which should be a priority for research moving forward. In particular, there is an urgent need for studies focusing on low-income countries, and on vulnerable population subgroups such as agricultural and other outdoor workers [20], for whom adaptation options will be much more limited. New study designs and data sources could also help advance the science of heat adaptation, taking advantages of new health and exposure sensors, citizen science, GIS, and big data.

Funding: This work was partially supported by contract # EP-D-14-031 from the U.S. Environmental Protection Agency.

Conflicts of Interest: The authors declare no conflict of interest.

References

1. Wang, Y.; Bobb, J.F.; Papi, B.; Wang, Y.; Kosheleva, A.; Di, Q.; Schwartz, J.D.; Dominici, F. Heat stroke admissions during heat waves in 1916 US counties for the period from 1999 to 2010 and their effect modifiers. *Environ. Health* **2016**, *15*, 83. [CrossRef] [PubMed]
2. Madrigano, J.; McCormick, S.; Kinney, P.L. The Two Ways of Assessing Heat-Related Mortality and Vulnerability. *Am. J. Public Health* **2015**, *105*, 2212–2213. [CrossRef] [PubMed]
3. Luber, G.E.; Sanchez, C.A. *Heat-Related Deaths—United States, 1999–2003*; US CDC: Atlanta, GA, USA, 2006; pp. 796–798.
4. Matte, T.D.; Lane, K.; Ito, K. Excess mortality attributable to extreme heat in new york city, 1997–2013. *Health Secur.* **2016**, *14*, 64–70. [CrossRef] [PubMed]
5. Semenza, J.C.; Rubin, C.H.; Falter, K.H.; Selanikio, J.D.; Flanders, W.D.; Howe, H.L.; Wilhelm, J.L. Heat-Related Deaths during the July 1995 Heat Wave in Chicago. *N. Eng. J. Med.* **1996**, *335*, 84–90. [CrossRef] [PubMed]
6. Vandentorren, S.; Suzan, F.; Medina, S.; Pascal, M.; Maulpoix, A.; Cohen, J.-C.; Ledrans, M. Mortality in 13 French cities during the August 2003 heat wave. *Am. J. Public Health* **2004**, *94*, 1518–1520. [CrossRef] [PubMed]
7. Curriero, F.C.; Heiner, K.S.; Samet, J.M.; Zeger, S.L.; Strug, L.; Patz, J.A. Temperature and mortality in 11 cities of the eastern United States. *Am. J. Epidemiol.* **2002**, *155*, 80–87. [CrossRef] [PubMed]
8. Anderson, B.G.; Bell, M.L. Weather-related mortality: How heat, cold, and heat waves affect mortality in the United States. *Epidemiology* **2009**, *20*, 205–213. [CrossRef] [PubMed]
9. Kinney, P.L.; Schwartz, J.; Pascal, M.; Petkova, E.; Le Tertre, A.; Medina, S.; Vautard, R. Winter season mortality: Will climate warming bring benefits? *Environ. Res. Lett.* **2015**, *10*, 064016. [CrossRef] [PubMed]

10. Ebi, K.L.; Mills, D. Winter mortality in a warming climate: A reassessment. *WIREs Clim. Chang.* **2013**, *4*, 203–212. [CrossRef]

11. Medina-Ramón, M.; Schwartz, J. Temperature, temperature extremes, and mortality: A study of acclimatization and effect modification in 50 United States cities. *Occup. Environ. Med.* **2007**, *64*, 827–833. [CrossRef] [PubMed]

12. McMichael, A.J.; Wilkinson, P.; Kovats, R.S.; Pattenden, S.; Hajat, S.; Armstrong, B.; Vajanapoom, N.; Niciu, E.M.; Mahomed, H.; Kingkeow, C.; et al. International study of temperature, heat and urban mortality: The 'ISOTHURM' project. *Int. J. Epidemiol.* **2008**, *37*, 1121–1131. [CrossRef] [PubMed]

13. Gasparrini, A.; Guo, Y.; Hashizume, M.; Lavigne, E.; Zanobetti, A.; Schwartz, J.; Tobias, A.; Tong, S.; Rocklöv, J.; Forsberg, B.; et al. Mortality risk attributable to high and low ambient temperature: A multicountry study observational study. *Lancet* **2015**. [CrossRef]

14. Guo, Y.M.; Gasparrini, A.; Armstrong, B.; Li, S.S.; Tawatsupa, B.; Tobias, A.; Lavigne, E.; Coelho, M.; Leone, M.; Pan, X.C.; et al. Global Variation in the Effects of Ambient Temperature on Mortality A Systematic Evaluation. *Epidemiology* **2014**, *25*, 781–789. [CrossRef] [PubMed]

15. Hondula, D.M.; Balling, R.C.J.; Vanos, J.K.; Georgescu, M. Rising temperatures, human health, and the role of adaptation. *Curr. Clim. Chang. Rep.* **2015**, *1*, 144–154. [CrossRef]

16. Barreca, A.; Clay, K.; Deschenes, O.; Greenstone, M.; Shapiro, J.S. Convergence in Adaptation to Climate Change: Evidence from High Temperatures and Mortality, 1900–2004. *Am. Econ. Rev.* **2015**, *105*, 247–251. [CrossRef]

17. Lee, M.; Nordio, F.; Zanobetti, A.; Kinney, P.; Vautard, R.; Schwartz, J. Acclimatization across space and time in the effects of temperature on mortality: A time-series analysis. *Environ. Health* **2014**, *13*, 89. [CrossRef] [PubMed]

18. Todd, N.; Valleron, A.J. Space-Time Covariation of Mortality with Temperature: A Systematic Study of Deaths in France, 1968–2009. *Environ. Health Perspect.* **2015**, *123*, 659–664. [CrossRef] [PubMed]

19. Kinney, P.L.; O'Neill, M.S.; Bell, M.L.; Schwartz, J. Approaches for estimating effects of climate change on heat-related deaths: Challenges and opportunities. *Environ. Sci. Policy* **2008**, *11*, 87–96. [CrossRef]

20. Sheridan, S.C.; Allen, M.J. Temporal trends in human vulnerability to excessive heat. *Environ. Res. Lett.* **2018**, *13*, 043001. [CrossRef]

21. Arbuthnott, K.; Hajat, S.; Heaviside, C.; Vardoulakis, S. Changes in population susceptibility to heat and cold over time: Assessing adaptation to climate change. *Environ. Health* **2016**, *15*, S33. [CrossRef] [PubMed]

22. Boeckmann, M.; Rohn, I. Is planned adaptation to heat reducing heat-related mortality and illness? A systematic review. *BMC Public Health* **2014**, *14*, 1112. [CrossRef] [PubMed]

23. Davis, R.E.; Knappenberger, P.C.; Michaels, P.J.; Novicoff, W.M. Changing heat-related mortality in the United States. *Environ. Health Perspect.* **2003**, *111*, 1712–1718. [CrossRef] [PubMed]

24. Bobb, J.F.; Peng, R.D.; Bell, M.L.; Dominici, F. Heat-Related Mortality and Adaptation to Heat in the United States. *Environ. Health Perspect.* **2014**, *122*, 811–816. [CrossRef] [PubMed]

25. Barreca, A.; Clay, K.; Deschenes, O.; Greenstone, M.; Shapiro, J.S. *Adapting to Climate Change: The Remarkable Decline in the U.S. Temperature-Mortality Relationship over the 20th Century*; Working Paper 18692; National Bureau of Economic Research: Cambridge, MA, USA, 2013.

26. Deschenes, O.; Greenstone, M. Climate Change, Mortality, and Adaptation: Evidence from Annual Fluctuations in Weather in the US. *Am. Econ. J.-Appl. Econ.* **2011**, *3*, 152–185. [CrossRef]

27. Deschenes, O. Temperature, human health, and adaptation: A review of the empirical literature. *Energy Econ.* **2014**, *46*, 606–619. [CrossRef]

28. Petkova, E.P.; Gasparrini, A.; Kinney, P.L. Heat and Mortality in New York City Since the Beginning of the 20th Century. *Epidemiology* **2014**, *25*, 554–560. [CrossRef] [PubMed]

29. Nordio, F.; Zanobetti, A.; Colicino, E.; Kloog, I.; Schwartz, J. Changing patterns of the temperature-mortality association by time and location in the US, and implications for climate change. *Environ. Int.* **2015**, *81*, 80–86. [CrossRef] [PubMed]

30. Astrom, D.O.; Forsberg, B.; Edvinsson, S.; Rocklov, J. Acute Fatal Effects of Short-Lasting Extreme Temperatures in Stockholm, Sweden Evidence Across a Century of Change. *Epidemiology* **2013**, *24*, 820–829. [CrossRef] [PubMed]

31. Astrom, D.O.; Tornevi, A.; Ebi, K.L.; Rocklov, J.; Forsberg, B. Evolution of Minimum Mortality Temperature in Stockholm, Sweden, 1901-2009. *Environ. Health Perspect.* **2016**, *124*, 740–744. [CrossRef] [PubMed]

32. Carson, C.; Hajat, S.; Armstrong, B.; Wilkinson, P. Declining vulnerability to temperature-related mortality in London over the 20th century. *Am. J. Epidemiol.* **2006**, *164*, 77–84. [CrossRef] [PubMed]

33. Donaldson, G.C.; Keatinge, W.R.; Nayha, S. Changes in summer temperature and heat-related mortality since 1971 in North Carolina, South Finland, and Southeast England. *Environ. Res.* **2003**, *91*, 1–7. [CrossRef]

34. Gasparrini, A.; Guo, Y.; Hashizume, M.; Kinney, P.L.; Petkova, E.P.; Lavigne, E.; Zanobetti, A.; Schwartz, J.D.; Tobias, A.; Leone, M.; et al. Temporal Variation in Heat-Mortality Associations: A Multicountry Study. *Environ. Health Perspect.* **2015**, *123*, 1200–1207. [CrossRef] [PubMed]

35. Ha, J.; Kim, H. Changes in the association between summer temperature and mortality in Seoul, South Korea. *Int. J. Biometeorol.* **2013**, *57*, 535–544. [CrossRef] [PubMed]

36. Onozuka, D.; Hagihara, A. Variation in vulnerability to extreme-temperature-related mortality in Japan: A 40-year time-series analysis. *Environ. Res.* **2015**, *140*, 177–184. [CrossRef] [PubMed]

37. Matzarakis, A.; Muthers, S.; Koch, E. Human biometeorological evaluation of heat-related mortality in Vienna. *Theor. Appl. Climatol.* **2011**, *105*, 1–10. [CrossRef]

38. Chung, Y.; Noh, H.; Honda, Y.; Hashizume, M.; Bell, M.L.; Guo, Y.L.L.; Kim, H. Temporal Changes in Mortality Related to Extreme Temperatures for 15 Cities in Northeast Asia: Adaptation to Heat and Maladaptation to Cold. *Am. J. Epidemiol.* **2017**, *185*, 907–913. [CrossRef] [PubMed]

39. Lee, W.; Choi, H.M.; Kim, D.; Honda, Y.; Guo, Y.L.L.; Kim, H. Temporal changes in mortality attributed to heat extremes for 57 cities in Northeast Asia. *Sci. Total Environ.* **2018**, *616*, 703–709. [CrossRef] [PubMed]

40. Yang, C.Y.; Meng, X.; Chen, R.J.; Cai, J.; Zhao, Z.H.; Wan, Y.; Kan, H.D. Long-term variations in the association between ambient temperature and daily cardiovascular mortality in Shanghai, China. *Sci. Total Environ.* **2015**, *538*, 524–530. [CrossRef] [PubMed]

41. de' Donato, F.K.; Leone, M.; Scortichini, M.; De Sario, M.; Katsouyanni, K.; Lanki, T.; Basagana, X.; Ballester, F.; Astrom, C.; Paldy, A.; et al. Changes in the Effect of Heat on Mortality in the Last 20 Years in Nine European Cities. Results from the PHASE Project. *Int. J. Environ. Res. Public Health* **2015**, *12*, 15567–15583. [CrossRef] [PubMed]

42. Ng, C.F.S.; Boeckmann, M.; Ueda, K.; Zeeb, H.; Nitta, H.; Watanabe, C.; Honda, Y. Heat-related mortality: Effect modification and adaptation in Japan from 1972 to 2010. *Glob. Environ. Chang.-Hum. Policy Dimens.* **2016**, *39*, 234–243. [CrossRef]

43. Christidis, N.; Donaldson, G.C.; Stott, P.A. Causes for the recent changes in cold- and heat-related mortality in England and Wales. *Clim. Chang.* **2010**, *102*, 539–553. [CrossRef]

44. Sheridan, S.C.; Dixon, P.G. Spatiotemporal trends in human vulnerability and adaptation to heat across the United States. *Anthropocene* **2017**, *20*, 61–73. [CrossRef]

45. Sanderson, M.; Arbuthnott, K.; Kovats, S.; Hajat, S.; Falloon, P. The use of climate information to estimate future mortality from high ambient temperature: A systematic literature review. *PLoS ONE* **2017**, *12*, e0180369. [CrossRef] [PubMed]

46. Knowlton, K.; Lynn, B.; Goldberg, R.A.; Rosenzweig, C.; Hogrefe, C.; Rosenthal, J.K.; Kinney, P.L. Projecting heat-related mortality impacts under a changing climate in the New York City region. *Am. J. Public Health* **2007**, *97*, 2028–2034. [CrossRef] [PubMed]

47. Huang, C.R.; Barnett, A.G.; Wang, X.M.; Vaneckova, P.; FitzGerald, G.; Tong, S.L. Projecting Future Heat-Related Mortality under Climate Change Scenarios: A Systematic Review. *Environ. Health Perspect.* **2011**, *119*, 1681–1690. [CrossRef] [PubMed]

48. Hayhoe, K.; Cayan, D.; Field, C.B.; Frumhoff, P.C.; Maurer, E.P.; Miller, N.L.; Moser, S.C.; Schneider, S.H.; Cahill, K.N.; Cleland, E.E.; et al. Emissions pathways, climate change, and impacts on California. *Proc. Natl. Acad. Sci. USA* **2004**, *101*, 12422–12427. [CrossRef] [PubMed]

49. Anderson, G.B.; Oleson, K.W.; Jones, B.; Peng, R.D. Projected trends in high-mortality heatwaves under different scenarios of climate, population, and adaptation in 82 US communities. *Clim. Chang.* **2018**, *146*, 455–470. [CrossRef] [PubMed]

50. Gosling, S.; McGregor, G.; Lowe, J. Climate change and heat-related mortality in six cities Part 2: Climate model evaluation and projected impacts from changes in the mean and variability of temperature with climate change. *Int. J. Biometeorol.* **2009**, *53*, 31–51. [CrossRef] [PubMed]

51. Honda, Y.; Kondo, M.; McGregor, G.; Kim, H.; Guo, Y.-L.; Hijioka, Y.; Yoshikawa, M.; Oka, K.; Takano, S.; Hales, S.; et al. Heat-related mortality risk model for climate change impact projection. *Environ. Health Prev. Med.* **2014**, *19*, 56–63. [CrossRef] [PubMed]

52. Mills, D.; Schwartz, J.; Lee, M.; Sarofim, M.; Jones, R.; Lawson, M.; Duckworth, M.; Deck, L. Climate change impacts on extreme temperature mortality in select metropolitan areas in the United States. *Clim. Chang.* **2015**, *131*, 83–95. [CrossRef]
53. Petkova, E.P.; Vink, J.K.; Horton, R.M.; Gasparrini, A.; Bader, D.A.; Francis, J.D.; Kinney, P.L. Towards More Comprehensive Projections of Urban Heat-Related Mortality: Estimates for New York City under Multiple Population, Adaptation, and Climate Scenarios. *Environ. Health Perspect.* **2017**, *125*, 47–55. [CrossRef] [PubMed]
54. Gosling, S.N.; Hondula, D.M.; Bunker, A.; Ibarreta, D.; Liu, J.G.; Zhang, X.X.; Sauerborn, R. Adaptation to Climate Change: A Comparative Analysis of Modeling Methods for Heat-Related Mortality. *Environ. Health Perspect.* **2017**, *125*, 087008. [CrossRef] [PubMed]
55. Barraclough, K.A.; Blashki, G.A.; Holt, S.G.; Agar, J.W.M. Climate change and kidney disease-threats and opportunities. *Kidney Int.* **2017**, *92*, 526–530. [CrossRef] [PubMed]
56. Li, T.T.; Horton, R.M.; Bader, D.A.; Zhou, M.G.; Liang, X.D.; Ban, J.; Sun, Q.H.; Kinney, P.L. Aging Will Amplify the Heat-related Mortality Risk under a Changing Climate: Projection for the Elderly in Beijing, China. *Sci. Rep.* **2016**, *6*, 28161. [CrossRef] [PubMed]
57. Wang, Y.; Nordio, F.; Nairn, J.; Zanobetti, A.; Schwartz, J.D. Accounting for adaptation and intensity in projecting heat wave-related mortality. *Environ. Res.* **2018**, *161*, 464–471. [CrossRef] [PubMed]

Review

Climate Change and Water-Related Infectious Diseases

Gordon Nichols [1,2,3,*], **Iain Lake** [3] **and Clare Heaviside** [1,4,5]

1   Centre for Radiation Chemicals and Environmental Hazards, Public Health England, Chilton, Oxon OX11 0RQ, UK; clare.heaviside@gmail.com

2   European Centre for Environment and Human Health, University of Exeter Medical School, C/O Knowledge Spa RCHT, Truro, Cornwall TR1 3HD, UK

3   School of Environmental Sciences, UEA, Norwich NR4 7TJ, UK; i.lake@uea.ac.uk

4   School of Geography, Earth and Environmental Sciences, University of Birmingham, Edgbaston, Birmingham B15 2TT, UK

5   Department of Social and Environmental Health Research, London School of Hygiene and Tropical Medicine, 15-17 Tavistock Place, London WC1H 9SH, UK

*   Correspondence: Gordon.nichols@me.com

Received: 31 May 2018; Accepted: 10 September 2018; Published: 2 October 2018

Abstract: Background: Water-related, including waterborne, diseases remain important sources of morbidity and mortality worldwide, but particularly in developing countries. The potential for changes in disease associated with predicted anthropogenic climate changes make water-related diseases a target for prevention. **Methods**: We provide an overview of evidence on potential future changes in water-related disease associated with climate change. **Results**: A number of pathogens are likely to present risks to public health, including cholera, typhoid, dysentery, leptospirosis, diarrhoeal diseases and harmful algal blooms (HABS). The risks are greatest where the climate effects drive population movements, conflict and disruption, and where drinking water supply infrastructure is poor. The quality of evidence for water-related disease has been documented. **Conclusions**: We highlight the need to maintain and develop timely surveillance and rapid epidemiological responses to outbreaks and emergence of new waterborne pathogens in all countries. While the main burden of waterborne diseases is in developing countries, there needs to be both technical and financial mechanisms to ensure adequate quantities of good quality water, sewage disposal and hygiene for all. This will be essential in preventing excess morbidity and mortality in areas that will suffer from substantial changes in climate in the future.

Keywords: climate change; waterborne disease; natural environment; risks; public health; cryptosporidiosis; cholera; leptospirosis; Legionnaires' disease

---

1. Introduction

The effect of human activity on observed changes to the climate system over recent decades is widely acknowledged and is a global cause for concern. Anthropogenic (man-made) climate change has led to a rise in annual global mean temperatures since pre-industrial times, with more rapid increases since the mid-1900s [1]. As well as changing weather patterns, increasing average temperatures and, potentially of more concern, is the increase in the frequency of extreme weather events which can have enormous human cost [2]. Climate change is seen as an example of a tragedy of the Commons [3], whereby it is in the interests of individuals to benefit from human activity but the overall impact on all people collectively will be negative unless there is an agreed intervention. It is generally the case that the largest impacts on health are realised in developing regions of the world such as the tropics, whereas the greatest contributors to greenhouse gas emissions are often developed countries which do not suffer the consequences of extreme events to the same extent [4]. Due to the lifetime of greenhouse gases such as carbon dioxide in the atmosphere, and the timescales associated with ocean warming, even if global CO_2 emissions were curtailed immediately, the effects on the earth's climate, including increasing temperatures and sea level rise would continue for a number of decades before starting to plateau [2]. However, this should be seen as a general call to action to reduce emissions as soon as possible, given that the impacts are likely to extend beyond current conditions and there are indications that economic investment now will be likely to reduce costs later [5]. There is also a need to develop adaptations to cope with changes in climate. Recent research highlighting the health benefits of limiting future temperature rises to the more ambitious target of 1.5 °C rather than 2 °C, in line with the Paris agreement of 2016, further emphasises the need to limit emissions [6].

Climate change affects health in a number of ways, and the impacts vary both geographically and between different populations. A growing and ageing population in much of the world means that the proportion of the population who are vulnerable to the effects of climate change will increase in the future [7]. The most direct impacts from climate change are from the effects of high temperatures, and from acute impacts relating to extreme events such as storms, floods and heatwaves. These physical or meteorological stressors can produce direct health effects, such as physical injury, illness, or mental health impacts due to the consequences of the aftermath. In places where infrastructure or adaptation measures are poor, the impacts will be more severe [2]. In addition to these types of impact on health, climate change is likely to modify or mediate existing health effects and exacerbate inequalities through a number of indirect pathways. These more indirect effects on health occur through climate interactions with ecosystems, water, biodiversity and land use changes. Climate change can lead to environmental degradation; can affect food and water availability and quality; and increase risks to health from pathogens, vectors and infectious diseases [8]. Civil conflict or mass movement of people may be partly driven by environmental degradation and can further increase risks to health. There is evidence to suggest that climate change can be a driver for civil war [9,10].

Waterborne and water-related diseases are sensitive to environmental conditions, some or all of which are likely to be affected by climate change. For example, climate change is likely to lead to changes in the frequency of heavy rainfall events, storms and drought periods [2], melting of polar ice and glaciers, warming and thermal expansion of the oceans causing sea level rise [11], and melting of permafrost, which may contribute to further warming [12]. Changes in interactions between the water cycle and the climate system will modify the risk from waterborne diseases from these physical impacts, as well as from the resulting risk of famine, water shortages, decreased water quality, increasing habitat for mosquitoes, alterations to seasonality of diseases and contaminated recreational waters. However, health impacts of waterborne disease over the longer term may be secondary to other health effects associated with other water issues (e.g., shortage, flooding, famine, the economy, sea level rise and war).

Tackling the climate change problem has focussed on mitigating the effects of greenhouse gas emissions through the Intergovernmental Panel on Climate Change (IPCC), through cross government international agreements on reducing carbon emissions and by providing reliable scientific evidence and reports. While the approach has generally been to reduce worldwide greenhouse gas emissions, the reductions are likely to be slow and work on adaptation strategies to deal with the climate change associated with overall increases in temperature is also being undertaken.

2. Climate Change and the Water Cycle

Even cursory examination of the evidence shows the importance of water in relation to climate change. The impact of water in the decline of civilisations has been examined and reviewed [13]. Climate change is likely to affect the water cycle across the globe [14] with potential influences on surface and groundwater quality; it will lead to changes in atmospheric water vapour content, changes in cloud types and cover, and changes in the frequency of severe storms. Over time, it is likely that there will be increased melting of glaciers and icecaps [15] and ocean warming (and associated thermal expansion) that will cause sea levels to rise [16]. It remains possible that the release of permafrost methane may contribute to further warming [12,17]. These changes in atmospheric conditions and moisture content are expected to result in an increase in floods, famine, hurricanes and tropical storms, drought [18], wild fires, chronic water shortages, decreased water quality, periods with increased mosquito vectors, alterations to the seasonality of diseases, contaminated recreational waters and rising sea levels. Most assessments of the risks to health from climate change are conducted over time periods ranging from the present to 80 years. However, there are potentially much larger risks over longer timescales, particularly if there is a substantial sea level rise, and prolonged periods with much-increased average and peak temperatures [19].

Although scientific understanding of how climate may affect weather patterns has increased enormously over the past few years, especially in relation to extreme weather, the consequences for, and influences on, water quality (e.g., microbiological quality) are far less studied [20]. Changes in water arise from interactions with the weather, affecting ecosystems, changing the flow of nutrients and pathogens in catchments, and influencing water quality. Climate change may influence the microbiological quality of river water, which may present risks to bathers [21]. A modelling approach involving Quantitative Microbial Risk Assessment found that although climate change increased the fluxes of a number of pathogens, the overall change in risk was limited. Part of this was due to the dilution effect of increasing rainfall. As risks are likely to be location specific, wider generalisations are difficult. The approach used in such studies could be transferred to other locations. In summary the IPCC states that in overall terms, climate change is projected to reduce the quality of raw waters [22].

3. Climate Change and Drinking Water-Related Infectious Diseases

The history of water-related disease dates back to the classical period when Empedocles is reported to have introduced drainage of a swamp (a public health intervention) to reduce disease [23]. The ability of waterborne diseases to cause large outbreaks that, in the case of typhoid and cholera, have a high morbidity and mortality, makes them important historically. A classification of water-related infections was drawn up by Bradley (Bradley's classification [24]) and has been adapted by others [25–27]. A modified-version groups diseases into waterborne; water access related; water based; water-related insect vectors; and engineered water systems, although this revision still excludes some water-related diseases (Table 1).

Drinking water is the most important waterborne disease risk because the contamination of large mains supplies can cause large outbreaks [28–32], and because small rural supplies are commonly contaminated [33,34] which, in developing countries, can lead to substantial infant mortality. Bottled water is usually safe, particularly natural mineral waters that are derived from deep wells or other secure sources, but outbreaks may occasionally occur. The water supplies on ships, trains and aircraft can be subject to contamination where the supply chain breaks down [35]. There are always potential risks when people drink untreated natural waters. The supply of clean water can become critical in areas of war, disaster, famine, drought, water shortage and flooding, and refugee supplies often need to be established rapidly to prevent outbreaks [36–41].

While climate change may affect the microbiological quality of water and water-related diseases, the arguments indicate notable uncertainty over what the specific impacts will be, and when and where they will be most acute. However, to assess the likely impacts upon the health of humans, there is a need to examine the resilience of society to changing water quality. Developing an understanding of these capabilities and adaptation potentials is key to assessing the likely influence of climate change.

Table 1. Criteria for water-related diseases [24–26,42].

Class	Sub-Class	Examples
Waterborne infection	Infectious through drinking water	*Cryptosporidium* spp., *Giardia* spp., *Vibrio cholerae*, *Dracunculus medinensis*
	Infectious through recreational water	*Cryptosporidium*, Adenovirus, *Leptospira* spp., *Naegleria fowleri*
	Infectious through inhalation (engineered systems)	*Legionella pneumophila*
	Infectious through contact	*Pseudomonas aeruginosa*
	Infectious through contamination of wounds	*Mycobacterium marinum*, *Vibrio vulnificus*, *Aeromonas hydrophila*
	Infectious through growth in equipment and water systems	*Mycobacterium avium* complex
	Infectious through growth in soil or water	*Acanthamoeba* spp., *Burkholderia pseudomallei*
	Infectious through growth in coastal waters	*Vibrio* spp.
	Infectious from food contamination with water/soil	*Clostridium botulinum*
	Infectious through water contamination of food	*Cyclospora cayetanensis*, *Cryptosporidium* spp., *Salmonella* Typhi
	Infections through near-drowning	*Aeromonas hydrophila*
	Infectious through injection of non-sterile water	*Clostridium botulinum*, *Clostridium novyi*
Waterborne chemical or toxin	Potent toxins through inhalation	*Ostreopsis* spp. (HABS)
	Potent toxins through seafood	Dinoflagellate and diatom fish and shellfish poisoning (HABS)
	Potent toxins through drinking	Cyanobacterial blooms (HABS)
	Potent toxins through dialysis	Cyanobacterial blooms (HABS)
	Potent toxins through recreational exposure	Cyanobacterial blooms (HABS)
Water washed (poor access)	Hygiene related	*Shigella* spp., *Chlamydia trachomatis*, scabies, pneumonia
	Infections related to drought	*Coccidioides immitis*
	Infections related to flooding	*Leptospira* spp.
Water based	Parasite lifecycle requiring water and transmitted by water	*Schistosoma* spp., *Dracunculus medinensis*
	Parasite lifecycle requiring water and transmitted by food	*Fasciola* spp., *Opisthorchis sinensis*, *Heterophyes heterophyes*
	Parasite lifecycle transmitted by waterborne route	*Dracunculus medinensis*, *Spirometra* spp., *Echinococcus* spp., *Sarcocystis* spp., *Toxoplasma gondii*
Water-related insect vectors	Vector breeding in water	Dengue virus, *Onchocerca* spp., *Trypanosoma* spp.
Waste water related	Parasites maturing in waste water	*Ascaris lumbricoides*, *Cyclospora cayetanensis*
Diseases related to damp	Toxicosis related to food stored damp	Mycotoxins (aflatoxin, patulin, ochratoxin)
	Disease related to living in damp conditions	Mycotoxins

One of the major pathways through which contaminated water affects individuals is though drinking water. In terms of management, these supplies range from unimproved sources where the individual is effectively consuming raw water, to large, managed supplies where multiple barriers exist to prevent microbiological contamination of water supplies. Climate change has a number of potential influences upon water treatment, and higher temperatures are known to enhance biological methods of water treatment [43]. However, countering this effect are a number of other factors linked to more extreme weather, such as increased rainfall and water turbidity [44,45] which may increase risks to microbiological water quality in some locations. A review of the impacts of climate change on surface water contamination concluded that it was likely to increase the risk associated with drinking water supplied mainly during extreme climatic events [46]. Pathogen risk was argued to rise mainly due to elevated temperatures and extreme rainfall, especially in temperate countries.

Ensuring appropriate water infrastructure, regular monitoring and appropriate management techniques, such as Water Safety Plans, are likely to be increasingly important to address changing risks. In the future rapid testing (e.g., PCR based) [47] and new treatment technologies (e.g., nanomaterials) [48] may play an increasing role in addressing climate change challenges. The ability to respond to changing risks will vary according to the resources available. In terms of supplies less able to adapt to a changing water quality, we highlight small water supplies, private water supplies as well as supplies in lower income areas as potentially being at greatest risk.

4. Settings Other Than Drinking Water, and the Range of Water-Related Diseases

Drinking water is not the only route through which potentially contaminated water may affect individuals; bathing water is another important pathway. As argued above, modelling is challenging, and this is especially the case given the multiple number of potential exposure points. Water quality modelling is currently possible, and could be extended to include climate change. Warning the public about water risks in relation to bathing water is one way to address changing risks associated with climate change [49].

Infections can be related to exposure to natural and man-made recreational waters. These include thermal waters (amoebae, *Legionella*) [50], inland recreational fresh waters, ponds and lakes (cyanobacteria, *Pseudomonas aeruginosa* [51], enteric pathogens, *Leptospira* spp., *Trichobilharzia* spp. [52], *Schistosoma* spp. [53], *Vibrio* spp.) [54], wild swimming (enteric pathogens), coastal waters including sea sports, sea water pools, bathing beaches (dinoflagellates and diatoms, jelly-fish larvae, toxic seaweed, enteric pathogens) and the beach environment, including run-off from fields and sewers, beach sand, and so forth. Recreational exposure to man-made fresh water pools includes treated swimming pools (*Cryptosporidium*), natural pools (enteric pathogens), spa baths (*P. aeruginosa* [55], *Legionella* spp.), water parks (*Cryptosporidium* spp.), foot wash and foot spas (*Mycobacterium* spp. [56] and *P. aeruginosa* [57]) and inflatables (*P. aeruginosa*, *Aeromonas hydrophila* [58]).

Infections from working in water can include *Schistosoma* spp., *Burkholderia pseudomallei* [59] and wound infections from water (*Vibrio vulnificus* [60], *Mycobacterium* spp., [61] *B. pseudomallei* [62]). Water transmission in man-made systems and equipment includes hospital/medical uses of water, water for dialysis and hydration (cyanobacteria), water for washing and decontamination, hospital water systems, water transmission in intensive care (*P. aeruginosa*) [63–66], contaminated equipment including endoscope washers (*Mycobacterium* spp.) [67], humidifiers, taps and wash basins, showers (*Legionella* spp.) and water births [68–71]. A variety of industrial waters can contribute to respiratory infection, including cooling towers and thermally polluted waters (*Legionella* spp.).

The disposal of waste can contribute to water contamination, particularly chemical contamination, but also pathogens. Sewage disposal is the main source of human faecal contamination in developed countries and sewage treatment is designed to reduce this to a minimum. However, animal waste probably represents a larger input to the natural environment as a result of defecation on fields and run-off. In developing countries, human faeces are commonly deposited in the natural environment at defaecation sites or middens where the 'night soil' matures over time and some pathogens require this to become infectious (*Ascaris lumbricoides* [72], *Cyclospora cayetanensis*). There are also potential risks associated with water passing through waste burial sites, particularly mass graves associated with plague, smallpox or anthrax, and from water running from leather processing sites (*Bacillus anthracis*).

Water is important in agriculture and food production, and irrigation may be conducted with water that is not of potable quality. Where this is done for salad items and soft fruit that are eaten without further treatment, then outbreaks can occur [73,74]. Contaminated water used for washing and food processing can also cause outbreaks. Food retailers require water on the premises in order for staff to wash their hands. Water-related foodborne disease (water-based) includes helminths and other macro-parasites, ciguatera, shellfish dinoflagellate toxins and similar toxins [75]. As shellfish filter large volumes of water, there are common outbreaks associated with faecal pathogens (especially norovirus and *Vibrio cholerae* [76], *Vibrio parahaemolyticus* [76] and *Vibrio vulnificus* [76]). Many processed drinks and foods contain water, and infections are prevented by source water protection, filtration, heat treatment or preservative treatment and a matrix that prevents pathogenic organisms from multiplying.

Where water is in short supply, water washed diseases may occur [14]. This can be in desert areas where lack of water and chronic water shortages, together with flies resulting from poor waste disposal, can allow the transmission of enteric (shigellosis) and eye infections (*Chlamydia trachomatis*).

Vector borne diseases (mosquito-borne, tick-borne, fly-borne, triatomid bug) can be strongly influenced by weather and geographic parameters, are likely to change in distribution as a result of climate change and are difficult to predict accurately [77], but are not examined further in this paper.

5. Methods and Reviews Related to Water Quality and Health

Systematic review (SR) and meta-analysis (MA) are methods which have been used extensively to elucidate the relationships between microbiological contamination and ill health. As climate change will impact water contamination, an examination of systematic reviews relating to water quality has been undertaken. A SR of the effectiveness of interventions to reduce the health impact of climate change around the world, found a weak evidence base and gaps in dealing with extreme weather events [78]. This follows a previous SR [79]. The seasonal differences in faecal contamination of source waters used for supplying drinking water have been examined in a SR [80]. They found increased contamination in the wet season. A general review of wider evidence for pathogens being waterborne and the impacts of climate are included in the review of pathogens (Table A1, Appendix A).

6. Flooding

Flooding causes local or widespread disruption of normal life and makes waste disposal difficult. A SR of waterborne infections and climate change found evidence of outbreaks and increased sporadic disease following flooding [37], and included cholera and other diarrhoeal diseases. Other reviews have found similar results [36] and the health impacts of flooding showed that monitoring, mitigation and communication had the potential to reduce loss of life [81]. Tsunami related flooding can have a disproportionate impact on women, children and the elderly [82]. The impact of flooding is generally greater in developing rural countries with poor infrastructure compared to developed countries where people in flooded properties can be readily moved to non-flooded areas for a temporary period and provided with potable water. This is particularly true of coastal areas [83]. After Hurricane Katrina, there was an outbreak of norovirus deriving from populations being held in a stadium [84] and an increase in *Vibrio* infections resulting from coastal waters [85]. An increase in flood related disease might be expected under climate change with altered weather patterns and more severe weather events. Early warning systems with effective disease surveillance, prevention and response are important in preventing infectious diseases following flooding [86].

7. Drought

Drought can be a hidden risk with the potential to cause a public health disaster [87]. A SR of the health effects of drought found the main categories included: Nutrition-related effects; water-related disease (including *Escherichia coli*, cholera and algal bloom); airborne and dust-related disease (including coccidioidomycosis); vector borne disease (including malaria, dengue and West Nile Virus); mental health effects; and other health effects [88]. A review in Canada found that drought can affect respiratory and mental health, with illnesses related to exposure to toxins, food, water and vector-borne diseases [89]. Drought-related health impacts vary widely and depend upon drought severity, population vulnerability, health and sanitation infrastructure, and available resources with which to mitigate impacts as they occur. Population resilience is affected by socio-economic environment. A SR has examined the likely impacts of climate change on water quality and disease in the Mekong delta basin [90].

8. Disasters

The water needs for disaster recovery identified relations between the amount of water provided and the diarrhoea and mortality rates, but emphasised the inadequacy of the data [91]. As responses to climate related emergencies influence subsequent morbidity and mortality, it is important to understand the impact of water, sanitation and hygiene (WASH) interventions on health outcomes in humanitarian crises. The current evidence base is limited, and is unsuitable for determining associative or causal relationships [92]. People living with HIV have a higher morbidity where there is contaminated water, poor hygiene and sanitation. Programs to improve these also improved morbidity [93].

9. Climate Impacts on Water, Sanitation and Hygiene (WASH)

Understanding the climate change impacts on water-related diseases requires an understanding of water, sanitation and hygiene. Much of the burden of waterborne disease appears to result from small and rural supplies, although there may be under-ascertainment of outbreaks in urban areas with mains supplies due to limited disease surveillance. This is important in an examination of the burden of waterborne disease and future changes as a result of altered weather patterns. Waterborne disease burden methodologies used in developed countries to attribute acute gastrointestinal infection (AGI) to drinking water, include simple point estimates, quantitative microbial risk assessment, Monte Carlo simulations based on assumptions and epidemiological data from the literature [94]. In developing countries, inadequate water and sanitation are associated with risk of diarrhoeal disease [95], particularly in young children, and raised maternal mortality occurs in households with poor sanitation and a poor water environment [96]. The microbiological quality of household water correlates with health outcomes (diarrhoea and trachoma) [97], although improved sources do not always provide water that is completely free of faecal contamination. Point of use devices can be effective [98], however, contamination of water between source and point of use remains a continuing problem [99]. An examination of interventions to reduce diarrhoea in less developed countries found, that while interventions were generally effective, the heterogeneity between studies made the exact conditions causing disease reductions difficult to assess [100], and others found a dearth of methodologically sound studies [101]. There were also inadequacies in behavioural models and frameworks for intervening in WASH specific interventions [102]. The impacts of WASH on child diarrhoeal morbidity has been examined in a number of studies [103] and there was a general lack of good quality studies of diarrhoea morbidity in children in India [104]. While water treatment (e.g., chlorine water treatment at point of use) can be effective, most studies are short term [105]. An examination of water distribution system deficiencies demonstrated that study blinding can be important [106]. Expecting to obtain good information on the risks from water-related infections under climate change, in the absence of reliable, experimental evidence for effectiveness of interventions, seems naïve.

Of 293 outbreaks linked to water supplies in Canada and the US, failure of existing treatment and lack of water treatment were the leading causes [107]. Temporary water outages and chronic outages in intermittently operated systems can be associated with gastrointestinal infection [106]. The Walkerton outbreak of *E. coli* O157 and *Campylobacter* in Canada highlighted the role that heavy rainfall can play in outbreaks [108,109]. However, this was also linked to poor management. There are many studies that have examined the role of rainfall before an outbreak. It is generally seen that heavy rain is more common before many outbreaks, suggesting the source water is compromised [32,36,110–112]. The association between waterborne outbreaks with a period of prolonged low rainfall in the four weeks before an outbreak may also indicate a vulnerability to weather [32], although this was not seen in other studies. There is a strong need for water utilities to build water safety plans that factor in likely changes in climate over future scenarios, while retaining an understanding of historical weather events. Behaviour change interventions for water and sanitation in developing countries have looked at risk factors, attitudinal factors, normative factors, ability factors, and self-regulation factors [113].

Social marketing for water and sanitation products showed this improves health threat awareness and provides a solution to reducing disease burden [114]. Good water sourcing, treatment, distribution, storage and clean point of use remain the key to community health. Water and sanitation in schools is an important area where improvements can facilitate improved educational achievement [115].

There is a need for robust epidemiological studies that quantify the health risks associated with both small, private water systems, and large community supplies. More information is needed on pathogen quantification, susceptibility of vulnerable sub-populations, the influence of extreme weather events, the proportions of the population served by different water sources and the treatment level, source water quality and condition of the distribution system infrastructure. The exposure to faecal contamination in potable water has been estimated, suggesting that there may be a substantial under-estimate of disease burden [116].

10. Which Water-Related Pathogens Are Important?

A wide range of pathogens are known to be transmitted through water (Table A1, Appendix A). Historically, some diseases, such as cholera, dysentery and typhoid have been very important from a public health perspective, causing extensive morbidity and mortality. Many of the diseases are a problem where there is limited infrastructure, as in developing countries or rural regions of developed countries and can be sensitive to social disruption and infrastructure damage. It is important to understand the future health impacts of climate change and to understand where the most important disease burdens will be. Waterborne diseases can be sensitive to emergence, but this is generally related to the discovery of new infectious agents or identification methods (e.g., cryptosporidiosis, microsporidiosis, legionellosis, hepatitis E) rather than newly emergent diseases. However, some classical waterborne pathogens may emerge in new areas with climate change (e.g., cholera), and Legionnaires' disease, which is predominately derived from contaminated water systems in the built environment, may increase as a result of raised temperatures.

An examination of 24 analytical studies assessing the association between extreme precipitation or temperature and drinking water-related waterborne infections, found that most studies showed a positive association with increased precipitation or temperature. A few studies showed an association with decreased precipitation and several in which there was no association [117]. Infections included cholera, typhoid, paratyphoid, campylobacter, shigella, hepatitis A, cryptosporidiosis, giardiasis and waterborne outbreaks.

10.1. Schistosomiasis

Schistosoma are transmitted as a result of cercaria burrowing through the skin of people working or bathing in contaminated waters. The cercaria develop in infected snails. Although temperature, precipitation and humidity are known to influence the development of schistosome parasites, as well as their intermediate snail hosts (*Biomphalaria* spp.; *Oncomelania* spp.) [118–121] and their internal defence system [122], modelling climate change impacts on disease can have mixed results [123,124]. Some scenarios show predicted increases and decreases [125], with degrees of uncertainty [126,127]. Increases in parasite growth related to temperature can be offset by increased snail mortality at higher temperatures [127]. Schistosoma eradication campaigns may be impacted by changes in snail distribution, migrant workers and weather [128]. The impact of water and sanitation on schistosomiasis has shown that access to safe water and adequate sanitation is important in reducing schistosomiasis [129,130].

10.2. Guinea Worm

Dracunculus medinensis is transmitted by drinking water contaminated with copepods that contain the parasite larvae, and infections are easily controlled with simple measures. This disease is moving rapidly towards eradication under the supervision of the Carter Centre, with the World Health Organisation, UNICEF and CDC [131,132]. Although climate is likely to influence disease, the biggest problems in elimination have been organising surveillance in conflict areas [133], and the recent demonstration of a transmission cycle in dogs, particularly in Chad [134].

10.3. Nematodes

Ascaris, *Trichuris* and hookworms are soil transmitted nematodes, and transmission is related to inadequate methods for faecal/wastewater disposal. Risk of *Necator* infections were linked to rainfall in East Timor [135]. For *Ascaris lumbricoides*, higher temperatures in the coolest quarter of the year resulted in reduced risk. There is little evidence to suggest that changes in climate might greatly increase disease transmission and evidence that helminth control measures, if effectively undertaken, might cause substantial reductions in disease. A SR of spatial and temporal distribution of soil transmitted helminths was used to determine changing disease burden and predict treatment needs for eradication [136]. Another soil helminth SR and MA identified reductions in disease associated with WASH activities, but emphasised the limited nature of the studies, which were mostly observational [137].

10.4. Protozoa

Waterborne cryptosporidiosis is common, even in developed countries because the parasite's oocysts are resistant to chlorine, which is the main chemical treatment used for disinfecting both drinking water and indoor recreational waters, including swimming pools [138]. The oocysts are common in natural waters and can be present in source waters. A SR and MA of the impacts of weather on surface waters indicated that contamination with *Cryptosporidium* and *Giardia* was 2.61 and 2.87 higher during and after heavy rainfall [139]. A SR was used to examine the risks of endemic cryptosporidiosis, and found increased risk associated with the unsafe use of water [140]. This makes the need for risk assessment necessary. Cryptosporidiosis outbreaks may well increase if there are more frequent incidents of severe weather. In many countries, surveillance involving the laboratory identification of isolates is absent or suboptimal. Without good surveillance, outbreaks will be missed, and the prevention of these not dealt with. A SR of *Toxoplasma* outbreaks showed longer incubation following exposure to contaminated water compared to meat related exposures [141]. A SR of intestinal protozoal infections found significantly lower odds of infection in people with treated water [142]. *Giardia* transmission is not as well understood as *Cryptosporidium*, mostly because typing approaches have been suboptimal and routine surveillance in developed countries has focussed on the link to travel, rather than indigenous sources of infection [143]. Hygiene linked to water is important and occurrence in developing countries shows no relationship to rainfall [144]. However, the infection is common, and determining any change due to climate may be difficult. Both *Giardia* and *Cryptosporidium* are common in developing countries, and include a variety of species and types, and it has been suggested that climate change will increase malnutrition and contamination of water sources [145].

Naegleria fowleri infections are commonly associated with thermally polluted waters and cases might increase with raised temperatures. However, many of the reported infections are associated with geothermal waters that will not be greatly impacted by climate change.

10.5. Cholera and Other Diarrhoeal Diseases

Cholera remains one of the major diarrhoeal disease concerns with climate change. There has been a long history of associations between seasonality and weather as contributing factors in outbreaks. Cholera is also readily inserted into disaster situations and can contribute to morbidity and mortality. Examination of the weather relationship in El Niño years to the occurrence of cholera suggests that weather might be able to be used for predicting outbreaks in East Africa [146]. However, the assumptions in models are not necessarily always proven in practice [147–149]. As pandemics of cholera have occurred in the past, there is a strong need to reduce the risks of another pandemic arising out of a disaster situation. Changes in seawater temperature may contribute to changes in disease incidence [150]. The impact of water, sanitation and hygiene interventions are important in the control of cholera [151]. Home water treatment and storage interventions can reduce diarrhoea and cholera [152]. This study demonstrated the dearth of good evidence. A SR looked at the environmental determinants of cholera in inland Africa [147]. They found that spread was linked to displaced populations and the poor water and sanitation associated with these settings. A similar review of coastal cholera by the same authors found that cholera seasonality is driven by rainfall-induced contamination of unprotected water sources [148].

The impact of a changing climate is most important when it comes to the quality of drinking water and diarrhoeal diseases, particularly in developing countries [37]. There are relationships between many of the bacterial infections and temperature, and evidence of links to heavy rainfall and flooding [32,110,117]. The sources of exposure may be agricultural [153] or from sewage, and the different sources contain different pathogens, with human viruses rarely being detected in animal waste [154]. Engagement with the water sector is important in reducing risks.

10.6. Legionella

Legionnaires' disease is commonly associated with thermally polluted waters [155]. Due to the relationship between temperature and the occurrence of *Legionella* spp. in water, one might expect an increase in cases or outbreaks associated with climate change. Studies indicate that temperature and humidity [156,157], or vapour pressure [158], may play a role in the occurrence of sporadic disease. Due to concerns about the multiplication of *Legionella* spp. at the consumer endpoint (taps and showers) the European Drinking Water Directive has been modified to include a risk assessment for this pathogen. A SR on cooling towers related to Legionnaires' disease provided quite detailed advice on risks and their mitigation [159]. As climate changes becomes more apparent, there may be improvements in our understanding of previously unrecognised transmission routes from the natural environment.

10.7. Leptospira spp.

Leptospirosis outbreaks are commonly linked to flooding in rural developing countries. This relates to the dislocation of small mammal populations [160–162]. It is expected that an increase in flooding will contribute to more leptospirosis cases. Floods are one of the most important drivers of leptospirosis on islands and in Asia [163].

10.8. Pseudomonas aeruginosa

The ability of *Pseudomonas* to grow in water is dependent on being able to extract nutrients from the surfaces it is growing on, and to bloom within the body of the water. This includes outbreaks of folliculitis associated with spa pools and is dependent on temperature. One must therefore expect some increase associated with increased temperatures. A SR of *Pseudomonas aeruginosa* in hospitals found evidence that the water systems can be the source of this organism, and that contamination mostly occurs at the distal end [164]. Water resource management is important in control of this [165]. *P. aeruginosa* infections are common, and numbers related to water use may increase with a changing climate.

10.9. Harmful Algal Blooms (HABs)

Harmful algal blooms have the potential to increase with changes in climate [75]. The neurotoxin producing dinoflagellate and diatom plankton mainly cause disease through shellfish poisoning and Ciguatera poisoning in fish, but can also be responsible for the mass deaths of fish. Monitoring of shellfish waters can facilitate shellfish bed closures as a means of controlling exposure through food, while ciguatera is mostly controlled by not eating carnivorous reef fish. Blooms can cause respiratory symptoms in people on affected bathing beaches. Freshwater cyanobacteria are a potential risk through untreated potable water, and the toxic products can cause liver disease. In developed countries, these problems should be dealt with by water utilities.

10.10. Shiga-Toxin Producing E. coli (STEC)

As water is commonly contaminated with the faeces of agricultural animals, ways of reducing *E. coli* O157 excretion by cattle have been examined [166]. Further work on preventing haemolytic uremic syndrome found that vaccine use on animals reduced carriage, but the study emphasised the importance of public health measures [167]. Patients were consistently found to have exposure to rodents, behavioural and sanitation related risk factors.

10.11. Norovirus

An examination of norovirus transmission routes showed higher attack rates in foodborne than waterborne transmission, and higher rates in surface derived than groundwater derived waters [168]. Given the very common occurrence of this pathogen, any change will possibly depend more on progress in vaccinations rather than an impact from climate change. However, outbreaks may occur when large populations are kept in confined areas during an emergency [84].

10.12. Trachoma

Chlamydia trachomatis is the cause of trachoma, a disease of the eyes related to water shortage. Water can be important in reducing trachoma, where the F and E of the SAFE strategy (surgery, antibiotics, facial cleanliness and environmental improvement) were strongly linked to disease reduction [169]. Another trachoma SR found the studies too limited to demonstrate the impact of F and E [170]. The SAFE interventions are dramatically reducing disease but could be disrupted in chaotic situations associated with climate change.

11. Preparing for Climate Change

The principal reason for drinking water infections is the absence of reliable infrastructure. For rural communities worldwide, this includes the source waters used for drinking, the methods used to transport it from source to household, safe storage within the household, any treatment used and hygiene and sewage disposal changes. The principal underlying factor is poverty. Improving water treatment needs to be accompanied by improvements in sewage disposal and hygiene training. Climate change will not alter these underlying principles. What could change is the frequency of water shortages, population movements, conflict, refugee camps and so on, and a deterioration in the hygienic quality of water that people receive. Good planning associated with disasters can reduce the risks of waterborne diseases [41]. Any increase in disease associated with disasters in developing countries can have knock-on effects in developed ones.

With climate change, it is important to build knowledge of the mechanisms by which changes in weather can influence individual pathogens, and how these may subsequently affect human health. Not all problems associated with diarrhoeal diseases are as a result of outbreaks and it is important to understand the drivers for sporadic disease, and the disease burdens associated with the range of common pathogens. Good surveillance is necessary in order to detect clusters and outbreaks at an early stage, so that they can be investigated and controlled in a timely way, and so that future outbreaks can be prevented.

Most enteric pathogens are very seasonal, with bacterial pathogens (e.g., *Salmonella* spp., *Campylobacter* spp., STEC) predominating in the summer months, while viral gastroenteritis predominantly peaks in the winter (e.g., norovirus, rotavirus) [171]. The analytical examination of weather drivers has relied on time series approaches, that have limitations due to the collinearity between, for example temperature and seasonality, making it difficult to get definitive evidence of the causal mechanisms and to predict the impact of changes in climate. A review of mathematical approaches to demonstrating weather influences on waterborne infections has been undertaken [172]. The studies grouped into two clusters: Process-based models (PBM) and time series and spatial epidemiology (TS-SE). A review of analytical epidemiology studies looked at the quality of evidence for associations with rainfall, temperature, and so on, and waterborne disease identified the difficulties with developing optimum approaches [117]. A more systematic examination of the seasonality of a full range of pathogens [171] may widen our understanding of those that are linked to climate fluctuations. There are also a range of less analytical approaches that can contribute to a better understanding of waterborne pathogens and future risks (Table A1, Appendix A). Further development of tools to separate the effects of weather from other influences on seasonality needs to be undertaken, utilising surveillance data from a wider geographic area and linking local cases to local weather.

12. Conclusions

There have been a number of previous reviews and publications highlighting potential impacts of climate change and water-related illness in water-related or waterborne disease [37,38,117,172–182]. Some approaches examine the technological means to best represent the data [42,175,177,178,183], some examine outbreaks [28,32] and others focus on effects of the hydrological cycle in particular regions [28,110,174,179], while others have adopted a systematic review approach [37]. This paper has examined a full range of water-related pathogens, providing an evidence base for regarding them as water-related and indicating which studies have provided evidence. In addition to a changing climate, the world will continue to experience an increasing human population, with ever greater inter-connectivity through travel and information technology. Animal infections will continue to contribute to zoonotic disease and we should expect new pathogens to continue to arise. With climate change we need to watch for the rise in cholera worldwide. Although this is predominantly a disease of the poor and is not generally a problem in developed countries that have safe drinking water, it still has the potential to cause pandemic disease if there is general chaos, as can be found in disaster situations. The growth of *Vibrio cholera* in coastal waters means there can be changes in exposure with alterations in rainfall and temperature [150]. There remains a strong need for timely surveillance and rapid epidemiological response to outbreaks and new waterborne pathogens. The needs of developing countries in relation to waterborne diseases are both technical and financial, and ensuring adequate quantities of good quality water will be essential in preventing excess morbidity and mortality in areas that will suffer from substantial changes in climate in the future.

Author Contributions: G.N. initiated the paper and reviewed the evidence for waterborne disease and climate change. C.H. provided the climate change overview and I.L. the interface between waterborne diseases and climate change. All three contributed to the manuscript production and producing a final text.

Acknowledgments: The research was funded in part by the National Institute for Health Research Health Protection Research Unit (NIHR HPRU) in Environmental Change and Health at the London School of Hygiene and Tropical Medicine in partnership with Public Health England (PHE), and in collaboration with the University of Exeter, University College London, and the Met Office; the UK Medical Research Council (MRC) and UK Natural Environment Research Council (NERC) for the MEDMI Project (https://www.data-mashup.org.uk). Some of the work was conducted while seconded from HPA/PHE to the European Centre for Disease Prevention and Control (ECDC). The views expressed are those of the authors not necessarily those of the NHS, the NIHR, the Department of Health and Social Care or Public Health England. IL is part funded by the National Institute for Health Research Health Protection Research Unit (NIHR HPRU) in Gastrointestinal Infections at University of Liverpool in partnership with Public Health England (PHE), in collaboration with University of East Anglia, University of Oxford and the Institute of Food Research.

Conflicts of Interest: The authors declare no conflicts of interest in respect of this work.

Abbreviations

CA	Cost analysis
CEO	Circumstantial evidence only
CSS	Cross sectional survey
EACI	Ecological association between climate and infections
EACO	Ecological association between climate and outbreaks
EACWQ	Ecological association between climate and water quality
FTA	Fault Tree Analysis
GAMTS	Generalised additive model time series

GDSE	Gastrointestinal and dermatological symptoms and exposure
GLM	Generalised linear model
HSM	Hindsight suitability model
LR	Literature review
MLM	Multi-level modelling
MM	Mathematical modelling
MMF	Microbiological monitoring of flooding
MMO	Microbiological monitoring of outbreaks
MMST	Microbiological monitoring with salinity and temperature
NBM	Negative binomial model
MSRA	Multiple stepwise regression analysis
NE	No published evidence
OI	Outbreak investigation
PMCC	pairwise-matched case-control study
POCS	Prospective observational cohort study
PORA	Post outbreak rainfall analysis
POTA	Post outbreak temperature analysis
POWE	Post outbreak water examination
PRA	Poisson regression analysis
RAI	Review of animal infections
RCS	Retrospective cohort study
RCCS	Retrospective case-crossover study
RILO	Rodent investigation linked to outbreak
ROS	Retrospective outbreak surveillance
RRM	Rainfall runoff model
RSA	Rainy season association
RSE	Recreational swimming exposure
SA	Spatial analysis
SCS	Sporadic case series
SFA	Seasonal factor analysis
TSAT	Time series analysis of temperature
QMRA	Quantitative microbial risk assessment
WMCS	Water microbiology and case series
WMR	Water microbiology and rainfall

Appendix A

Table A1. Evidence of water related, waterborne, water-based, water contamination of equipment, water washed, water-based food contamination, water toxicosis, respiratory aerosol transmitted organisms, those that might be subject to climate change and the types of evidence.

Pathogen	How Climate Might Affect Disease Occurrence	Strength of Evidence for Water Related Infections	Human Waterborne Outbreaks or Infections	Type of Infection Route	Study Linking Weather/Climate and Infection
Acanthamoeba	*A. polyphaga* linked to contact lens washing and hygiene important. Infections linked to flooding (presumed contamination of potable water). Water contamination links.	Strong	[183–194]	Waterborne	PMCC [195]; LR [196]
Adenovirus	Subgroups A–E cause upper respiratory infections, conjunctivitis, febrile illness, sore throat and swollen glands. The enteric subgroup F adenoviruses Ad40 and Ad41 cause gastroenteritis in children. Contamination of groundwater used as a drinking water source and from faecal or respiratory contamination of untreated recreational waters. Swimming pool outbreaks.	Moderate	[184,197–201]	Waterborne	POWE [202]; OI [203]
Aeromonas spp.	Can cause wound infections and are thought to be a cause of diarrhoea. Point source diarrhoea outbreaks do not occur and evidence of their role in diarrhoeal disease is equivocal. They are commonly found in raw foods, sewage and source waters, and can bloom in drinking water distribution systems. Seasonality similar to *Salmonella*.	Weak	No waterborne outbreaks [204].	Waterborne	NE
Arcobacter spp.	*A. butzleri* is a Campylobacter-like bacteria that is associated with watery diarrhoea. *A. skirrowii* can cause persistent diarrhoea. *A. cryaerophilus* is a third species found in human infections. *Arcobacter* spp. can be isolated from animals, birds, food, water and the environment. Arcobacter infections are uncommon and there is no evidence of the impact of weather on infections. Three waterborne outbreaks but no evidence Arcobacter was the cause.	Weak	[202,205]	Waterborne	NE
Ascaris	*A. lumbricoides* ova in 'night soil' need to mature at ambient temperatures before they are infectious. Rain can wash infectious oocysts into water sources used for irrigating salad vegetables. Transmission through irrigation of salad vegetables.	Weak	[206–208]	Water based	NE
Astrovirus	Astroviruses cause diarrhoea in children under five years old. Viruses are excreted in faeces, and they will be present in sewage. Contact with contaminated recreational waters may be a risk factor. Outbreaks often mixed. Outbreak linked to flood water contamination of shellfish with several viruses.	Weak	[31,209–212]	Waterborne	POWE [213]
Burkholderia pseudomallei	Waterborne outbreaks in Australia with circumstantial & epidemiological evidence for water as the source. Infections and outbreaks linked to tropical storms, rainfall and drought.	Strong	[214–224]	Waterborne	POWE [216,222]
Burkholderia cepacia	Contamination of indoor water systems, predominantly in hospitals; contaminated sterile solutions, [225–229] disinfectant solution and taps can occur. Clinical circumstantial & epidemiological evidence but no link to weather parameters.	Strong	[225,227,230–245]	Waterborne	NE

Table A1. *Cont.*

Pathogen	How Climate Might Affect Disease Occurrence	Strength of Evidence for Water Related Infections	Human Waterborne Outbreaks or Infections	Type of Infection Route	Study Linking Weather/Climate and Infection
Balamuthia mandrillaris	Occurs in soil and causes amoebic meningoencephalitis. Limited evidence of occurrence in thermal water sources and hot water systems [246–248].	Weak	[185,246,249]	Waterborne	NE
Balantidium coli	A ciliated protozoan organism found in wild and domestic pigs. It causes dysentery in man and other primates. The routes of transmission are not clear. Rare infection thought to be drinking water-borne but little supporting evidence.	Weak	[187,250–252]	Waterborne	NE
Blastocystis hominis	An amoeba that can be detected in patients with diarrhoea. Limited evidence of family outbreak [253] but no waterborne outbreaks. Its significance as a human pathogen is unclear [254–256]. Transmission is probably by contaminated food and water. Little evidence of waterborne transmission. No weather links.	Weak	[187,190,257–259]	Waterborne	NE
Bocavirus	Human bocavirus (HBoV) is comparatively newly discovered and is thought to be spread from person to person by the respiratory and faecal-oral routes. The virus has been detected in respiratory samples, faeces and blood. It has been commonly detected in children with upper or lower respiratory infections. Infection has been seen worldwide. No evidence of waterborne infection or infections linked to water, although it has been detected in sewage [260].	None	None	Unknown	NE
Campylobacter spp.	The commonest bacterial cause of diarrhoea. Most infections are sporadic, but waterborne outbreaks linked to camp sites, travelling abroad, hospitals and large communities. Infection is commonly derived from contaminated poultry and water for the chicken flocks may be one source of contamination. *Campylobacter* spp. are spiral/curved organisms when isolated from patients, and change to a more resistant coccal stage when present in water. Most human infections are caused by *C. jejuni*, *C. coli* and *C. lari*. *Campylobacter fetus* subsp. *fetus*, can cause human infections, with septicaemia and gall bladder infection being more common than with the other species. *C. upsaliensis*, *C. hyointestinalis* subsp. *lawsonii* and *C. hyointestinalis* subsp. *hyointestinalis* are occasionally isolated from diarrhoeal patients. Infection through contaminated drinking water—heavy rainfall	Strong, outbreaks represent a small percentage of cases	[261–268]	Waterborne	PORA [109,269]; EACI [270]
Chilomastix mesnili	A flagellated protozoan that is thought to have limited pathogenicity. Parasite present in wastewater. It is probably transmitted by contaminated food and water, and can be found in patients with diarrhoea. Weak circumstantial evidence of water transmission [271–274]. No link to climate.	Weak	NE	Unknown	NE
Chlamydia trachomatis	Trachoma, caused by *C. trachomatis*, is associated with a lack of water. Transmission by flies is reduced by face washing.	Strong	NE	Water washed	NE
Chryseobacterium/ Elizabethkingia/ Flavobacterium	*E. meningoseptica*, (syn. *C. meningoseptica*, *F. meningoseptica*). Contamination of washed wounds. Water aerator colonisation and ICU infection—no identified climate association.	Weak, circumstantial	[275–277]	Waterborne	NE

Table A1. *Cont.*

Pathogen	How Climate Might Affect Disease Occurrence	Strength of Evidence for Water Related Infections	Human Waterborne Outbreaks or Infections	Type of Infection Route	Study Linking Weather/Climate and Infection
Coronavirus (SARS)	Has caused severe respiratory illness and can be excreted in faeces. Survives in sewage for 2 days at 20 °C and 14 days at 4 °C [278]. Possible contamination of bathing waters during outbreak. Possible aerosolisation in buildings.	Weak	NE	Unknown	NE
Cryptosporidium spp.	Cause diarrhoea in young mammals and in humans but cannot grow in the environment. Large waterborne outbreaks have been reported throughout the world. Oocysts are excreted in faeces and sewage. Many species (*C. hominis*, *C. parvum*, *C. meleagridis*, *C. cuniculus*, *C. ubiquitum*, *C. viatorum*, *C. canis*, *C. felis*, *C. suis*, *C. scrofarum*, *C. bovis*, and *C. muris*) and genotypes of *C. parvum*) can cause human disease. Rainfall can contribute to drinking water contamination from both human and animal faeces [279].	Strong	[28,138,187,190, 271–274,280–285]	Waterborne	SA [286] SFA [287]; OI [288–292]; POWE [202]; FTA [293]; RSA [294]
Cyanobacteria	Grow as blooms or mats, mostly within fresh water bodies. There are a large variety of species, many producing potent toxins that can cause acute and chronic disease in mammals, including man. The toxins include microcystins, nodularins, anatoxins, Saxitoxins, aplysiatoxins, cylindrospremopsins, beta-methyl-amino-L-alanine (BMAA) and lipopolysaccharides. Algal blooms are more commonly found in eutrophic (eutrophic waters have a high concentration of nutrients) inland waters. Human health risks arise if the water is consumed untreated, if people bathe or participate in water contact sports in waters with a scum or heavy bloom and if contaminated water is used in renal dialysis. There have been some notable outbreaks associated with cyanobacterial toxins with a high mortality rate in dialysis patients. There are also associations between exposure to cyanobacterial toxins and long-term health risks including cancer. The risks from BMAA linked to neurological disease are unclear. Climate influence on algal blooms. Human recreational and drinking water exposures.	Strong for outbreaks linked to peritoneal dialysis.	[295,296]	Water toxicosis	CEO
Cyanobacteria—*Anabena* spp.	Found in eutrophic waters. Has caused liver damage and deaths in waterfowl, sheep, cattle and other agricultural animals. *Vibrio cholerae* has been found to survive for long periods inside *A. variabilis*. Toxin producing species include *A. flos-aquae*, *A. lemmermannii*, *A. circinalis*, *A. millerii* and *A. planctonica*.	Weak	NE	Potential water toxicosis	NE
Cyanobacteria—*Aphanisomenon* spp.	Produce the toxins anatoxin-a, saxitoxins and cylindrospermopsin. Toxin producing species includes *A. ovalisporum*.	None	NE	Potential water toxicosis	NE
Cyanobacteria—*Anacystis* spp.	*A. nidulans* can produce lipopolysaccharide toxin.	None	NE	Potential water toxicosis	NE

Table A1. *Cont.*

Pathogen	How Climate Might Affect Disease Occurrence	Strength of Evidence for Water Related Infections	Human Waterborne Outbreaks or Infections	Type of Infection Route	Study Linking Weather/Climate and Infection
Cyanobacteria—*Cylindrospermopsis raciborskii*	Found in eutrophic waters and produces anatoxin-a. Palm Island Mystery Disease in Queensland, Australia caused 140 children to be hospitalised with a variety of symptoms including malaise, anorexia, vomiting, headache, liver enlargement, bloody diarrhoea and kidney damage. Water from a dam contaminated with *C. raciborskii* was thought to have been responsible.	Strong	[297,298]	Water toxicosis	NE
Cyanobacteria—*Lyngbya majuscule*	A marine cyanobacteria that produces a variety of toxic metabolites including apratoxin A, Palau'amide, 15-norlyngbyapeptin A and lyngbyabellin D, a quinoline alkaloid, malyngamide T and the potent neurotoxins antillatoxin, antillatoxin B, and kalkitoxin. Debromoaplysiatoxin has been isolated from *Lyngbya gracilis* also has dermonecrotic activity and may be the dermatitis-producing substance in *L. majuscula*, the causative agent of "swimmers' itch" outbreaks in Hawaiian waters.	Moderate	[299]	Water toxicosis	NE
Cyanobacteria—*Microcystis* spp.	*M. aeruginosa* is a common cyanobacteria found in eutrophic waters. It can cause hepatic failure and diarrhoea in man and other animals. An association was found between drinking water from a reservoir contaminated with *M. aeruginosa* and raised liver enzymes in a population in New South Wales, Australia. Other toxic species include *M. viridis* and *M. botrys*.	Strong	[300–302]	Water toxicosis	GDSE [303]
Cyanobacteria—*Nodularia* spp.	*N. spumigena* can cause poisoning of cattle and sheep which drink contaminated water. The nodularin toxin is hepatotoxic.	Weak	NE	Potential water toxicosis	NE
Cyanobacteria—*Nostoc* spp.	Produce microcystins and the beta-methyl-amino-L-alanine (BMAA) that has been implicated as the cause of neurodegenerative disease among the Chamorro people of Guam (Guam disease). Evidence not conclusive.	Weak	NE	Potential water toxicosis	NE
Cyanobacteria—*Oscillatoria* spp.	Produce anatoxin-a which can cause neurotoxicity leading to respiratory failure. Consumption of water contaminated with the toxin has been associated with liver cancer in China and acute respiratory failure and death in dogs. Includes *O. agardhii*, *O. nitroviridis* and *O. limosa*.	Moderate	NE	Potential water toxicosis	NE
Cyanobacteria—Planktothrix spp.	Produces microcystins, anatoxin-a and homoanatoxin-a. The main toxin producing species are *P. agardhii*, *P. mougeotii* and *P. formosa*.	None	NE	Potential water toxicosis	NE
Cyanobacteria—*Phormidium* spp.	*P. favosum* is a cyanobacterium that produces anatoxin-a and has been associated with neurotoxicosis in dogs. Toxin producing species include *P. bijugatum*, *P. molle*, *P. papyraceum*, *P. uncinatum* and *P. autumnale*. Toxin producing strains have been found in water from reservoirs in Australia.	None	NE	Potential water toxicosis	NE

Table A1. *Cont.*

Pathogen	How Climate Might Affect Disease Occurrence	Strength of Evidence for Water Related Infections	Human Waterborne Outbreaks or Infections	Type of Infection Route	Study Linking Weather/Climate and Infection
Cyanobacteria—*Raphidiopsis*	*R. mediterranea* can produce the toxins homoanatoxin-a, anatoxin-a, and non-toxic 4-hydroxyhomoanatoxin-a.	None	NE	Potential water toxicosis	NE
Cyanobacteria—*Trichodesmium*	*T. erythraeum* is a marine cyanobacteria that is reported to produce a neurotoxin. This organism contributes to some ciguatera toxin accumulation in fish.	Moderate	NE	Potential water toxicosis	NE
Cyanobacteria—*Umezakia natans*	Produces the toxin cylindrospermopsin.	None	NE	Potential water toxicosis	NE
Cyclospora cayetanensis	Contaminated water applied to salads and soft fruits through irrigation/spraying. Repeated outbreaks linked to Mexico.	Strong	[189,190,271–274,304,305]	Waterborne, Water contaminated food	NE
Dinoflagellates and diatoms	These are protozoan organisms that can produce a range of potent toxins. They occur predominantly in saltwater and, under the right conditions, can produce blooms that cause 'red tides' that can cause toxic effects in fish and other sea-life. The toxins can accumulate within shellfish, causing paralytic shellfish poisoning (PSP), diarrhoretic shellfish poisoning (DSP), neurotoxic shellfish poisoning (NSP), Amnesic Shellfish Poisoning (ASP). Some of the toxins can also accumulate through passing up the food chain to give carnivorous fish that are toxic (ciguatera toxin). Coastal blooms causing respiratory symptoms, ciguatera and shellfish poisoning. Blooms of dinoflagellates are linked to weather and nutrients.	Strong	[306–324]	Toxin contamination of marine foods	CEO [325]
Diatom—*Nitzschia* spp., *Pseudo-nitzschia* spp. & *Amphora* spp. Amnesic Shellfish Poisoning	*Pseudo-nitzschia* spp. And *Nitzschia navis-varingica* are diatoms that cause Amnesic Shellfish Poisoning through the production of domoic acids. Symptoms include dizziness, nausea, vomiting, cramps, diarrhoea, dizziness, headache, seizures, disorientation, short-term memory loss, respiratory difficulty and coma. The toxins are regularly detected in UK waters and shellfish. Blooms of *Pseudo-nitzschia australis* have been implicated in deaths of fish and sea lions. Other species include P. *calliantha*, P. *delicatissima*, P. *fraudulenta*, P. *galaxiae*, P. *multiseries*, P. *multistriata*, P. *pungens*, P. *seriata*, P. *turgidula*. *Amphora coffeaeformis* is a diatom that produces domoic acid, but its role in human disease is unclear.	Moderate	[326–329]	Toxin contamination of marine foods	NE

Table A1. *Cont.*

Pathogen	How Climate Might Affect Disease Occurrence	Strength of Evidence for Water Related Infections	Human Waterborne Outbreaks or Infections	Type of Infection Route	Study Linking Weather/Climate and Infection
Dinoflagellate—*Alexandrium* spp., *Pyrodinium* spp. & *Gymnodinium* spp. Paralytic Shellfish Poisoning	Cause Paralytic Shellfish Poisoning (PSP) through toxin accumulation in bivalve molluscs. They can also cause mass fish kills. Species include *A. acatenella, A. andersonii, A. balechii, A. catenella, A. fundyense, A. hiranoi, A. minutum, A. monilatum, A. ostenfeldi* and *A. tamiyavanichii*. The toxins include saxitoxin and neosaxitoxin, spirolide, goniodomin A. Change in algal blooms with climate. Transmission through shellfish. *Gymnodinium catenatum* produces red tides and high mortality in fishes and can also cause Paralytic Shellfish Poisoning (PSP). *Pyrodinium bahamense* var. *compressa* can produce red tides has been implicated in PSP.	Strong	[189]	Toxin contamination of marine foods	NE
Dinoflagellate—*Dinophysis* spp. & *Procentrum* spp. Diarrhoretic Shellfish Poisoning	Cause Diarrhoretic Shellfish Poisoning (DSP) through accumulation of the toxins in bivalve molluscs. The toxins include okadaic acid, dinophysistoxin-1 (DTX1), dinophysistoxin-2 (DTX2) and pectenotoxin-2 (PTX2). The implicated species include *D. acuminata, D. acuta, D. caudata, D. fortii, D. miles, D. mitra, D. norvegica, D. rapa, D. rotundata, D. sacculus* and *D. tripos. Procentrum arenarium, P. belizeanum, P. cassubicum, P. faustiae, P. hoffmannianum, P. lima, P. maculosum* produce one or more toxins including okadaic acid, DTX1, DTX-2, prorocentrolide and fast acting toxins (FAT). Other species with less evidence of toxicity include *P. borbonicum, P. concavum, P. emarginatum, P. minimum, P. emarginatum, P. minimum, P. mexicanum, P. arabianum.*	Strong	[318,319,330, 331]	Toxin contamination of marine foods	NE
Dinoflagellate—*Karenia* spp. Neurotoxic Shellfish Poisoning	The species *K. bicuneiformis, K. brevis, K. brevisulcata, K. concordia, K. cristata, K. mikimotoi, K. papilionacea, K. selliformis, K. umbella* produce red tides and can cause Neurotoxic Shellfish Poisoning (NSP). They also cause high mortality in fishes, invertebrates, marine animals and plants, and human respiratory distress, eye and skin irritations. Most species produce brevetoxin in culture. *K. brevis* is synonymous with *Gymnodinium brevis* and *Ptychodiscus brevis.*	Moderate	[309]	Toxin contamination of marine foods	NE
Dinoflagellate—*Protoperidinium* Azaspiracid Shellfish Poisoning	*P. crassipes* produces azaspiracid, which gives DSP-like symptoms of Azaspiracid Shellfish Poisoning in humans but a mixture of DSP-and neurotoxin-like effects in mice.	None	[332]	Toxin contamination of marine foods	NE
Dinoflagellate—*Gambierdiscus* spp. & *Gonyaulax tamarensis.* Ciguatera food poisoning	Produce ciguatoxin or maitotoxin that can pass up the food chain and accumulate in dangerous amounts in certain species of carnivorous fish and shellfish accumulate the toxins as they pass up the food chain. Ciguatera food poisoning produces gastrointestinal, neurological, and cardiovascular symptoms including diarrhoea, vomiting, abdominal pain, reversal of temperature sensation, muscular aches, dizziness, anxiety, sweating, and a numbness and tingling of the mouth and digits. Toxic species of *Gambierdiscus* include *G. pacificus, G. australes, G. yasumotoi* and *G. polynesiensis* and *Gonyaulax tamarensis.*	Strong	[333–344]	Toxin contamination of marine foods	NE

Table A1. *Cont.*

Pathogen	How Climate Might Affect Disease Occurrence	Strength of Evidence for Water Related Infections	Human Waterborne Outbreaks or Infections	Type of Infection Route	Study Linking Weather/Climate and Infection
Dinoflagellate— *Ostreopsis* spp.	*Ostreopsis lenticularis* is the presumed vector of ciguatera poisoning in the Caribbean. The toxin accumulates through the food chain and can be present in toxic amounts in carnivorous fish. *O. siamensis* produces ostreocin D, a palytoxin analogue and has been implicated in clupeotoxism, a fatal toxicosis through ingestion of clupeoid fish (sardines, anchovies and herring). *O. lenticularis* produces ostreotoxin-1 and -3, palytoxin analogues (polyethers), mascarenotoxin-A and -B. *O. mascarenensis* is possibly responsible for palytoxin poisoning, which in humans results in cramps, nausea, diarrhoea, etc, after eating of crabs and certain fish. *O. ovata* is also be a toxin producer and has been associated with toxic effects of bathing in a bloom of this organism (nausea, breathing difficulties, high fever, stomach cramps, irritation of the eyes, vomiting and diarrhea) in Italy.	Strong	[308,345]	Toxin contamination of marine foods; Potential water toxicosis	NE
Dinoflagellate— *Hematodinium* spp.	Cause Bitter Crab Disease and Pink Crab Disease in crabs, making them unpalatable. There is little evidence for a health risk associated with the consumption of infected crabs.	None	NE	Potential water/food toxicosis	NE
Dinoflagellate— *Heterocapsa* spp.	*H. circularisquama* forms red tides and can cause mass fish kills. Their role in human disease remains unclear.		NE	Potential water/food toxicosis	NE
Dinoflagellate— *Pfiesteria* spp.	*Pfiesteria piscicida* is a dinoflagellate that resides in estuarine waters and is responsible for mass fish kills. It has been associated with skin and neurological problems in humans exposed to the toxins, but this work has been challenged. *Pfiesteria shumwayae* causes mass fish kills in estuarine waters. No human illness has yet been associated with this organism.	Weak	NE	Unknown	NE
Dinoflagellate— *Protoceratium reticulatum*	*P. reticulatum* produces yessotoxin which may accumulate in bivalves and is toxic to mice. Its role in human disease is unclear.	None	NE	Unknown	NE
Dracunculus medinensis	*Dracunculus medinensis* life cycle involves the water flea Cyclops. It is the cause of dracontiasis. Human infection results from the consumption of water contaminated with infected water fleas. The adult worm emerges on the foot or leg, and rhabditoid larvae are released into the water where they re-infect water fleas. There is a WHO lead worldwide programme to eradicate Guinea Worm. Rainfall contamination of source waters. Infection associated with water scarcity and start of rainy season.	Strong	[346–351]	Waterborne	RSA [352,353]
Entamoeba histolytica	Causes invasive amoebic dysentery, with abscess formation in the liver and other body sites. Colonisation may occur through water contaminated with the cysts, often in developing countries where it is endemic. People infected with this organism have usually been travelling abroad recently. *E. histolytica* is visually similar but genetically distinct from *E. dispar*. A few waterborne outbreaks.	Moderate	[187,189,190, 258,271–274, 354,355]	Waterborne	NE

Table A1. *Cont.*

Pathogen	How Climate Might Affect Disease Occurrence	Strength of Evidence for Water Related Infections	Human Waterborne Outbreaks or Infections	Type of Infection Route	Study Linking Weather/Climate and Infection
Other *Entamoeba* spp.	*Entamoeba coli*, *E. dispar* and *E. moshnikovii* can colonise the human intestinal tract and is incapable of causing the invasive form of amoebic dysentery, and is not thought to be an enteric pathogen. *Endolimax nana*, *Enteromonas hominis*, *Retortomonas intestinalis*, *Iodamoeba butschlii*, *Pentatrichomonas hominis*, *Trichomonas hominis* are thought to be non-pathogenic. They are probably transmitted by contaminated food and water, and can be found in patients with diarrhoea. No evidence of rainfall impact.	Weak	NE	Unknown	NE
Echinococcus spp.	Main species *E. granulosus* and *E.multilocularis* and rarer species *E. vogeli* and *E.oligarthrus*. *E. granulosus* forms hydatids (fluid filled vacuoles) in wild ruminants, sheep, pigs, cattle and man, *E. g. equinus* in horses but not man, *E. g. canadensis* in caribou, reindeer and man and *E. g. borealis* cervids and man. *E. multilocularis* has subspecies including *E. m. multilocularis* in Europe and *E. m. sibiricencis* in North America. *E. vogeli* in South American Bush dogs with hydatids in rodents. *E. oligarthrus* in wild cats with hydatids in rodents. Infection results from consumption of *Echinococcus* ova and has been linked to the consumption of contaminated drinking water and direct contamination from dogs or foxes. Strong epidemiological association with drinking water contamination. Rainfall may contribute to contamination.	Strong	[356–363]	Waterborne	NE
Enterobius vermicularis	No waterborne outbreaks but sporadic infection related to drinking water quality. No weather associations.	Weak	[364–366]	Waterborne	NE
Enteroviruses including coxsackie, ECHO, polio	Gross contamination of drinking water leading to enterovirus outbreaks.	Moderate	[197,367–373]	Waterborne	OI [373], QMRA [374]
Enterovirus—Coxsackievirus A and B	Cause mycocarditis, pericarditis, skin rash, meningitis, respiratory infections and fever and are transmitted through the faecal-oral route. Coxsackie A can cause hand, foot and mouth disease, herpangina (fever, sore throat, pain on swallowing, anorexia and vomiting) and conjunctivitis. Coxsackie B can cause pleurodynia (chest pain) and severe infections in neonates. Outbreaks of Bornholm disease do not show evidence of being waterborne.	Weak	NE	Not waterborne	NE
Enterovirus—Echovirus	These enteric viruses usually manifest themselves as a maculopapular skin rash but can also cause meningitis. There are a number of serotypes.	Weak	[375–377]	Waterborne	NE
Enterovirus—Parechovirus	Parechovirus type 1 (PeV1) and type 2 (PeV2) were previously known as echovirus 22 and echovirus 23, but there are six types. Infections can include meningoencephalitis, fever and myositis and may play a role in intra-uterine fetal deaths. Parechoviruses have been isolated from some patients with gastroenteritis. PeV1 has been associated with otitis media and cough.	Weak	NE	Unknown	NE

Table A1. *Cont.*

Pathogen	How Climate Might Affect Disease Occurrence	Strength of Evidence for Water Related Infections	Human Waterborne Outbreaks or Infections	Type of Infection Route	Study Linking Weather/Climate and Infection
Enteroviruses—Poliovirus	There are three poliovirus types that have been eradicated from the UK. The WHO is organising the worldwide eradication of poliovirus. Wildtype WPV2 poliovirus was last observed in October 1999. Strategies for eradication include expanded use of type 1 monovalent oral poliovirus vaccine (mOPV1) to eliminate WPV1 transmission before WPV3 and targeted use of type 3 monovalent vaccine mOPV3 in selected areas.	Weak	NE	Unknown	NE
Escherichia coli	*Escherichia coli* contain both non-pathogenic organisms and strains that are an important cause of diarrhoeal disease (termed enterovirulent *E. coli*). Several classes have been defined based on the possession of distinct virulence factors. These are the enteropathogenic (EPEC), enterotoxigenic (ETEC), enteroinvasive (EIEC) and Vero cytotoxigenic (VTEC), enteroaggregative (EAggEC) and diffusely adherent (DAEC) strains of *E. coli*. In developing countries food and contaminated water are important vehicles for transmission of *E. coli*. These are less common in developed countries where there are higher standards of general hygiene.	Strong	[35,378–390]	Waterborne	NE
Escherichia coli— enteropathogenic (EPEC)	The classical enteropathogenic *E. coli* (EPEC) belong to a small number of serotypes and cause sporadic cases and outbreaks of diarrhoea in children, usually under the age of 2 years. A wider range of EPEC strains have caused sporadic cases and outbreaks affecting adults.	Weak	[378]	Waterborne	NE
Escherichia coli— enterotoxigenic (ETEC)	Common cause of traveller's diarrhoea that can be transmitted by water.	Strong	[35,379–383]	Waterborne	NE
Escherichia coli— enteroinvasive (EIEC)	Transmission through poor hygiene or contaminated food or water.	Moderate	[384]	Waterborne	NE
Escherichia coli— Shiga cytotoxigenic (STEC; VTEC; EHEC)	Infection through contaminated drinking water—heavy rainfall.	Strong	[385–390]	Waterborne	CA [391]; OI [109]
Escherichia coli— enteroaggregative (EAggEC)	Enteroaggregative *E. coli* (EAggEC) cause sporadic cases and outbreaks affecting all ages and like Enterotoxigenic *E. coli* (ETEC) are a major cause of travellers' diarrhoea.	Moderate	[392]	Waterborne	NE
Escherichia coli— diffusely adherent (DAEC)	Transmission poorly understood.	NE	NE	Unknown	NE

Table A1. *Cont.*

Pathogen	How Climate Might Affect Disease Occurrence	Strength of Evidence for Water Related Infections	Human Waterborne Outbreaks or Infections	Type of Infection Route	Study Linking Weather/Climate and Infection
Fasciola gigantica	A liver fluke (helminth) that is common in cattle in the Middle East, Africa and Asia. The parasite requires a snail as an intermediate host, and man is infected through the consumption of aquatic plants contaminated with the metacercaria.	Moderate	[393]	Water based	NE
Fasciola hepatica	A liver fluke (helminth) that is common in herbivores that graze in wet pasture. The parasite requires a snail as an intermediate host, and man is occasionally infected through the consumption of aquatic plants, particularly watercress, contaminated with the metacercaria.	Strong	[394]	Water based	RAI [395–499]
Fasciolopsis buski	An intestinal fluke (helminth) that is common in areas of Indonesia and Northern Thailand and the Far East. The parasite requires a snail as an intermediate host, and man is infected through the consumption of aquatic plants, particularly water chestnuts and water caltrop, contaminated with the metacercaria.	Moderate	[400]	Water based	NE
Giardia spp.	Grows attached to the small intestinal lining and causes malabsorption in people. The parasite can be isolated from the faeces of wild and domestic animals, and waterborne outbreaks are usually associated with recreational water use. The parasite cyst, which is found in faeces, is moderately resistant to chlorine. The modes of transmission remain unclear. Giardia can be transmitted through recreational and drinking water, although hygiene is also important.	Strong	[28,143,271–274,281,283,401,402]	Waterborne	CSS [403]
Gongylonema pulchrum	Occurs commonly in domestic cattle and other vertebrates, but gongylonemiasis is very rare in humans. Only 48 cases have been described in the literature since 1864. An infection in Hungary was linked to ingestion of contaminated water from an open draw well.	Moderate	NE	Water based	NE
Helicobacter pylori	*H. pylori* is associated with gastritis, and with gastric and duodenal ulcers. The mode of transmission is not entirely understood. The reservoir of H. pylori is the digestive tracts of humans and some primates, and transmission is considered to be from person-to-person with the most common route being oral-oral. *H. pylori* is shed in the faeces after turnover of the gastric mucosa, and has been detected by PCR in sewage in Peru. Sewage pollution is therefore a possible though not proven route of transmission.	Moderate	[404–415]	Waterborne	NE
Helicobacter spp.	Is larger and more tightly helical than *H. pylori* and is associated with a small percentage of patients with gastritis. *Helicobacter bilis* is a Gram negative spiral bacterium that is regarded as an enteric helicobacter and has been implicated in liver disease, particularly biliary tract cancer. Other enteric helicobacters that may be implicated in liver or intestinal disease include *H. hepaticus*, *H. rappani* and *H. pullorum*. *Helicobacter winghamensis* is a *Helicobacter* sp. that have been isolated from patients with gastroenteritis. It's role as a cause of diarrhoea has not been demonstrated. *H. pylori* can contaminate drinking water sources and infection is associated with drinking water source. No evidence of climate drivers.	None	NE	Unknown	NE

Table A1. *Cont.*

Pathogen	How Climate Might Affect Disease Occurrence	Strength of Evidence for Water Related Infections	Human Waterborne Outbreaks or Infections	Type of Infection Route	Study Linking Weather/Climate and Infection
Hepatitis A	*Hepatitis A virus* causes hepatitis and can be acquired by person-to-person, through contaminated water, shellfish, and foods eaten raw or washed in contaminated water and waterborne routes. Infection resulting from sewage contamination of source waters and shellfish. Some rainfall associations.	Strong	[198,200,416–424]	Waterborne	PORA = [425]; SFA [426,427]
Hepatitis E	Hepatitis E virus has a genome of single stranded RNA. Epidemiological evidence suggests that the disease can be transmitted by drinking water contaminated with faeces or contact with an environment contaminated with faeces. Pigs may be an important reservoir of infection. Infections in the UK are associated with overseas travel. Large waterborne outbreaks	Strong	[198,209,428–436]	Waterborne	OI [428]
Isospora belli	A coccidian (protozoan) parasite related to *Cryptosporidium* and *Cyclospora*. Infection with *I. belli* may cause colicky pain, diarrhoea, malabsorption and fever. Transmission of *Isospora belli* to humans from contaminated water and infected animals is suspected but has not been clearly demonstrated. Close contact in barracks or institutions, the presence of other intestinal disease, and poor hygiene appear to favour transmission. Limited evidence of waterborne transmission	Weak	[187]	Unknown	NE
Laribacter hongkongensis	An aerobic, spiral shaped gram-negative bacteria isolated from a few people suffering from travel related diarrhoea. Cases were first identified in Hong Kong but have since occurred in other parts of the World. It has also been isolated from freshwater fish in China. The causative role of this organism in human gastroenteritis is still unproven.	None	NE	Unknown	NE
Legionella pneumophila	Legionellosis linked to climate (humidity/vapour pressure)	Strong	[275,437–443]	Aerosol transmission	PRA, RCCS [158,444,445]; POCS [446]; OI [447]
Leptospira spp.	Tightly coiled spiral bacteria that cause Weil's Disease (jaundice) in people. Infection is from rodents and agricultural and domestic animals, usually through exposure to contaminated water or urine. Drinking or exposing wounds or mucous membranes to contaminated water can result in infection. Infection through natural water contaminated by rodent urine and occasionally through unchlorinated drinking water. Outbreaks follow heavy rainfall and flooding and occasionally abnormally low rainfall.	Strong	[36,448–455]	Waterborne	[456]; OI [160–162,457–468]; RILO [469]; SA [470]; RCS [471]; NBM [472]; CSS [473]
Microsporidia	*Enterocytozoon bieneusi* infection linked to transmission through food and water. *Encephalitozoon hellem* keratoconjunctivitis possibly related to water or mud. Link to rainy season in Singapore.	Weak	[187,189,474–481]	Waterborne	RSA [478]

Table A1. *Cont.*

Pathogen	How Climate Might Affect Disease Occurrence	Strength of Evidence for Water Related Infections	Human Waterborne Outbreaks or Infections	Type of Infection Route	Study Linking Weather/Climate and Infection
Microsporidia—*Encephalitozoon* spp.	*E. intestinalis* is a microsporidian (protozoan) parasite that is associated with diarrhoeal disease in immunodeficient patients, particularly AIDS. Its transmissibility by food and water is not known. *E. cuniculi* is associated with disseminated infection involving the kidneys, sinuses, lungs, brain and conjunctiva in immunodeficient patients, particularly AIDS. Its transmissibility by food and water is not known. *E. hellem* is associated with nasal and ocular disease in immunodeficient patients, particularly AIDS. Its transmissibility by food and water is not known.	Weak	NE	Unknown	NE
Microsporidia—*Enterocytozoon bieneusi*	A parasite associated with diarrhoeal disease in immunodeficient patients, particularly AIDS. Its transmissibility by food and water is not known. *E. bieneusi* is usually confined to the intestines and is not associated with disseminated infection. *E. bieneusi* has been detected in pigs, cats, dogs and a rhesus monkey. *E. bieneusi* is transmitted through the usual faecal-oral pathways including food and water. Outbreaks have been linked to water and food.	Weak	[189,474]	Waterborne	NE
Mycobacterium spp. (not TB)	Infections related to buildings and the built environment.	Strong	[275,482–492]	Waterborne	LR [493]
Mycobacterium avium	Contains three subspecies (*M. avium* subsp. *avium*, *M. avium* subsp. *sylvaticum* and *M. avium* subsp. *paratuberculosis*). Organisms belonging to the *Mycobacterium avium* complex (MAC) include *M. avium*, *M. intracellulare* and *M. scrofulacium*. They can cause intestinal infection, septicaemia and wasting in AIDS patients, up to 50% of whom may develop MAC bacteraemia. Where the isolates have been speciated they have been found to be predominantly *M. avium*. Isolates recovered from water are probably a source of human infection. Organisms of the MAC group are able to grow in both hot and cold-water systems.	Moderate	[275,494,495]	Waterborne	NE
Mycobacterium paratuberculosis	A slow-growing bacteria that causes Johne's disease in agricultural animals. It has been implicated in the causation of Crohn's disease in humans. *M. paratuberculosis* (or *M. avium* subsp. *paratuberculosis* (Map)) has been detected in water using PCR but has not to date been isolated from potable water supplies.	Weak	NE	Unknown	NE
Naegleria fowleri	Colonises thermally polluted waters. Infections in Southern US are seasonal, with more in the summer. Infections in cattle are also seasonal. Infections may increase in some countries with warmer temperatures. Runoff from heavy rains introduces this organism into lakes, ponds, and surface waters [189,496].	Strong, links to water contamination	[187,189,497]	Waterborne	SCS [498]
Norovirus	Is mostly transmitted person-to-person. Transmission has also been indicated via contaminated ice, stored water on cruise ships, borehole water and contaminated recreational bathing waters. Municipal drinking water supplies have been implicated in outbreaks of gastroenteritis, usually following contamination by sewage. Strongly seasonal. Link to shellfish contaminated from infected faeces. Coastal water contamination linked to rainfall.	Strong	[416,499–508]	Waterborne	OI [509]

Table A1. *Cont.*

Pathogen	How Climate Might Affect Disease Occurrence	Strength of Evidence for Water Related Infections	Human Waterborne Outbreaks or Infections	Type of Infection Route	Study Linking Weather/Climate and Infection
Plesiomonas shigelloides	*Plesiomonas shigelloides* is a bacterial pathogen that has been implicated as a cause of diarrhoeal disease. It is common in the natural waters in tropical countries. No well documented waterborne outbreaks (two reported in 1978). No associations with rainfall or temperature.	None	[51,296,497, 510–513]	Waterborne	NE
Pseudomonas aeruginosa	Contamination of natural and man-made bathing waters. Most folliculitis infections are related to spa pools. Swimming in natural waters can cause otitis externa. No evidence of rain or temperature link.	Strong	[497,514]	Waterborne	NE
Rhinosporidium seeberi	Infections linked to lake, river and well water in India. Outbreak linked to bathing in stagnant water in Serbia.	Moderate	[515–518]	Waterborne	NE
Rotavirus	Rotavirus. Rotaviruses are part of the Reovirus family and have a double stranded RNA genome. Exposure is by contact with infected individuals or contaminated water or other materials. Group C rotaviruses have been identified throughout the world. Group B rotaviruses have caused large outbreaks of diarrhoeal illness in mainland China. The virus entered the population as a result of faecal contamination of water supplies drawn from rivers, and then spread through the population by person-to-person contact. Waterborne outbreaks in developing countries	Weak	[197,519–525]	Waterborne	[526]
Stenotrophomonas maltophilia	Hospital outbreaks linked to contaminated water systems.	Moderate	[527–531]	Waterborne	NE
Salmonella spp.	Are strongly associated with foodborne disease but can also be transmitted through waterborne and other routes. There is a particular problem with the occurrence of Salmonellas in agricultural animals and the transmission to people through poor slaughter hygiene and kitchen practices. This particularly results from their ability to grow within foods to numbers that can cause disease. Outbreaks and sporadic disease related to contaminated drinking water and recreational water. No clear links to rain.	Moderate	[518,532–545]	Waterborne	GAMTS [546]; GLM [546]; EACWQ [547]
Salmonella Paratyphi	Causes paratyphoid fever. Infections can be acquired from travel overseas, and food and waterborne infection can also occur.	Moderate	[548]	Waterborne	NE
Salmonella Typhi	Causes typhoid (fever and diarrhoea). The infectious dose is low, and large waterborne outbreaks were common in the first half of the 20th century. Waterborne infection can arise through sewage contamination of drinking water and through typhoid carriers who are water workers. There is no natural animal host of this pathogen. Most infections are in people returning from less developed countries.	Strong	[535,541,549]	Waterborne	RSA [550]
Sappinia diploidea	A rare cause of amoebic cerebral abscess.	None	[551,552]	Waterborne	NE

Table A1. *Cont.*

Pathogen	How Climate Might Affect Disease Occurrence	Strength of Evidence for Water Related Infections	Human Waterborne Outbreaks or Infections	Type of Infection Route	Study Linking Weather/Climate and Infection
Sapovirus	A calicivirus, formerly called "Sapporo-like virus" (SLV), classic or typical calicivirus and are associated with relatively mild gastroenteritis in children. Outbreak linked to flood water contamination of shellfish with several viruses.	Weak	[183]		POWE [213]
Sarcocystis hominis	Sporocysts are infectious, causing muscle infections through drinking water.	Strong	[553,554]	Waterborne	NE
Schistosoma spp.	These are flukes (helminth) which are transmitted through the contamination of water with faeces containing the ova. Cases linked to flooding and land surface temperature.	Strong	[555]	Water based	OI [556]; SA [557–560]; RRM [561]; MM [562]; MLM [563]
Schistosoma intercalatum	The life cycle involves the ova hatching and infecting specific snail species, and the cercaria infect people occupationally or recreationally exposed to contaminated water through the skin.	Strong	[564]	Water based	NE
Schistosoma haematobium	The life cycle involves the ova hatching and infecting specific snail species, and the cercaria infect people occupationally or recreationally exposed to contaminated water through the skin. Infection is found in Africa and the Middle East.	Strong	[564,565]	Water based	NE
Schistosoma japonicum	Infection is found in Eastern Asia including Japan and Korea. Links to rainfall and temperature.	Strong	[566]	Water based	CSS [563]
Schistosoma mansoni	The life cycle involves the ova hatching and infecting specific snail species, and the cercaria infect people occupationally or recreationally exposed to contaminated water through the skin.	Strong	[564,567–570]	Water based	OI [556]; SA [560]
Schistosoma mekongi	The life cycle involves the ova hatching and infecting specific snail species, and the cercaria infect people occupationally or recreationally exposed to contaminated water (usually river water) through the skin. Infection is restricted to South East Asia, particularly the Mekong River basin in Kampuchea.	Strong	[571–574]	Water based	NE
Schistosoma spindale	A fluke (helminth) that commonly causes cercarial dermatitis of paddy field workers in Assam, India. It is unclear which species of animal the schistosomes derive from although ducks are likely.	Strong	[575]	Water based	NE
Spirometra spp.	These are tapeworms (helminth) and occur in amphibious animals including frogs. Ingestion of the first stage larvae which are present in the copepod *Cyclops* results in human sparganosis, with 'Sparganum' larvae forming in nodules. Infection is thought to derive from contaminated drinking water or the consumption of uncooked frogs or snakes. *S. mansonoides* (*S. erinacei-europaei*) is the organism most commonly diagnosed.	Weak	[576–579]	Waterborne	NE

Table A1. *Cont.*

Pathogen	How Climate Might Affect Disease Occurrence	Strength of Evidence for Water Related Infections	Human Waterborne Outbreaks or Infections	Type of Infection Route	Study Linking Weather/Climate and Infection
Shigella spp.	Causes dysentery and can be readily passed between children and adults in unsanitary environments. They can occasionally be transmitted through contaminated food. Waterborne outbreaks of shigellosis have been common in the past, but are now uncommon and are usually the result of a combination of inadequate treatment, post treatment contamination and chlorination breakdown. *Shigella* spp. will not grow within water distribution systems. There are four species. *Shigella sonnei* is commonly transmitted between children and can be a problem in schools in the UK. *Shigella dysenteriae* is a more severe infection than *S. sonnei* and causes severe dysentery. Most *S. boydii, S. flexneri* and *S. dysenteriae* infections within England and Wales are usually acquired abroad. Lake water recreational outbreaks.	Strong	[355,580–584]	Waterborne	OI [284]
Torovirus	Toroviruses are enveloped viruses that have been linked to enteric infections in horses (Berne virus), cattle (Breda virus), pigs, and humans and transmission is thought to be via the faecal-oral route. In humans toroviruses have been found in infants with necrotising enterocolitis but their role in gastroenteritis remains unproved.	Weak	NE	Unknown	NE
Toxoplasma gondii	A protozoan parasite which occurs in a wide range of warm-blooded animals. The only definitive host in which the full sexual cycle has been observed is members of the cat family (Felidae), which excrete the oocysts which contaminate the environment and source waters. People can be infected from consuming food or water that is contaminated with oocysts or the consumption of undercooked meat which contains tissue cysts. Infection can be a particular problem for pregnant women and immunocompromised patients. Some evidence that heavy rainfall can precede outbreaks.	Strong	[187,190,585–589]	Waterborne	OI [590]; SA [591];
Trichobilharzia regent	A schistosome of birds and the cercaria can give rise to an itchy rash (known as cercarial dermatitis or swimmer's itch) in people who have had contact with the water. They are otherwise thought to be non-pathogenic to humans. The parasites mature and lay eggs in the nasal cavities of waterfowl. The lifecycle involves snails.	Strong	[592–599]	Water based	NE
Vibrio cholerae	Causes cholera, a disease that is characterised by acute and life-threatening diarrhoea and dehydration usually in epidemic outbreaks. Cholera is transmitted through drinking water, shellfish and contaminated food. The disease is usually restricted to less developed countries where drinking water and waste disposal are poor, and to migrant populations associated with drought, flood, famine and war. Evidence of links to rainfall over the last century.	Strong	[600–602]	Waterborne	[603,604]; GAMTS [605]; [606]; EACO [607]; POWE [608]
Vibrio parahaemolyticus	Inhabits estuarine and marine environments. It can cause food-poisoning through the contamination of seafood. *V. parahaemolyticus* associated with raised water temperature.	Moderate	[609]	Foodborne through seafood	RSE [610]

Table A1. *Cont.*

Pathogen	How Climate Might Affect Disease Occurrence	Strength of Evidence for Water Related Infections	Human Waterborne Outbreaks or Infections	Type of Infection Route	Study Linking Weather/Climate and Infection
Vibrio vulnificus	*Vibrio vulnificus* can cause severe, soft tissue infections, septicaemia, and deaths. Infection is through the consumption of contaminated seafood (particularly raw oysters). *V. vulnificus* infection increased following hurricane Katrina.	Strong	[609,611,612]	Waterborne; Foodborne through seafood	MMF [613]; HSM [614]; MMST [615]; OI [611]
Vibrio spp. (other than *V. cholerae*)	A variety of Vibrio spp. can cause human disease, including the halophilic *V. parahaemolyticus, V. fluvialis, V. hollisae* and the non-halophilic vibrios non-O1 *V. cholerae* and *V. mimicus*. Cholera is a classical waterborne disease, and the water route is still important in developing countries. There is no evidence that vibrios are able to cause human disease by growing within water distribution systems. *Vibrio* spp. are part of normal marine flora and can be found in marine, estuarine and river water. These organisms and proliferate during the summer months. People are infected through the consumption of raw or undercooked contaminated shellfish, other foods and faecally contaminated water. A large infective dose is required to initiate infection and person-to-person transmission does not occur. Infections in the United Kingdom tend to be in travellers returning from developing countries. Non-cholera *V. cholera* in warmer Baltic waters.	Strong	[36,609]	Waterborne	MMO = [616]; OI [617]; TSAT [611]; WMR [615]; POTA [617]
Yersinia spp.	*Yersinia* spp. are bacteria that can cause diarrhoea, arthritis and mesenteric lymphadenitis, and small waterborne outbreaks of infection can occur. Infection usually derives from an animal source, particularly pigs. Non-pathogenic types can be detected in people and are not thought to be involved as a cause of diarrhoea. *Y. pseudotuberculosis* causes fever, enlargement of the mesenteric lymph nodes, pseudo-appendicitis, septicaemia and diarrhoea. *Yersinia enterocolitica* can cause diarrhoea, arthritis and mesenteric lymphadenitis, and small waterborne outbreaks of infection can occur. *Y. fredriksenii* and *Y. kristensenii* have been isolated from the faecal samples of people with diarrhoea, but their pathogenic role is unclear.	Moderate	[618–620]	Waterborne	NE
Mixed causes	Climate change and waterborne illness [180,181] Relationship between recreational water outbreaks and temperature [54]. Review of waterborne outbreaks [621]	NE	[187,190,257, 258]	Mixed	ROS [36,622]; FTA [293]; RCCS [32,269]; QMRA [374,623]

References

1. IPCC. Climate Change 2013: The Physical Science Basis. In *Contribution of Working Group I to the Fifth Assessment Report of the Intergovernmental Panel on Climate Change*; Stocker, T.F.Q.D., Plattner, G.K., Tignor, M., Allen, S.K., Boschung, J., Nauels, A., Xia, Y., Bex, V., Midgley, P.M., Eds.; Cambridge University Press: Cambridge, UK; New York, NY, USA, 2013; pp. 1–1535.
2. IPCC. Managing the Risks of Extreme Events and Disasters to Advance Climate Change Adaptation. In *A Special Report of Working Groups I and II of the Intergovernmental Panel on Climate Change*; Field, C.B.B.V., Stocker, T.F., Qin, D., Dokken, D.J., Ebi, K.L., Mastrandrea, M.D., Mach, K.J., Plattner, G.K., Allen, S.K., Tignor, M., et al., Eds.; Cambridge University Press: Cambridge, UK; New York, NY, USA, 2012; pp. 1–582.
3. Hardin, G. The tragedy of the commons. The population problem has no technical solution; it requires a fundamental extension in morality. *Science* **1968**, *162*, 1243–1248. [PubMed]
4. Patz, J.A.; Campbell-Lendrum, D.; Holloway, T.; Foley, J.A. Impact of regional climate change on human health. *Nature* **2005**, *438*, 310–317. [CrossRef] [PubMed]
5. Stern, N. *Stern Review: The Economics of Climate Change*; Edward Elgar Publishing: Cheltenham, UK, 2008.
6. Gasparrini, A.; Guo, Y.; Sera, F.; Vicedo-Cabrera, A.M.; Huber, V.; Tong, S.; de Sousa Zanotti Stagliorio Coelho, M.; Nascimento Saldiva, P.H.; Lavigne, E.; Matus Correa, P.; et al. Projections of temperature-related excess mortality under climate change scenarios. *Lancet Planet Health* **2017**, *1*, e360–e367. [CrossRef]
7. Melorose, J.P.R.; Careas, S. *World Population Prospects*; United Nations: New York, NY, USA, 2015; pp. 587–592.
8. McMichael, A.J.; Wilcox, B.A. Climate change, human health, and integrative research: A transformative imperative. *EcoHealth* **2009**, *6*, 163–164. [CrossRef] [PubMed]
9. Hsiang, S.M.; Burke, M.; Miguel, E. Quantifying the influence of climate on human conflict. *Science* **2013**, *341*, 1235367. [CrossRef] [PubMed]
10. Hsiang, S.M.; Meng, K.C.; Cane, M.A. Civil conflicts are associated with the global climate. *Nature* **2011**, *476*, 438–441. [CrossRef] [PubMed]
11. Dangendorf, F.; Herbst, S.; Reintjes, R.; Kistemann, T. Spatial patterns of diarrhoeal illnesses with regard to water supply structures—A GIS analysis. *Int. J. Hyg. Environ. Health* **2002**, *205*, 183–191. [CrossRef] [PubMed]
12. Schuur, E.A.; McGuire, A.D.; Schadel, C.; Grosse, G.; Harden, J.W.; Hayes, D.J.; Hugelius, G.; Koven, C.D.; Kuhry, P.; Lawrence, D.M.; et al. Climate change and the permafrost carbon feedback. *Nature* **2015**, *520*, 171–179. [CrossRef] [PubMed]
13. McMichael, A.J.; Woodward, A.; Muir, C. *Climate Change and the Health of Nations: Famines, Fevers, and the Fate of Populations*; Oxford University Press: New York, NY, USA, 2017; 370p.
14. Patz, J.A.; Grabow, M.L.; Limaye, V.S. When it rains, it pours: Future climate extremes and health. *Ann. Glob. Health* **2014**, *80*, 332–344. [CrossRef] [PubMed]
15. Raper, S.C.; Braithwaite, R.J. Low sea level rise projections from mountain glaciers and icecaps under global warming. *Nature* **2006**, *439*, 311–313. [CrossRef] [PubMed]
16. Jevrejeva, S.; Jackson, L.P.; Riva, R.E.; Grinsted, A.; Moore, J.C. Coastal sea level rise with warming above 2 degrees C. *Proc. Natl. Acad. Sci. USA* **2016**, *113*, 13342–13347. [CrossRef] [PubMed]
17. Koven, C.D.; Schuur, E.A.; Schadel, C.; Bohn, T.J.; Burke, E.J.; Chen, G.; Chen, X.; Ciais, P.; Grosse, G.; Harden, J.W.; et al. A simplified, data-constrained approach to estimate the permafrost carbon-climate feedback. *Philos. Trans. Ser. A Math. Phys. Eng. Sci.* **2015**, *373*. [CrossRef] [PubMed]
18. Orth, R.; Zscheischler, J.; Seneviratne, S.I. Record dry summer in 2015 challenges precipitation projections in Central Europe. *Sci. Rep.* **2016**, *6*, 28334. [CrossRef] [PubMed]
19. Steffen, W.; Rockström, J.; Richardson, K.; Lenton, T.M.; Folke, C.; Liverman, D.; Summerhayes, C.P.; Barnosky, A.D.; Cornell, S.E.; Crucifix, M.; et al. Trajectories of the Earth System in the Anthropocene. *Proc. Natl. Acad. Sci. USA* **2018**, *115*, 8252–8259. [CrossRef] [PubMed]
20. Michalak, A.M. Study role of climate change in extreme threats to water quality. *Nature* **2016**, *535*, 349–350. [CrossRef] [PubMed]
21. Sterk, A.; de Man, H.; Schijven, J.F.; de Nijs, T.; de Roda Husman, A.M. Climate change impact on infection risks during bathing downstream of sewage emissions from CSOs or WWTPs. *Water Res.* **2016**, *105*, 11–21. [CrossRef] [PubMed]

22. Jiménez Cisneros, B.E.O.T.; Arnell, N.W.; Benito, G.; Cogley, J.G.; Döll, P.; Jiang, T.; Mwakalila, S.S. *Freshwater Resources*; Cambridge University Press: Cambridge, UK; New York, NY, USA, 2014; pp. 229–269.

23. Stathakou, N.P.; Stathakou, G.P.; Damianaki, S.G.; Ioannou, E.T.; Stavrianeas, N.G. Empedocles' bio-medical Comments: A Precursor of Modern Science. Available online: https://www.google.com/search?client=safari&rls=en&q=N.P.+Stathakou,MD,+G.P.+Stathakou,MD,+S.G.+Damianaki,MD,+E.+Toumbis+Ioannou,MD+and+N.G.+Stavrianeas,MD+Empedocles%E2%80%99+bio-medical+comments:+A+precursor+of+modern+science+Copyright+%C2%A9+Priory+Lodge+Education+Limited+2007&ie=UTF-8&oe=UTF-8 (accessed on 19 September 2018).

24. White, G.B.D.; White, A. *Drawers of Water*; University of Chicago Press: Chicago, IL, USA, 1972.

25. Bartram, J.H.P. Bradley Classification of disease transmission routes for water-related hazards. In *Routledge Handbook of Water and Health*; Bartram, J.B.R., Coclanis, P.A., Gute, D.M., Kay, D., Pond, K., Robertson, W., Rouse, M.J., Eds.; Routledge: London, UK; New York, NY, USA, 2015; pp. 20–37.

26. Nichols, G.L.; Mølbak, K. The sources and environmental origins of gastrointestinal pathogens. In *Environmental Medicine*; Ayres, J.G.H., Nichols, G.L., Maynard, R.L., Eds.; Hodder Arnold: London, UK, 2010; pp. 359–372.

27. Sobsey, M.D. Waterborne and water-washed disease. In *Routledge Handbook of Water and Health*; Bartram, J.B.R., Coclanis, P.A., Gute, D.M., Kay, D., Pond, K., Robertson, W., Rouse, M.J., Eds.; Routledge: London, UK; New York, NY, USA, 2015; pp. 38–50.

28. Guzman-Herrador, B.; Carlander, A.; Ethelberg, S.; Freiesleben de Blasio, B.; Kuusi, M.; Lund, V.; Lofdahl, M.; MacDonald, E.; Nichols, G.; Schonning, C.; et al. Waterborne outbreaks in the Nordic countries, 1998 to 2012. *Euro Surveill.* **2015**, *20*, 21160. [CrossRef] [PubMed]

29. Richardson, H.Y.; Nichols, G.; Lane, C.; Lake, I.R.; Hunter, P.R. Microbiological surveillance of private water supplies in England: The impact of environmental and climate factors on water quality. *Water Res.* **2009**, *43*, 2159–2168. [CrossRef] [PubMed]

30. Semenza, J.C.; Nichols, G. Cryptosporidiosis surveillance and water-borne outbreaks in Europe. *Euro Surveill.* **2007**, *12*, E13–E14. [CrossRef] [PubMed]

31. Smith, A.; Reacher, M.; Smerdon, W.; Adak, G.K.; Nichols, G.; Chalmers, R.M. Outbreaks of waterborne infectious intestinal disease in England and Wales, 1992–2003. *Epidemiol. Infect.* **2006**, *134*, 1141–1149. [CrossRef] [PubMed]

32. Nichols, G.; Lane, C.; Asgari, N.; Verlander, N.Q.; Charlett, A. Rainfall and outbreaks of drinking water related disease and in England and Wales. *J. Water Health* **2009**, *7*, 1–8. [CrossRef] [PubMed]

33. Said, B.; Wright, F.; Nichols, G.L.; Reacher, M.; Rutter, M. Outbreaks of infectious disease associated with private drinking water supplies in England and Wales 1970–2000. *Epidemiol. Infect.* **2003**, *130*, 469–479. [PubMed]

34. Rutter, M.; Nichols, G.L.; Swan, A.; De Louvois, J. A survey of the microbiological quality of private water supplies in England. *Epidemiol. Infect.* **2000**, *124*, 417–425. [CrossRef] [PubMed]

35. Rooney, R.M.; Bartram, J.K.; Cramer, E.H.; Mantha, S.; Nichols, G.; Suraj, R.; Todd, E.C. A review of outbreaks of waterborne disease associated with ships: Evidence for risk management. *Public Health Rep.* **2004**, *119*, 435–442. [CrossRef] [PubMed]

36. Cann, K.F.; Thomas, D.R.; Salmon, R.L.; Wyn-Jones, A.P.; Kay, D. Extreme water-related weather events and waterborne disease. *Epidemiol. Infect.* **2013**, *141*, 671–686. [CrossRef] [PubMed]

37. Levy, K.; Woster, A.P.; Goldstein, R.S.; Carlton, E.J. Untangling the Impacts of Climate Change on Waterborne Diseases: A Systematic Review of Relationships between Diarrheal Diseases and Temperature, Rainfall, Flooding, and Drought. *Environ. Sci. Technol.* **2016**, *50*, 4905–4922. [CrossRef] [PubMed]

38. Funari, E.; Manganelli, M.; Sinisi, L. Impact of climate change on waterborne diseases. *Ann. dell'Istituto Super. Sanita* **2012**, *48*, 473–487. [CrossRef]

39. Dale, K.; Kirk, M.; Sinclair, M.; Hall, R.; Leder, K. Reported waterborne outbreaks of gastrointestinal disease in Australia are predominantly associated with recreational exposure. *Aust. N. Z. J. Public Health* **2010**, *34*, 527–530. [CrossRef] [PubMed]

40. Nicole, W. The WASH approach: Fighting waterborne diseases in emergency situations. *Environ. Health Perspect.* **2015**, *123*, A6–A15. [CrossRef] [PubMed]

41. Lim, J.H.; Yoon, D.; Jung, G.; Joo Kim, W.; Lee, H.C. Medical needs of tsunami disaster refugee camps. *Fam. Med.* **2005**, *37*, 422–428. [PubMed]

42. Nichols, G. Water supply, Waterborne infection and sewage/sludge disposal. In *Hobbs's Food Poisoning and Food Hygiene*; McLauchlin, J.L.C., Ed.; Hodder Arnold: London, UK, 2007; pp. 146–165.

43. Burton, F.L.; Tchobanoglous, G.; Tsuchihashi, R.; Stensel, H.D. *Wastewater Engineering: Treatment and Reuse*; McGraw-Hill Education: Columbus, OH, USA, 2003; p. 1819.

44. Schwartz, J.; Levin, R.; Goldstein, R. Drinking water turbidity and gastrointestinal illness in the elderly of Philadelphia. *J. Epidemiol. Community Health* **2000**, *54*, 45–51. [CrossRef] [PubMed]

45. Lake, I.R.; Bentham, G.; Kovats, R.S.; Nichols, G.L. Effects of weather and river flow on cryptosporidiosis. *J. Water Health* **2005**, *3*, 469–474. [CrossRef] [PubMed]

46. Delpla, I.; Jung, A.V.; Baures, E.; Clement, M.; Thomas, O. Impacts of climate change on surface water quality in relation to drinking water production. *Environ. Int.* **2009**, *35*, 1225–1233. [CrossRef] [PubMed]

47. Jiang, Y.S.; Riedel, T.E.; Popoola, J.A.; Morrow, B.R.; Cai, S.; Ellington, A.D.; Bhadra, S. Portable platform for rapid in-field identification of human fecal pollution in water. *Water Res.* **2017**, *131*, 186–195. [CrossRef] [PubMed]

48. Voisin, H.; Bergstrom, L.; Liu, P.; Mathew, A.P. Nanocellulose-Based Materials for Water Purification. *Nanomaterials* **2017**, *7*, E57. [CrossRef] [PubMed]

49. McPhail, C.D.; Stidson, R.T. Aquatic Bathing water signage and predictive water quality models in Scotland. *Ecosyst. Health Manag.* **2009**, *12*, 183–186. [CrossRef]

50. Whiley, H.; Bentham, R.; Brown, M.H. Legionella Persistence in Manufactured Water Systems: Pasteurization Potentially Selecting for Thermal Tolerance. *Front. Microbiol.* **2017**, *8*, 1330. [CrossRef] [PubMed]

51. Craun, G.F.; Calderon, R.L.; Craun, M.F. Outbreaks associated with recreational water in the United States. *Int. J. Environ. Health Res.* **2005**, *15*, 243–262. [CrossRef] [PubMed]

52. Soldanova, M.; Selbach, C.; Kalbe, M.; Kostadinova, A.; Sures, B. Swimmer's itch: Etiology, impact, and risk factors in Europe. *Trends Parasitol.* **2013**, *29*, 65–74. [CrossRef] [PubMed]

53. Xu, J.F.; Xu, J.; Li, S.Z.; Jia, T.W.; Huang, X.B.; Zhang, H.M.; Chen, M.; Yang, G.J.; Gao, S.J.; Wang, Q.Y.; et al. Transmission risks of schistosomiasis japonica: Extraction from back-propagation artificial neural network and logistic regression model. *PLoS Negl. Trop. Dis.* **2013**, *7*, e2123. [CrossRef] [PubMed]

54. Schets, F.M.; De Roda Husman, A.M.; Havelaar, A.H. Disease outbreaks associated with untreated recreational water use. *Epidemiol. Infect.* **2011**, *139*, 1114–1125. [CrossRef] [PubMed]

55. Roser, D.J.; van den Akker, B.; Boase, S.; Haas, C.N.; Ashbolt, N.J.; Rice, S.A. Pseudomonas aeruginosa dose response and bathing water infection. *Epidemiol. Infect.* **2014**, *142*, 449–462. [CrossRef] [PubMed]

56. Redbord, K.P.; Shearer, D.A.; Gloster, H.; Younger, B.; Connelly, B.L.; Kindel, S.E.; Lucky, A.W. Atypical Mycobacterium furunculosis occurring after pedicures. *J. Am. Acad. Dermatol.* **2006**, *54*, 520–524. [CrossRef] [PubMed]

57. Fiorillo, L.; Zucker, M.; Sawyer, D.; Lin, A.N. The pseudomonas hot-foot syndrome. *N. Engl. J. Med.* **2001**, *345*, 335–338. [CrossRef] [PubMed]

58. Julia Manresa, M.; Vicente Villa, A.; Gene Giralt, A.; Gonzalez-Ensenat, M.A. Aeromonas hydrophila folliculitis associated with an inflatable swimming pool: Mimicking Pseudomonas aeruginosa infection. *Pediatr. Dermatol.* **2009**, *26*, 601–603. [CrossRef] [PubMed]

59. Gauthier, J.; Gerome, P.; Defez, M.; Neulat-Ripoll, F.; Foucher, B.; Vitry, T.; Crevon, L.; Valade, E.; Thibault, F.M.; Biot, F.V. Melioidosis in Travelers Returning from Vietnam to France. *Emerg. Infect. Dis.* **2016**, *22*, 1671–1673. [CrossRef] [PubMed]

60. Jones, E.H.; Feldman, K.A.; Palmer, A.; Butler, E.; Blythe, D.; Mitchell, C.S. Vibrio infections and surveillance in Maryland, 2002–2008. *Public Health Rep.* **2013**, *128*, 537–545. [CrossRef] [PubMed]

61. Ozluer, S.M.; De'Ambrosis, B.J. Mycobacterium abscessus wound infection. *Australas. J. Dermatol.* **2001**, *42*, 26–29. [CrossRef] [PubMed]

62. Merritt, A.J.; Peck, M.; Gayle, D.; Levy, A.; Ler, Y.H.; Raby, E.; Gibbs, T.M.; Inglis, T.J. Cutaneous Melioidosis Cluster Caused by Contaminated Wound Irrigation Fluid. *Emerg. Infect. Dis.* **2016**, *22*. [CrossRef] [PubMed]

63. Trautmann, M.; Halder, S.; Lepper, P.M.; Exner, M. Reservoirs of Pseudomonas aeruginosa in the intensive care unit. The role of tap water as a source of infection. *Bundesgesundheitsblatt Gesundheitsforschung Gesundheitsschutz* **2009**, *52*, 339–344. [CrossRef] [PubMed]

64. Cohen, R.; Babushkin, F.; Shimoni, Z.; Cohen, S.; Litig, E.; Shapiro, M.; Adler, A.; Paikin, S. Water faucets as a source of Pseudomonas aeruginosa infection and colonization in neonatal and adult intensive care unit patients. *Am. J. Infect. Control* **2017**, *45*, 206–209. [CrossRef] [PubMed]

65. Walker, J.T.; Jhutty, A.; Parks, S.; Willis, C.; Copley, V.; Turton, J.F.; Hoffman, P.N.; Bennett, A.M. Investigation of healthcare-acquired infections associated with Pseudomonas aeruginosa biofilms in taps in neonatal units in Northern Ireland. *J. Hosp. Infect.* **2014**, *86*, 16–23. [CrossRef] [PubMed]

66. Jefferies, J.M.; Cooper, T.; Yam, T.; Clarke, S.C. Pseudomonas aeruginosa outbreaks in the neonatal intensive care unit—A systematic review of risk factors and environmental sources. *J. Med. Microbiol.* **2012**, *61*, 1052–1061. [CrossRef] [PubMed]

67. Maloney, S.; Welbel, S.; Daves, B.; Adams, K.; Becker, S.; Bland, L.; Arduino, M.; Wallace, R., Jr.; Zhang, Y.; Buck, G.; et al. Mycobacterium abscessus pseudoinfection traced to an automated endoscope washer: Utility of epidemiologic and laboratory investigation. *J. Infect. Dis.* **1994**, *169*, 1166–1169. [CrossRef] [PubMed]

68. Al-Assaf, N.; Moore, H.; Leifso, K.; Ben Fadel, N.; Ferretti, E. Disseminated Neonatal Herpes Simplex Virus Type 1 After a Water Birth. *J. Pediatr. Infect. Dis. Soc.* **2017**, *6*, e169–e172. [CrossRef] [PubMed]

69. Fritschel, E.; Sanyal, K.; Threadgill, H.; Cervantes, D. Fatal legionellosis after water birth, Texas, USA, 2014. *Emerg. Infect. Dis.* **2015**, *21*, 130–132. [CrossRef] [PubMed]

70. Soileau, S.L.; Schneider, E.; Erdman, D.D.; Lu, X.; Ryan, W.D.; McAdams, R.M. Case report: Severe disseminated adenovirus infection in a neonate following water birth delivery. *J. Med. Virol.* **2013**, *85*, 667–669. [CrossRef] [PubMed]

71. Franzin, L.; Cabodi, D.; Scolfaro, C.; Gioannini, P. Microbiological investigations on a nosocomial case of Legionella pneumophila pneumonia associated with water birth and review of neonatal cases. *InfezMed* **2004**, *12*, 69–75.

72. Berendes, D.; Levy, K.; Knee, J.; Handzel, T.; Hill, V.R. Ascaris and Escherichia coli Inactivation in an Ecological Sanitation System in Port-au-Prince, Haiti. *PLoS ONE* **2015**, *10*, e0125336. [CrossRef] [PubMed]

73. McKerr, C.; Adak, G.K.; Nichols, G.; Gorton, R.; Chalmers, R.M.; Kafatos, G.; Cosford, P.; Charlett, A.; Reacher, M.; Pollock, K.G.; et al. An Outbreak of Cryptosporidium parvum across England & Scotland Associated with Consumption of Fresh Pre-Cut Salad Leaves, May 2012. *PLoS ONE* **2015**, *10*, e0125955. [CrossRef]

74. Nichols, G.L.; Freedman, J.; Pollock, K.G.; Rumble, C.; Chalmers, R.M.; Chiodini, P.; Hawkins, G.; Alexander, C.L.; Godbole, G.; Williams, C.; et al. Cyclospora infection linked to travel to Mexico, June to September 2015. *Euro Surveill.* **2015**, *20*. [CrossRef] [PubMed]

75. Nichols, G.L. Human health risks from toxic cyanobacteria, dinoflagellates and diatoms. In *Environmental Medicine*; Ayers, J., Harrison, R., Maynard, R., Nichols, G.L., Eds.; Hodder Arnold: London, UK, 2010.

76. Dance, D.A. Melioidosis. *Curr. Opin. Infect. Dis.* **2002**, *15*, 127–132. [CrossRef] [PubMed]

77. Ogden, N.H. Climate change and vector-borne diseases of public health significance. *FEMS Microbiol. Lett.* **2017**, *364*. [CrossRef] [PubMed]

78. Bouzid, M.; Hooper, L.; Hunter, P.R. The effectiveness of public health interventions to reduce the health impact of climate change: A systematic review of systematic reviews. *PLoS ONE* **2013**, *8*, e62041. [CrossRef] [PubMed]

79. Sargeant, J.M.; Amezcua, M.R.; Rajic, A.; Waddell, L. Pre-harvest interventions to reduce the shedding of *E. coli* O157 in the faeces of weaned domestic ruminants: A systematic review. *Zoonoses Public Health* **2007**, *54*, 260–277. [CrossRef] [PubMed]

80. Kostyla, C.; Bain, R.; Cronk, R.; Bartram, J. Seasonal variation of fecal contamination in drinking water sources in developing countries: A systematic review. *Sci. Total Environ.* **2015**, *514*, 333–343. [CrossRef] [PubMed]

81. Doocy, S.; Daniels, A.; Murray, S.; Kirsch, T.D. The human impact of floods: A historical review of events 1980–2009 and systematic literature review. *PLoS Curr.* **2013**, *5*. [CrossRef] [PubMed]

82. Doocy, S.; Daniels, A.; Dick, A.; Kirsch, T.D. The human impact of tsunamis: A historical review of events 1900–2009 and systematic literature review. *PLoS Curr.* **2013**, *5*. [CrossRef] [PubMed]

83. Barbier, E.B. Policy: Hurricane Katrina's lessons for the world. *Nature* **2015**, *524*, 285–287. [CrossRef] [PubMed]

84. Yee, E.L.; Palacio, H.; Atmar, R.L.; Shah, U.; Kilborn, C.; Faul, M.; Gavagan, T.E.; Feigin, R.D.; Versalovic, J.; Neill, F.H.; et al. Widespread outbreak of norovirus gastroenteritis among evacuees of Hurricane Katrina residing in a large "megashelter" in Houston, Texas: Lessons learned for prevention. *Clin. Infect. Dis.* **2007**, *44*, 1032–1039. [CrossRef] [PubMed]

85. CDC. *Vibrio* illnesses after Hurricane Katrina—Multiple states, August–September 2005. *MMWR Morb. Mortal. Wkly. Rep.* **2005**, *54*, 928–931.

86. Brown, L.; Murray, V. Examining the relationship between infectious diseases and flooding in Europe: A systematic literature review and summary of possible public health interventions. *Disaster Health* **2013**, *1*, 117–127. [CrossRef] [PubMed]

87. Sena, A.; Barcellos, C.; Freitas, C.; Corvalan, C. Managing the health impacts of drought in Brazil. *Int. J. Environ. Res. Public Health* **2014**, *11*, 10737–10751. [CrossRef] [PubMed]

88. Stanke, C.; Kerac, M.; Prudhomme, C.; Medlock, J.; Murray, V. Health effects of drought: A systematic review of the evidence. *PLoS Curr.* **2013**, *5*. [CrossRef] [PubMed]

89. Yusa, A.; Berry, P.; Cheng J, J.; Ogden, N.; Bonsal, B.; Stewart, R.; Waldick, R. Climate Change, Drought and Human Health in Canada. *Int. J. Environ. Res. Public Health* **2015**, *12*, 8359–8412. [CrossRef] [PubMed]

90. Phung, D.; Huang, C.; Rutherford, S.; Chu, C.; Wang, X.; Nguyen, M. Climate change, water quality, and water-related diseases in the Mekong Delta Basin: A systematic review. *Asia Pac. J. Public Health* **2015**, *27*, 265–276. [CrossRef] [PubMed]

91. De Buck, E.; Borra, V.; De Weerdt, E.; Vande Veegaete, A.; Vandekerckhove, P. A systematic review of the amount of water per person per day needed to prevent morbidity and mortality in (post-)disaster settings. *PLoS ONE* **2015**, *10*, e0126395. [CrossRef] [PubMed]

92. Ramesh, A.; Blanchet, K.; Ensink, J.H.; Roberts, B. Evidence on the Effectiveness of Water, Sanitation, and Hygiene (WASH) Interventions on Health Outcomes in Humanitarian Crises: A Systematic Review. *PLoS ONE* **2015**, *10*, e0124688. [CrossRef] [PubMed]

93. Yates, T.; Lantagne, D.; Mintz, E.; Quick, R. The impact of water, sanitation, and hygiene interventions on the health and well-being of people living with HIV: A systematic review. *J. Acquir. Immune Defic. Syndr.* **2015**, *68* (Suppl. 3), S318–S330. [CrossRef] [PubMed]

94. Murphy, H.M.; Pintar, K.D.; McBean, E.A.; Thomas, M.K. A systematic review of waterborne disease burden methodologies from developed countries. *J. Water Health* **2014**, *12*, 634–655. [CrossRef] [PubMed]

95. Wolf, J.; Pruss-Ustun, A.; Cumming, O.; Bartram, J.; Bonjour, S.; Cairncross, S.; Clasen, T.; Colford, J.M., Jr.; Curtis, V.; De France, J.; et al. Assessing the impact of drinking water and sanitation on diarrhoeal disease in low- and middle-income settings: Systematic review and meta-regression. *Trop. Med. Int. Health* **2014**, *19*, 928–942. [CrossRef] [PubMed]

96. Benova, L.; Cumming, O.; Campbell, O.M. Systematic review and meta-analysis: Association between water and sanitation environment and maternal mortality. *Trop. Med. Int. Health* **2014**, *19*, 368–387. [CrossRef] [PubMed]

97. Stelmach, R.D.; Clasen, T. Household water quantity and health: A systematic review. *Int. J. Environ. Res. Public Health* **2015**, *12*, 5954–5974. [CrossRef] [PubMed]

98. Fewtrell, L.; Kaufmann, R.B.; Kay, D.; Enanoria, W.; Haller, L.; Colford, J.M., Jr. Water, sanitation, and hygiene interventions to reduce diarrhoea in less developed countries: A systematic review and meta-analysis. *Lancet Infect. Dis.* **2005**, *5*, 42–52. [CrossRef]

99. Wright, J.; Gundry, S.; Conroy, R. Household drinking water in developing countries: A systematic review of microbiological contamination between source and point-of-use. *Trop. Med. Int. Health* **2004**, *9*, 106–117. [CrossRef] [PubMed]

100. Clasen, T.; Schmidt, W.P.; Rabie, T.; Roberts, I.; Cairncross, S. Interventions to improve water quality for preventing diarrhoea: Systematic review and meta-analysis. *BMJ* **2007**, *334*, 782. [CrossRef] [PubMed]

101. Peletz, R.; Mahin, T.; Elliott, M.; Harris, M.S.; Chan, K.S.; Cohen, M.S.; Bartram, J.K.; Clasen, T.F. Water, sanitation, and hygiene interventions to improve health among people living with HIV/AIDS: A systematic review. *AIDS* **2013**, *27*, 2593–2601. [CrossRef] [PubMed]

102. Dreibelbis, R.; Winch, P.J.; Leontsini, E.; Hulland, K.R.; Ram, P.K.; Unicomb, L.; Luby, S.P. The Integrated Behavioural Model for Water, Sanitation, and Hygiene: A systematic review of behavioural models and a framework for designing and evaluating behaviour change interventions in infrastructure-restricted settings. *BMC Public Health* **2013**, *13*, 1015. [CrossRef] [PubMed]

103. Loevinsohn, M.; Mehta, L.; Cuming, K.; Nicol, A.; Cumming, O.; Ensink, J.H. The cost of a knowledge silo: A systematic re-review of water, sanitation and hygiene interventions. *Health Policy Plan.* **2015**, *30*, 660–674. [CrossRef] [PubMed]

104. Ganguly, E.; Sharma, P.K.; Bunker, C.H. Prevalence and risk factors of diarrhea morbidity among under-five children in India: A systematic review and meta-analysis. *Indian J. Child Health* **2015**, *2*, 152–160.

105. Arnold, B.F.; Colford, J.M., Jr. Treating water with chlorine at point-of-use to improve water quality and reduce child diarrhea in developing countries: A systematic review and meta-analysis. *Am. J. Trop. Med. Hyg.* **2007**, *76*, 354–364. [CrossRef] [PubMed]

106. Ercumen, A.; Gruber, J.S.; Colford, J.M., Jr. Water distribution system deficiencies and gastrointestinal illness: A systematic review and meta-analysis. *Environ. Health Perspect.* **2014**, *122*, 651–660. [CrossRef] [PubMed]

107. Pons, W.; Young, I.; Truong, J.; Jones-Bitton, A.; McEwen, S.; Pintar, K.; Papadopoulos, A. A Systematic Review of Waterborne Disease Outbreaks Associated with Small Non-Community Drinking Water Systems in Canada and the United States. *PLoS ONE* **2015**, *10*, e0141646. [CrossRef] [PubMed]

108. Waterborne outbreak of gastroenteritis associated with a contaminated municipal water supply, Walkerton, Ontario, May–June 2000. *Can. Commun. Dis. Rep.* **2000**, *26*, 170–173.

109. Auld, H.; MacIver, D.; Klaassen, J. Heavy rainfall and waterborne disease outbreaks: The Walkerton example. *J. Toxicol. Environ. Health A* **2004**, *67*, 1879–1887. [CrossRef] [PubMed]

110. Guzman Herrador, B.; de Blasio, B.F.; Carlander, A.; Ethelberg, S.; Hygen, H.O.; Kuusi, M.; Lund, V.; Lofdahl, M.; MacDonald, E.; Martinez-Urtaza, J.; et al. Association between heavy precipitation events and waterborne outbreaks in four Nordic countries, 1992–2012. *J. Water Health* **2016**, *14*, 1019–1027. [CrossRef] [PubMed]

111. O'Dwyer, J.; Morris Downes, M.; Adley, C.C. The impact of meteorology on the occurrence of waterborne outbreaks of vero cytotoxin-producing *Escherichia coli* (VTEC): A logistic regression approach. *J. Water Health* **2016**, *14*, 39–46. [CrossRef] [PubMed]

112. Drayna, P.; McLellan, S.L.; Simpson, P.; Li, S.H.; Gorelick, M.H. Association between rainfall and pediatric emergency department visits for acute gastrointestinal illness. *Environ. Health Perspect.* **2010**, *118*, 1439–1443. [CrossRef] [PubMed]

113. Mosler, H.J. A systematic approach to behavior change interventions for the water and sanitation sector in developing countries: A conceptual model, a review, and a guideline. *Int. J. Environ. Health Res.* **2012**, *22*, 431–449. [CrossRef] [PubMed]

114. Evans, W.D.; Pattanayak, S.K.; Young, S.; Buszin, J.; Rai, S.; Bihm, J.W. Social marketing of water and sanitation products: A systematic review of peer-reviewed literature. *Soc. Sci. Med.* **2014**, *110*, 18–25. [CrossRef] [PubMed]

115. Jasper, C.; Le, T.T.; Bartram, J. Water and sanitation in schools: A systematic review of the health and educational outcomes. *Int. J. Environ. Res. Public Health* **2012**, *9*, 2772–2787. [CrossRef] [PubMed]

116. Bain, R.; Cronk, R.; Hossain, R.; Bonjour, S.; Onda, K.; Wright, J.; Yang, H.; Slaymaker, T.; Hunter, P.; Pruss-Ustun, A.; et al. Global assessment of exposure to faecal contamination through drinking water based on a systematic review. *Trop. Med. Int. Health* **2014**, *19*, 917–927. [CrossRef] [PubMed]

117. Guzman Herrador, B.R.; de Blasio, B.F.; MacDonald, E.; Nichols, G.; Sudre, B.; Vold, L.; Semenza, J.C.; Nygard, K. Analytical studies assessing the association between extreme precipitation or temperature and drinking water-related waterborne infections: A review. *Environ. Health* **2015**, *14*, 29. [CrossRef] [PubMed]

118. Camargo, E.A.F.; Camargo, J.T.F.; Neves, M.F.; Simoes, L.F.; Bastos, L.A.D.; Magalhaes, L.A.; Zanotti-Magalhaes, E.M. Assessment of the impact of changes in temperature in Biomphalaria glabrata (Say, 1818) melanic and albino variants infected with *Schistosoma mansoni* (Sambon, 1907). *Braz. J. Boil.* **2017**, *77*, 490–494. [CrossRef] [PubMed]

119. McCreesh, N.; Nikulin, G.; Booth, M. Predicting the effects of climate change on *Schistosoma mansoni* transmission in eastern Africa. *Parasites Vectors* **2015**, *8*, 4. [CrossRef] [PubMed]

120. McCreesh, N.; Arinaitwe, M.; Arineitwe, W.; Tukahebwa, E.M.; Booth, M. Effect of water temperature and population density on the population dynamics of *Schistosoma mansoni* intermediate host snails. *Parasites Vectors* **2014**, *7*, 503. [CrossRef] [PubMed]

121. Zhou, X.N.; Yang, G.J.; Yang, K.; Wang, X.H.; Hong, Q.B.; Sun, L.P.; Malone, J.B.; Kristensen, T.K.; Bergquist, N.R.; Utzinger, J. Potential impact of climate change on schistosomiasis transmission in China. *Am. J. Trop. Med. Hyg.* **2008**, *78*, 188–194. [CrossRef] [PubMed]

122. Nelson, M.K.; Cruz, B.C.; Buena, K.L.; Nguyen, H.; Sullivan, J.T. Effects of abnormal temperature and starvation on the internal defense system of the schistosome-transmitting snail Biomphalaria glabrata. *J. Invertebr. Pathol.* **2016**, *138*, 18–23. [CrossRef] [PubMed]

123. Pedersen, U.B.; Karagiannis-Voules, D.A.; Midzi, N.; Mduluza, T.; Mukaratirwa, S.; Fensholt, R.; Vennervald, B.J.; Kristensen, T.K.; Vounatsou, P.; Stensgaard, A.S. Comparison of the spatial patterns of schistosomiasis in Zimbabwe at two points in time, spaced twenty-nine years apart: Is climate variability of importance? *Geospat. Health* **2017**, *12*, 505. [CrossRef] [PubMed]

124. Stensgaard, A.S.; Booth, M.; Nikulin, G.; McCreesh, N. Combining process-based and correlative models improves predictions of climate change effects on *Schistosoma mansoni* transmission in eastern Africa. *Geospat. Health* **2016**, *11*, 406. [CrossRef] [PubMed]

125. McCreesh, N.; Booth, M. The effect of simulating different intermediate host snail species on the link between water temperature and schistosomiasis risk. *PLoS ONE* **2014**, *9*, e87892. [CrossRef] [PubMed]

126. Moore, J.L.; Liang, S.; Akullian, A.; Remais, J.V. Cautioning the use of degree-day models for climate change projections in the presence of parametric uncertainty. *Ecol. Appl.* **2012**, *22*, 2237–2247. [CrossRef] [PubMed]

127. Mangal, T.D.; Paterson, S.; Fenton, A. Predicting the impact of long-term temperature changes on the epidemiology and control of schistosomiasis: A mechanistic model. *PLoS ONE* **2008**, *3*, e1438. [CrossRef] [PubMed]

128. Abou-El-Naga, I.F. Biomphalaria alexandrina in Egypt: Past, present and future. *J. Biosci.* **2013**, *38*, 665–672. [CrossRef] [PubMed]

129. Grimes, J.E.; Croll, D.; Harrison, W.E.; Utzinger, J.; Freeman, M.C.; Templeton, M.R. The relationship between water, sanitation and schistosomiasis: A systematic review and meta-analysis. *PLoS Negl. Trop. Dis.* **2014**, *8*, e3296. [CrossRef] [PubMed]

130. Steinmann, P.; Keiser, J.; Bos, R.; Tanner, M.; Utzinger, J. Schistosomiasis and water resources development: Systematic review, meta-analysis, and estimates of people at risk. *Lancet Infect. Dis.* **2006**, *6*, 411–425. [CrossRef]

131. Hopkins, D.R.; Ruiz-Tiben, E.; Eberhard, M.L.; Roy, S.L.; Weiss, A.J. Progress Toward Global Eradication of Dracunculiasis-January 2015–June 2016. *MMWR Morb. Mortal. Wkly. Rep.* **2016**, *65*, 1112–1116. [CrossRef] [PubMed]

132. Tayeh, A.; Cairncross, S.; Cox, F.E.G. Guinea worm: From Robert Leiper to eradication. *Parasitology* **2017**, *144*, 1643–1648. [CrossRef] [PubMed]

133. Enserink, M. Infectious diseases. Guinea worm eradication at risk in South Sudanese war. *Science* **2014**, *343*, 236. [CrossRef] [PubMed]

134. Callaway, E. Dogs thwart effort to eradicate Guinea worm. *Nature* **2016**, *529*, 10–11. [CrossRef] [PubMed]

135. Wardell, R.; Clements, A.C.A.; Lal, A.; Summers, D.; Llewellyn, S.; Campbell, S.J.; McCarthy, J.; Gray, D.J.; V Nery, S. An environmental assessment and risk map of Ascaris lumbricoides and Necator americanus distributions in Manufahi District, Timor-Leste. *PLoS Negl. Trop. Dis.* **2017**, *11*, e0005565. [CrossRef] [PubMed]

136. Karagiannis-Voules, D.A.; Biedermann, P.; Ekpo, U.F.; Garba, A.; Langer, E.; Mathieu, E.; Midzi, N.; Mwinzi, P.; Polderman, A.M.; Raso, G.; et al. Spatial and temporal distribution of soil-transmitted helminth infection in sub-Saharan Africa: A systematic review and geostatistical meta-analysis. *Lancet Infect. Dis.* **2015**, *15*, 74–84. [CrossRef]

137. Strunz, E.C.; Addiss, D.G.; Stocks, M.E.; Ogden, S.; Utzinger, J.; Freeman, M.C. Water, sanitation, hygiene, and soil-transmitted helminth infection: A systematic review and meta-analysis. *PLoS Med.* **2014**, *11*, e1001620. [CrossRef] [PubMed]

138. Chalmers, R.M. Waterborne outbreaks of cryptosporidiosis. *Ann. Ist. Super. Sanita* **2012**, *48*, 429–446. [CrossRef] [PubMed]

139. Young, I.; Smith, B.A.; Fazil, A. A systematic review and meta-analysis of the effects of extreme weather events and other weather-related variables on Cryptosporidium and Giardia in fresh surface waters. *J. Water Health* **2015**, *13*, 1–17. [CrossRef] [PubMed]

140. Gualberto, F.A.; Heller, L. Endemic Cryptosporidium infection and drinking water source: A systematic review and meta-analyses. *Water Sci. Technol.* **2006**, *54*, 231–238. [CrossRef] [PubMed]

141. Meireles, L.R.; Ekman, C.C.; de Andrade, H.F., Jr.; Luna, E.J. Human Toxoplasmosis Outbreaks and the Agent Infecting Form. Findings from a Systematic Review. *Rev. Inst. Med. Trop. Sao Paulo* **2015**, *57*, 369–376. [CrossRef] [PubMed]

142. Speich, B.; Croll, D.; Furst, T.; Utzinger, J.; Keiser, J. Effect of sanitation and water treatment on intestinal protozoa infection: A systematic review and meta-analysis. *Lancet Infect Dis.* **2016**, *16*, 87–99. [CrossRef]

143. Stuart, J.M.; Orr, H.J.; Warburton, F.G.; Jeyakanth, S.; Pugh, C.; Morris, I.; Sarangi, J.; Nichols, G. Risk factors for sporadic giardiasis: A case-control study in southwestern England. *Emerg. Infect. Dis.* **2003**, *9*, 229–233. [CrossRef] [PubMed]

144. Mukherjee, A.K.; Chowdhury, P.; Rajendran, K.; Nozaki, T.; Ganguly, S. Association between Giardia duodenalis and coinfection with other diarrhea-causing pathogens in India. *Biomed. Res. Int.* **2014**, *2014*, 786480. [CrossRef] [PubMed]

145. Squire, S.A.; Ryan, U. *Cryptosporidium* and *Giardia* in Africa: Current and future challenges. *Parasites Vectors* **2017**, *10*, 195. [CrossRef] [PubMed]

146. Moore, S.M.; Azman, A.S.; Zaitchik, B.F.; Mintz, E.D.; Brunkard, J.; Legros, D.; Hill, A.; McKay, H.; Luquero, F.J.; Olson, D.; et al. El Nino and the shifting geography of cholera in Africa. *Proc. Natl. Acad. Sci. USA* **2017**, *114*, 4436–4441. [CrossRef] [PubMed]

147. Rebaudet, S.; Sudre, B.; Faucher, B.; Piarroux, R. Environmental determinants of cholera outbreaks in inland Africa: A systematic review of main transmission foci and propagation routes. *J. Infect. Dis.* **2013**, *208* (Suppl. 1), S46–S54. [CrossRef] [PubMed]

148. Rebaudet, S.; Sudre, B.; Faucher, B.; Piarroux, R. Cholera in coastal Africa: A systematic review of its heterogeneous environmental determinants. *J. Infect. Dis.* **2013**, *208* (Suppl. 1), S98–S106. [CrossRef] [PubMed]

149. Martinez, P.P.; Reiner, R.C., Jr.; Cash, B.A.; Rodo, X.; Shahjahan Mondal, M.; Roy, M.; Yunus, M.; Faruque, A.S.; Huq, S.; King, A.A.; et al. Cholera forecast for Dhaka, Bangladesh, with the 2015–2016 El Nino: Lessons learned. *PLoS ONE* **2017**, *12*, e0172355. [CrossRef] [PubMed]

150. Semenza, J.C.; Trinanes, J.; Lohr, W.; Sudre, B.; Lofdahl, M.; Martinez-Urtaza, J.; Nichols, G.L.; Rocklov, J. Environmental Suitability of Vibrio Infections in a Warming Climate: An Early Warning System. *Environ. Health Perspect.* **2017**, *125*, 107004. [CrossRef] [PubMed]

151. Taylor, D.L.; Kahawita, T.M.; Cairncross, S.; Ensink, J.H. The Impact of Water, Sanitation and Hygiene Interventions to Control Cholera: A Systematic Review. *PLoS ONE* **2015**, *10*, e0135676. [CrossRef] [PubMed]

152. Gundry, S.; Wright, J.; Conroy, R. A systematic review of the health outcomes related to household water quality in developing countries. *J. Water Health* **2004**, *2*, 1–13. [CrossRef] [PubMed]

153. Boxall, A.B.; Hardy, A.; Beulke, S.; Boucard, T.; Burgin, L.; Falloon, P.D.; Haygarth, P.M.; Hutchinson, T.; Kovats, R.S.; Leonardi, G.; et al. Impacts of climate change on indirect human exposure to pathogens and chemicals from agriculture. *Environ. Health Perspect.* **2009**, *117*, 508–514. [CrossRef] [PubMed]

154. Bridge, J.W.; Oliver, D.M.; Chadwick, D.; Godfray, H.C.; Heathwaite, A.L.; Kay, D.; Maheswaran, R.; McGonigle, D.F.; Nichols, G.; Pickup, R.; et al. Engaging with the water sector for public health benefits: Waterborne pathogens and diseases in developed countries. *Bull. World Health Organ.* **2010**, *88*, 873–875. [CrossRef] [PubMed]

155. Lee, J.V.N. Legionnaires' disease. In *Environmental Medicine*; Ayres, J.G.H., Nichols, G.L., Maynard, R.L., Eds.; Hodder Arnold: London, UK, 2010; pp. 224–231.

156. Halsby, K.D.; Joseph, C.A.; Lee, J.V.; Wilkinson, P. The relationship between meteorological variables and sporadic cases of Legionnaires' disease in residents of England and Wales. *Epidemiol. Infect.* **2014**, *142*, 2352–2359. [CrossRef] [PubMed]

157. Ricketts, K.D.; Charlett, A.; Gelb, D.; Lane, C.; Lee, J.V.; Joseph, C.A. Weather patterns and Legionnaires' disease: A meteorological study. *Epidemiol. Infect.* **2009**, *137*, 1003–1012. [CrossRef] [PubMed]

158. Conza, L.; Casati, S.; Limoni, C.; Gaia, V. Meteorological factors and risk of community-acquired Legionnaires' disease in Switzerland: An epidemiological study. *BMJ Open* **2013**, *3*, e002428. [CrossRef] [PubMed]

159. Walser, S.M.; Gerstner, D.G.; Brenner, B.; Holler, C.; Liebl, B.; Herr, C.E. Assessing the environmental health relevance of cooling towers—A systematic review of legionellosis outbreaks. *Int. J. Hyg. Environ. Health* **2014**, *217*, 145–154. [CrossRef] [PubMed]

160. Smith, J.K.; Young, M.M.; Wilson, K.L.; Craig, S.B. Leptospirosis following a major flood in Central Queensland, Australia. *Epidemiol. Infect.* **2012**, *141*, 585–590. [CrossRef] [PubMed]

161. Dechet, A.M.; Parsons, M.; Rambaran, M.; Mohamed-Rambaran, P.; Florendo-Cumbermack, A.; Persaud, S.; Baboolal, S.; Ari, M.D.; Shadomy, S.V.; Zaki, S.R.; et al. Leptospirosis outbreak following severe flooding: A rapid assessment and mass prophylaxis campaign; Guyana, January–February 2005. *PLoS ONE* **2012**, *7*, e39672. [CrossRef] [PubMed]

162. Naranjo, M.; Suarez, M.; Fernandez, C.; Amador, N.; Gonzalez, M.; Batista, N.; Gonzalez, I.; Valdes, Y.; Infante, J.F.; Sierra, G. Study of a Leptospirosis Outbreak in Honduras Following Hurricane Mitch and Prophylactic Protection of the vax-SPIRAL(R) Vaccine. *MEDICC Rev.* **2008**, *10*, 38–42. [PubMed]

163. Mwachui, M.A.; Crump, L.; Hartskeerl, R.; Zinsstag, J.; Hattendorf, J. Environmental and Behavioural Determinants of Leptospirosis Transmission: A Systematic Review. *PLoS Negl. Trop. Dis.* **2015**, *9*, e0003843. [CrossRef] [PubMed]

164. Loveday, H.P.; Wilson, J.A.; Kerr, K.; Pitchers, R.; Walker, J.T.; Browne, J. Association between healthcare water systems and Pseudomonas aeruginosa infections: A rapid systematic review. *J. Hosp. Infect.* **2014**, *86*, 7–15. [CrossRef] [PubMed]

165. Wisener, L.V.; Sargeant, J.M.; O'Connor, A.M.; Faires, M.C.; Glass-Kaastra, S.K. The use of direct-fed microbials to reduce shedding of *Escherichia coli* O157 in beef cattle: A systematic review and meta-analysis. *Zoonoses Public Health* **2015**, *62*, 75–89. [CrossRef] [PubMed]

166. Thomas, D.E.; Elliott, E.J. Interventions for preventing diarrhea-associated hemolytic uremic syndrome: Systematic review. *BMC Public Health* **2013**, *13*, 799. [CrossRef] [PubMed]

167. Bitler, E.J.; Matthews, J.E.; Dickey, B.W.; Eisenberg, J.N.; Leon, J.S. Norovirus outbreaks: A systematic review of commonly implicated transmission routes and vehicles. *Epidemiol. Infect.* **2013**, *141*, 1563–1571. [CrossRef] [PubMed]

168. Stocks, M.E.; Ogden, S.; Haddad, D.; Addiss, D.G.; McGuire, C.; Freeman, M.C. Effect of water, sanitation, and hygiene on the prevention of trachoma: A systematic review and meta-analysis. *PLoS Med.* **2014**, *11*, e1001605. [CrossRef] [PubMed]

169. Travers, A.; Strasser, S.; Palmer, S.L.; Stauber, C. The added value of water, sanitation, and hygiene interventions to mass drug administration for reducing the prevalence of trachoma: A systematic review examining. *J. Environ. Public Health* **2013**, *2013*, 682093. [CrossRef] [PubMed]

170. Cherrie, M.P.C.; Nichols, G.; LoIacono, G.; Sarran, C.; Hajat, S.; Fleming, L. Pathogen seasonality and links with weather in England and Wales: A big data time series analysis. *BMC Public Health* **2017**, *18*, 1067. [CrossRef] [PubMed]

171. Lo Iacono, G.; Armstrong, B.; Fleming, L.E.; Elson, R.; Kovats, S.; Vardoulakis, S.; Nichols, G.L. Challenges in developing methods for quantifying the effects of weather and climate on water-associated diseases: A systematic review. *PLoS Negl. Trop. Dis.* **2017**, *11*, e0005659. [CrossRef] [PubMed]

172. Lo Iacono, G.; Nichols, G.L. Modeling the Impact of Environment on Infectious Diseases. In *Oxford Research Encyclopedia of Environmental Sciences*; Oxford University Press: Oxford, UK, 2017.

173. Hodges, M.; Belle, J.H.; Carlton, E.J.; Liang, S.; Li, H.; Luo, W.; Freeman, M.C.; Liu, Y.; Gao, Y.; Hess, J.J.; et al. Delays reducing waterborne and water-related infectious diseases in China under climate change. *Nat. Clim. Chang.* **2014**, *4*, 1109–1115. [CrossRef] [PubMed]

174. Schijven, J.; Bouwknegt, M.; de Roda Husman, A.M.; Rutjes, S.; Sudre, B.; Suk, J.E.; Semenza, J.C. A decision support tool to compare waterborne and foodborne infection and/or illness risks associated with climate change. *Risk Anal.* **2013**, *33*, 2154–2167. [CrossRef] [PubMed]

175. Moors, E.; Singh, T.; Siderius, C.; Balakrishnan, S.; Mishra, A. Climate change and waterborne diarrhoea in northern India: Impacts and adaptation strategies. *Sci. Total. Environ.* **2013**, *468–469*, S139–S151. [CrossRef] [PubMed]

176. Semenza, J.C.; Houser, C.; Herbst, S.; Rechenburg, A.; Suk, J.E.; Frechen, T.; Kistemann, T. Knowledge Mapping for Climate Change and Food- and Waterborne Diseases. *Crit. Rev. Environ. Sci. Technol.* **2012**, *42*, 378–411. [CrossRef] [PubMed]

177. Semenza, J.C.; Herbst, S.; Rechenburg, A.; Suk, J.E.; Hoser, C.; Schreiber, C.; Kistemann, T. Climate Change Impact Assessment of Food- and Waterborne Diseases. *Crit. Rev. Environ. Sci. Technol.* **2012**, *42*, 857–890. [CrossRef] [PubMed]

178. Patz, J.A.; Vavrus, S.J.; Uejio, C.K.; McLellan, S.L. Climate change and waterborne disease risk in the Great Lakes region of the U.S. *Am. J. Prev. Med.* **2008**, *35*, 451–458. [CrossRef] [PubMed]

179. Schijven, J.F.; de Roda Husman, A.M. Effect of climate changes on waterborne disease in The Netherlands. *Water Sci. Technol.* **2005**, *51*, 79–87. [CrossRef] [PubMed]

180. Charron, D.; Thomas, M.; Waltner-Toews, D.; Aramini, J.; Edge, T.; Kent, R.; Maarouf, A.; Wilson, J. Vulnerability of waterborne diseases to climate change in Canada: A review. *J. Toxicol. Environ. Health Part A* **2004**, *67*, 1667–1677. [CrossRef] [PubMed]

181. Hunter, P.R. Climate change and waterborne and vector-borne disease. *J. Appl. Microbiol.* **2003**, *94*, 37S–46S. [CrossRef] [PubMed]

182. Lo Iacono, G.; Cunningham, A.A.; Fichet-Calvet, E.; Garry, R.F.; Grant, D.S.; Leach, M.; Moses, L.M.; Nichols, G.; Schieffelin, J.S.; Shaffer, J.G.; et al. A Unified Framework for the Infection Dynamics of Zoonotic Spillover and Spread. *PLoS Negl. Trop. Dis.* **2016**, *10*, e0004957. [CrossRef] [PubMed]

183. Rasanen, S.; Lappalainen, S.; Kaikkonen, S.; Hamalainen, M.; Salminen, M.; Vesikari, T. Mixed viral infections causing acute gastroenteritis in children in a waterborne outbreak. *Epidemiol. Infect.* **2010**, *138*, 1227–1234. [CrossRef] [PubMed]

184. Mena, K.D.; Gerba, C.P. Waterborne adenovirus. *Rev. Environ. Contam. Toxicol.* **2009**, *198*, 133–167. [PubMed]

185. Schuster, F.L.; Visvesvara, G.S. Amebae and ciliated protozoa as causal agents of waterborne zoonotic disease. *Vet. Parasitol.* **2004**, *126*, 91–120. [CrossRef] [PubMed]

186. Lamoth, F.; Greub, G. Amoebal pathogens as emerging causal agents of pneumonia. *FEMS Microbiol. Rev.* **2010**, *34*, 260–280. [CrossRef] [PubMed]

187. Karanis, P.; Kourenti, C.; Smith, H. Waterborne transmission of protozoan parasites: A worldwide review of outbreaks and lessons learnt. *J. Water Health* **2007**, *5*, 1–38. [CrossRef] [PubMed]

188. Nwachuku, N.; Gerba, C.P. Health effects of *Acanthamoeba* spp. and its potential for waterborne transmission. *Rev. Environ. Contam. Toxicol.* **2004**, *180*, 93–131. [PubMed]

189. Marshall, M.M.; Naumovitz, D.; Ortega, Y.; Sterling, C.R. Waterborne protozoan pathogens. *Clin. Microbiol. Rev.* **1997**, *10*, 67–85. [PubMed]

190. Baldursson, S.; Karanis, P. Waterborne transmission of protozoan parasites: Review of worldwide outbreaks—An update 2004–2010. *Water Res.* **2011**, *45*, 6603–6614. [CrossRef] [PubMed]

191. Anger, C.; Lally, J.M. Acanthamoeba: A review of its potential to cause keratitis, current lens care solution disinfection standards and methodologies, and strategies to reduce patient risk. *Eye Contact Lens* **2008**, *34*, 247–253. [CrossRef] [PubMed]

192. Foulks, G.N. Acanthamoeba keratitis and contact lens wear: Static or increasing problem? *Eye Contact Lens* **2007**, *33*, 412–414. [CrossRef] [PubMed]

193. Joslin, C.E.; Tu, E.Y.; Shoff, M.E.; Booton, G.C.; Fuerst, P.A.; McMahon, T.T.; Anderson, R.J.; Dworkin, M.S.; Sugar, J.; Davis, F.G.; et al. The association of contact lens solution use and Acanthamoeba keratitis. *Am. J. Ophthalmol.* **2007**, *144*, 169–180. [CrossRef] [PubMed]

194. Tanhehco, T.; Colby, K. The clinical experience of Acanthamoeba keratitis at a tertiary care eye hospital. *Cornea* **2010**, *29*, 1005–1010. [CrossRef] [PubMed]

195. Meier, P.A.; Mathers, W.D.; Sutphin, J.E.; Folberg, R.; Hwang, T.; Wenzel, R.P. An epidemic of presumed Acanthamoeba keratitis that followed regional flooding. Results of a case-control investigation. *Arch. Ophthalmol.* **1998**, *116*, 1090–1094. [CrossRef] [PubMed]

196. Ibrahim, Y.W.; Boase, D.L.; Cree, I.A. Factors affecting the epidemiology of Acanthamoeba keratitis. *Ophthalmic Epidemiol.* **2007**, *14*, 53–60. [CrossRef] [PubMed]

197. Chitambar, S.; Gopalkrishna, V.; Chhabra, P.; Patil, P.; Verma, H.; Lahon, A.; Arora, R.; Tatte, V.; Ranshing, S.; Dhale, G.; et al. Diversity in the enteric viruses detected in outbreaks of gastroenteritis from Mumbai, Western India. *Int. J. Environ. Res. Public Health* **2012**, *9*, 895–915. [CrossRef] [PubMed]

198. Guerrero-Latorre, L.; Carratala, A.; Rodriguez-Manzano, J.; Calgua, B.; Hundesa, A.; Girones, R. Occurrence of water-borne enteric viruses in two settlements based in Eastern Chad: Analysis of hepatitis E virus, hepatitis A virus and human adenovirus in water sources. *J. Water Health* **2011**, *9*, 515–524. [CrossRef] [PubMed]

199. Mellou, K.; Sideroglou, T.; Potamiti-Komi, M.; Kokkinos, P.; Ziros, P.; Georgakopoulou, T.; Vantarakis, A. Epidemiological investigation of two parallel gastroenteritis outbreaks in school settings. *BMC Public Health* **2013**, *13*, 241. [CrossRef] [PubMed]

200. Sinclair, R.G.; Jones, E.L.; Gerba, C.P. Viruses in recreational water-borne disease outbreaks: A review. *J. Appl. Microbiol.* **2009**, *107*, 1769–1780. [CrossRef] [PubMed]

201. Artieda, J.; Pineiro, L.; Gonzalez, M.; Munoz, M.; Basterrechea, M.; Iturzaeta, A.; Cilla, G. A swimming pool-related outbreak of pharyngoconjunctival fever in children due to adenovirus type 4, Gipuzkoa, Spain, 2008. *Euro Surveill.* **2009**, *14*, 19125. [PubMed]

202. Fong, T.T.; Mansfield, L.S.; Wilson, D.L.; Schwab, D.J.; Molloy, S.L.; Rose, J.B. Massive microbiological groundwater contamination associated with a waterborne outbreak in Lake Erie, South Bass Island, Ohio. *Environ. Health Perspect.* **2007**, *115*, 856–864. [CrossRef] [PubMed]

203. Kukkula, M.; Arstila, P.; Klossner, M.L.; Maunula, L.; Bonsdorff, C.H.; Jaatinen, P. Waterborne outbreak of viral gastroenteritis. *Scand. J. Infect. Dis.* **1997**, *29*, 415–418. [CrossRef] [PubMed]

204. Wadstrom, T.; Ljungh, A. Aeromonas and Plesiomonas as food- and waterborne pathogens. *Int. J. Food Microbiol.* **1991**, *12*, 303–311. [CrossRef]

205. Collado, L.; Figueras, M.J. Taxonomy, epidemiology, and clinical relevance of the genus *Arcobacter*. *Clin. Microbiol. Rev.* **2011**, *24*, 174–192. [CrossRef] [PubMed]

206. Asaolu, S.O.; Ofoezie, I.E.; Odumuyiwa, P.A.; Sowemimo, O.A.; Ogunniyi, T.A. Effect of water supply and sanitation on the prevalence and intensity of Ascaris lumbricoides among pre-school-age children in Ajebandele and Ifewara, Osun State, Nigeria. *Trans. R. Soc. Trop. Med. Hyg.* **2002**, *96*, 600–604. [CrossRef]

207. Carneiro, F.F.; Cifuentes, E.; Tellez-Rojo, M.M.; Romieu, I. The risk of Ascaris lumbricoides infection in children as an environmental health indicator to guide preventive activities in Caparao and Alto Caparao, Brazil. *Bull. World Health Organ.* **2002**, *80*, 40–46. [PubMed]

208. Pham-Duc, P.; Nguyen-Viet, H.; Hattendorf, J.; Zinsstag, J.; Phung-Dac, C.; Zurbrugg, C.; Odermatt, P. Ascaris lumbricoides and Trichuris trichiura infections associated with wastewater and human excreta use in agriculture in Vietnam. *Parasitol. Int.* **2013**, *62*, 172–180. [CrossRef] [PubMed]

209. Divizia, M.; Gabrieli, R.; Donia, D.; Macaluso, A.; Bosch, A.; Guix, S.; Sanchez, G.; Villena, C.; Pinto, R.M.; Palombi, L.; et al. Waterborne gastroenteritis outbreak in Albania. *Water Sci. Technol.* **2004**, *50*, 57–61. [CrossRef] [PubMed]

210. Gofti-Laroche, L.; Gratacap-Cavallier, B.; Demanse, D.; Genoulaz, O.; Seigneurin, J.M.; Zmirou, D. Are waterborne astrovirus implicated in acute digestive morbidity (E.MI.R.A. study)? *J. Clin. Virol.* **2003**, *27*, 74–82. [CrossRef]

211. Sezen, F.; Aval, E.; Agkurt, T.; Yilmaz, S.; Temel, F.; Gulesen, R.; Korukluoglu, G.; Sucakli, M.B.; Torunoglu, M.A.; Zhu, B.P. A large multi-pathogen gastroenteritis outbreak caused by drinking contaminated water from antique neighbourhood fountains, Erzurum city, Turkey, December 2012. *Epidemiol. Infect.* **2015**, *143*, 704–710. [CrossRef] [PubMed]

212. Villena, C.; Gabrieli, R.; Pinto, R.M.; Guix, S.; Donia, D.; Buonomo, E.; Palombi, L.; Cenko, F.; Bino, S.; Bosch, A.; et al. A large infantile gastroenteritis outbreak in Albania caused by multiple emerging rotavirus genotypes. *Epidemiol. Infect.* **2003**, *131*, 1105–1110. [CrossRef] [PubMed]

213. Le Guyader, F.S.; Le Saux, J.C.; Ambert-Balay, K.; Krol, J.; Serais, O.; Parnaudeau, S.; Giraudon, H.; Delmas, G.; Pommepuy, M.; Pothier, P.; et al. Aichi virus, norovirus, astrovirus, enterovirus, and rotavirus involved in clinical cases from a French oyster-related gastroenteritis outbreak. *J. Clin. Microbiol.* **2008**, *46*, 4011–4017. [CrossRef] [PubMed]

214. Baker, A.; Tahani, D.; Gardiner, C.; Bristow, K.L.; Greenhill, A.R.; Warner, J. Groundwater seeps facilitate exposure to *Burkholderia pseudomallei*. *Appl. Environ. Microbiol.* **2011**, *77*, 7243–7246. [CrossRef] [PubMed]

215. Sapian, M.; Khair, M.T.; How, S.H.; Rajalingam, R.; Sahhir, K.; Norazah, A.; Khebir, V.; Jamalludin, A.R. Outbreak of melioidosis and leptospirosis co-infection following a rescue operation. *Med. J. Malays.* **2012**, *67*, 293–297.

216. Inglis, T.J.; O'Reilly, L.; Merritt, A.J.; Levy, A.; Heath, C.H. The aftermath of the Western Australian melioidosis outbreak. *Am. J. Trop. Med. Hyg.* **2011**, *84*, 851–857. [CrossRef] [PubMed]

217. Currie, B.J.; Haslem, A.; Pearson, T.; Hornstra, H.; Leadem, B.; Mayo, M.; Gal, D.; Ward, L.; Godoy, D.; Spratt, B.G.; et al. Identification of melioidosis outbreak by multilocus variable number tandem repeat analysis. *Emerg. Infect. Dis.* **2009**, *15*, 169–174. [CrossRef] [PubMed]

218. Cheng, A.C.; Jacups, S.P.; Gal, D.; Mayo, M.; Currie, B.J. Extreme weather events and environmental contamination are associated with case-clusters of melioidosis in the Northern Territory of Australia. *Int. J. Epidemiol.* **2006**, *35*, 323–329. [CrossRef] [PubMed]

219. Ashbolt, N.J. Microbial contamination of drinking water and disease outcomes in developing regions. *Toxicology* **2004**, *198*, 229–238. [CrossRef] [PubMed]

220. Currie, B.J.; Mayo, M.; Anstey, N.M.; Donohoe, P.; Haase, A.; Kemp, D.J. A cluster of melioidosis cases from an endemic region is clonal and is linked to the water supply using molecular typing of Burkholderia pseudomallei isolates. *Am. J. Trop. Med. Hyg.* **2001**, *65*, 177–179. [CrossRef] [PubMed]

221. Inglis, T.J.; Garrow, S.C.; Henderson, M.; Clair, A.; Sampson, J.; O'Reilly, L.; Cameron, B. Burkholderia pseudomallei traced to water treatment plant in Australia. *Emerg. Infect. Dis.* **2000**, *6*, 56–59. [PubMed]

222. Inglis, T.J.; Garrow, S.C.; Adams, C.; Henderson, M.; Mayo, M.; Currie, B.J. Acute melioidosis outbreak in Western Australia. *Epidemiol. Infect.* **1999**, *123*, 437–443. [CrossRef] [PubMed]

223. Inglis, T.J.; Garrow, S.C.; Adams, C.; Henderson, M.; Mayo, M. Dry-season outbreak of melioidosis in Western Australia. *Lancet* **1998**, *352*, 1600. [CrossRef]

224. Merianos, A.; Patel, M.; Lane, J.M.; Noonan, C.N.; Sharrock, D.; Mock, P.A.; Currie, B. The 1990–1991 outbreak of melioidosis in the Northern Territory of Australia: Epidemiology and environmental studies. *Southeast Asian J. Trop. Med. Public Health* **1993**, *24*, 425–435. [PubMed]

225. Nasser, R.M.; Rahi, A.C.; Haddad, M.F.; Daoud, Z.; Irani-Hakime, N.; Almawi, W.Y. Outbreak of Burkholderia cepacia bacteremia traced to contaminated hospital water used for dilution of an alcohol skin antiseptic. *Infect. Control Hosp. Epidemiol.* **2004**, *25*, 231–239. [CrossRef] [PubMed]

226. Roberts, L.A.; Collignon, P.J.; Cramp, V.B.; Alexander, S.; McFarlane, A.E.; Graham, E.; Fuller, A.; Sinickas, V.; Hellyar, A. An Australia-wide epidemic of Pseudomonas pickettii bacteraemia due to contaminated "sterile" water for injection. *Med. J. Aust.* **1990**, *152*, 652–655. [PubMed]

227. Romero-Gomez, M.P.; Quiles-Melero, M.I.; Pena Garcia, P.; Gutierrez Altes, A.; Garcia de Miguel, M.A.; Jimenez, C.; Valdezate, S.; Saez Nieto, J.A. Outbreak of *Burkholderia cepacia* bacteremia caused by contaminated chlorhexidine in a hemodialysis unit. *Infect. Control Hosp. Epidemiol.* **2008**, *29*, 377–378. [CrossRef] [PubMed]

228. Rutala, W.A.; Weber, D.J.; Thomann, C.A.; John, J.F.; Saviteer, S.M.; Sarubbi, F.A. An outbreak of Pseudomonas cepacia bacteremia associated with a contaminated intra-aortic balloon pump. *J. Thorac. Cardiovasc. Surg.* **1988**, *96*, 157–161. [PubMed]

229. Schaffner, W.; Reisig, G.; Verrall, R.A. Outbreak of Pseudomonas cepacia infection due to contaminated anaesthetics. *Lancet* **1973**, *1*, 1050–1051. [CrossRef]

230. Conly, J.M.; Klass, L.; Larson, L.; Kennedy, J.; Low, D.E.; Harding, G.K. Pseudomonas cepacia colonization and infection in intensive care units. *Can. Med. Assoc. J.* **1986**, *134*, 363–366.

231. Kotsanas, D.; Brett, J.; Kidd, T.J.; Stuart, R.L.; Korman, T.M. Disinfection of Burkholderia cepacia complex from non-touch taps in a neonatal nursery. *J. Perinat. Med.* **2008**, *36*, 235–239. [CrossRef] [PubMed]

232. Antony, B.; Cherian, E.V.; Boloor, R.; Shenoy, K.V. A sporadic outbreak of Burkholderia cepacia complex bacteremia in pediatric intensive care unit of a tertiary care hospital in coastal Karnataka, South India. *Indian J. Pathol. Microbiol.* **2016**, *59*, 197–199. [CrossRef] [PubMed]

233. Douce, R.W.; Zurita, J.; Sanchez, O.; Cardenas Aldaz, P. Investigation of an outbreak of central venous catheter-associated bloodstream infection due to contaminated water. *Infect. Control Hosp. Epidemiol.* **2008**, *29*, 364–366. [CrossRef] [PubMed]

234. Kaitwatcharachai, C.; Silpapojakul, K.; Jitsurong, S.; Kalnauwakul, S. An outbreak of Burkholderia cepacia bacteremia in hemodialysis patients: An epidemiologic and molecular study. *Am. J. Kidney Dis.* **2000**, *36*, 199–204. [CrossRef] [PubMed]

235. Lee, J.K. Two outbreaks of Burkholderia cepacia nosocomial infection in a neonatal intensive care unit. *J. Paediatr. Child Health* **2008**, *44*, 62–66. [CrossRef] [PubMed]

236. Lee, S.; Han, S.W.; Kim, G.; Song, D.Y.; Lee, J.C.; Kwon, K.T. An outbreak of Burkholderia cenocepacia associated with contaminated chlorhexidine solutions prepared in the hospital. *Am. J. Infect. Control* **2013**, *41*, e93–e96. [CrossRef] [PubMed]

237. Loukil, C.; Saizou, C.; Doit, C.; Bidet, P.; Mariani-Kurkdjian, P.; Aujard, Y.; Beaufils, F.; Bingen, E. Epidemiologic investigation of Burkholderia cepacia acquisition in two pediatric intensive care units. *Infect. Control Hosp. Epidemiol.* **2003**, *24*, 707–710. [CrossRef] [PubMed]

238. Lucero, C.A.; Cohen, A.L.; Trevino, I.; Rupp, A.H.; Harris, M.; Forkan-Kelly, S.; Noble-Wang, J.; Jensen, B.; Shams, A.; Arduino, M.J.; et al. Outbreak of Burkholderia cepacia complex among ventilated pediatric patients linked to hospital sinks. *Am. J. Infect. Control* **2011**, *39*, 775–778. [CrossRef] [PubMed]

239. Magalhaes, M.; Doherty, C.; Govan, J.R.; Vandamme, P. Polyclonal outbreak of Burkholderia cepacia complex bacteraemia in haemodialysis patients. *J. Hosp. Infect.* **2003**, *54*, 120–123. [CrossRef]

240. Moreira, B.M.; Leobons, M.B.; Pellegrino, F.L.; Santos, M.; Teixeira, L.M.; de Andrade Marques, E.; Sampaio, J.L.; Pessoa-Silva, C.L. *Ralstonia pickettii* and *Burkholderia cepacia* complex bloodstream infections related to infusion of contaminated water for injection. *J. Hosp. Infect.* **2005**, *60*, 51–55. [CrossRef] [PubMed]

241. Paul, L.M.; Hegde, A.; Pai, T.; Shetty, S.; Baliga, S.; Shenoy, S. An Outbreak of Burkholderia cepacia Bacteremia in a Neonatal Intensive Care Unit. *Indian J. Pediatr.* **2016**, *83*, 285–288. [CrossRef] [PubMed]
242. Peterson, A.E.; Chitnis, A.S.; Xiang, N.; Scaletta, J.M.; Geist, R.; Schwartz, J.; Dement, J.; Lawlor, E.; Lipuma, J.J.; O'Connell, H.; et al. Clonally related Burkholderia contaminans among ventilated patients without cystic fibrosis. *Am. J. Infect. Control* **2013**, *41*, 1298–1300. [CrossRef] [PubMed]
243. Rosengarten, D.; Block, C.; Hidalgo-Grass, C.; Temper, V.; Gross, I.; Budin-Mizrahi, A.; Berkman, N.; Benenson, S. Cluster of pseudoinfections with *Burkholderia cepacia* associated with a contaminated washer-disinfector in a bronchoscopy unit. *Infect. Control Hosp. Epidemiol.* **2010**, *31*, 769–771. [CrossRef] [PubMed]
244. Singhal, T.; Shah, S.; Naik, R. Outbreak of *Burkholderia cepacia* complex bacteremia in a chemotherapy day care unit due to intrinsic contamination of an antiemetic drug. *Indian J. Med. Microbiol.* **2015**, *33*, 117–119. [CrossRef] [PubMed]
245. Souza, A.V.; Moreira, C.R.; Pasternak, J.; Hirata Mde, L.; Saltini, D.A.; Caetano, V.C.; Ciosak, S.; Azevedo, F.M.; Severino, P.; Vandamme, P.; et al. Characterizing uncommon *Burkholderia cepacia* complex isolates from an outbreak in a haemodialysis unit. *J. Med. Microbiol.* **2004**, *53*, 999–1005. [CrossRef] [PubMed]
246. Latifi, A.R.; Niyyati, M.; Lorenzo-Morales, J.; Haghighi, A.; Seyyed Tabaei, S.J.; Lasjerdi, Z. Presence of *Balamuthia mandrillaris* in hot springs from Mazandaran province, northern Iran. *Epidemiol. Infect.* **2016**, *144*, 2456–2461. [CrossRef] [PubMed]
247. Retana-Moreira, L.; Abrahams-Sandi, E.; Cabello-Vilchez, A.M.; Reyes-Batlle, M.; Valladares, B.; Martinez-Carretero, E.; Pinero, J.E.; Lorenzo-Morales, J. Isolation and molecular characterization of *Acanthamoeba* and *Balamuthia mandrillaris* from combination shower units in Costa Rica. *Parasitol. Res.* **2014**, *113*, 4117–4122. [CrossRef] [PubMed]
248. Lares-Jimenez, L.F.; Booton, G.C.; Lares-Villa, F.; Velazquez-Contreras, C.A.; Fuerst, P.A. Genetic analysis among environmental strains of *Balamuthia mandrillaris* recovered from an artificial lagoon and from soil in Sonora, Mexico. *Exp. Parasitol.* **2014**, *145*, S57–S61. [CrossRef] [PubMed]
249. Pindyck, T.N.; Dvorscak, L.E.; Hart, B.L.; Palestine, M.D.; Gallant, J.E.; Allen, S.E.; SantaCruz, K.S. Fatal Granulomatous Amebic Encephalitis Due to Balamuthia mandrillaris in New Mexico: A Case Report. *Open Forum Infect. Dis.* **2014**, *1*, ofu062. [CrossRef] [PubMed]
250. Walzer, P.D.; Judson, F.N.; Murphy, K.B.; Healy, G.R.; English, D.K.; Schultz, M.G. Balantidiasis outbreak in Truk. *Am. J. Trop. Med. Hyg.* **1973**, *22*, 33–41. [CrossRef] [PubMed]
251. Schuster, F.L.; Ramirez-Avila, L. Current world status of *Balantidium coli*. *Clin. Microbiol. Rev.* **2008**, *21*, 626–638. [CrossRef] [PubMed]
252. Bellanger, A.P.; Scherer, E.; Cazorla, A.; Grenouillet, F. Dysenteric syndrome due to *Balantidium coli*: A case report. *New Microbiol.* **2013**, *36*, 203–205. [PubMed]
253. Guglielmetti, P.; Cellesi, C.; Figura, N.; Rossolini, A. Family outbreak of Blastocystis hominis associated gastroenteritis. *Lancet* **1989**, *2*, 1394. [CrossRef]
254. Kurt, O.; Dogruman Al, F.; Tanyuksel, M. Eradication of Blastocystis in humans: Really necessary for all? *Parasitol. Int.* **2016**, *65*, 797–801. [CrossRef] [PubMed]
255. Dagci, H.; Kurt, O.; Demirel, M.; Mandiracioglu, A.; Aydemir, S.; Saz, U.; Bart, A.; VAN Gool, T. Epidemiological and diagnostic features of blastocystis infection in symptomatic patients in izmir province, Turkey. *Iran. J. Parasitol.* **2014**, *9*, 519–529. [PubMed]
256. Chandramathi, S.; Suresh, K.; Anita, Z.B.; Kuppusamy, U.R. Infections of Blastocystis hominis and microsporidia in cancer patients: Are they opportunistic? *Trans. R. Soc. Trop. Med. Hyg.* **2012**, *106*, 267–269. [CrossRef] [PubMed]
257. Smith, H.; Nichols, R.A. Zoonotic protozoa—Food for thought. *Parassitologia* **2006**, *48*, 101–104. [PubMed]
258. Tuncay, S.; Delibas, S.; Inceboz, T.; Over, L.; Oral, A.M.; Akisu, C.; Aksoy, U. An outbreak of gastroenteritis associated with intestinal parasites. *Turk. Parazitol. Derg.* **2008**, *32*, 249–252.
259. Jimenez-Gonzalez, D.E.; Martinez-Flores, W.A.; Reyes-Gordillo, J.; Ramirez-Miranda, M.E.; Arroyo-Escalante, S.; Romero-Valdovinos, M.; Stark, D.; Souza-Saldivar, V.; Martinez-Hernandez, F.; Flisser, A.; et al. Blastocystis infection is associated with irritable bowel syndrome in a Mexican patient population. *Parasitol. Res.* **2012**, *110*, 1269–1275. [CrossRef] [PubMed]
260. Iaconelli, M.; Divizia, M.; Della Libera, S.; Di Bonito, P.; La Rosa, G. Frequent Detection and Genetic Diversity of Human Bocavirus in Urban Sewage Samples. *Food Environ. Virol.* **2016**, *8*, 289–295. [CrossRef] [PubMed]

261. Gubbels, S.M.; Kuhn, K.G.; Larsson, J.T.; Adelhardt, M.; Engberg, J.; Ingildsen, P.; Hollesen, L.W.; Muchitsch, S.; Molbak, K.; Ethelberg, S. A waterborne outbreak with a single clone of Campylobacter jejuni in the Danish town of Koge in May 2010. *Scand. J. Infect. Dis.* **2012**, *44*, 586–594. [CrossRef] [PubMed]

262. Karagiannis, I.; Sideroglou, T.; Gkolfinopoulou, K.; Tsouri, A.; Lampousaki, D.; Velonakis, E.N.; Scoulica, E.V.; Mellou, K.; Panagiotopoulos, T.; Bonovas, S. A waterborne Campylobacter jejuni outbreak on a Greek island. *Epidemiol. Infect.* **2010**, *138*, 1726–1734. [CrossRef] [PubMed]

263. Jakopanec, I.; Borgen, K.; Vold, L.; Lund, H.; Forseth, T.; Hannula, R.; Nygard, K. A large waterborne outbreak of campylobacteriosis in Norway: The need to focus on distribution system safety. *BMC Infect. Dis.* **2008**, *8*, 128. [CrossRef] [PubMed]

264. Richardson, G.; Thomas, D.R.; Smith, R.M.; Nehaul, L.; Ribeiro, C.D.; Brown, A.G.; Salmon, R.L. A community outbreak of *Campylobacter jejuni* infection from a chlorinated public water supply. *Epidemiol. Infect.* **2007**, *135*, 1151–1158. [CrossRef] [PubMed]

265. Martin, S.; Penttinen, P.; Hedin, G.; Ljungstrom, M.; Allestam, G.; Andersson, Y.; Giesecke, J. A case-cohort study to investigate concomitant waterborne outbreaks of Campylobacter and gastroenteritis in Soderhamn, Sweden, 2002–3. *J. Water Health* **2006**, *4*, 417–424. [CrossRef] [PubMed]

266. Kuusi, M.; Nuorti, J.P.; Hanninen, M.L.; Koskela, M.; Jussila, V.; Kela, E.; Miettinen, I.; Ruutu, P. A large outbreak of campylobacteriosis associated with a municipal water supply in Finland. *Epidemiol. Infect.* **2005**, *133*, 593–601. [CrossRef] [PubMed]

267. Merritt, A.; Miles, R.; Bates, J. An outbreak of Campylobacter enteritis on an island resort, north Queensland. *Commun. Dis. Intell.* **1999**, *23*, 215–219, discussion 220. [PubMed]

268. Engberg, J.; Gerner-Smidt, P.; Scheutz, F.; Moller Nielsen, E.; On, S.L.; Molbak, K. Water-borne Campylobacter jejuni infection in a Danish town—A 6-week continuous source outbreak. *Clin. Microbiol. Infect.* **1998**, *4*, 648–656. [CrossRef] [PubMed]

269. Thomas, K.M.; Charron, D.F.; Waltner-Toews, D.; Schuster, C.; Maarouf, A.R.; Holt, J.D. A role of high impact weather events in waterborne disease outbreaks in Canada, 1975–2001. *Int. J. Environ. Health Res.* **2006**, *16*, 167–180. [CrossRef] [PubMed]

270. Nichols, G.L. Fly transmission of Campylobacter. *Emerg. Infect. Dis.* **2005**, *11*, 361–364. [CrossRef] [PubMed]

271. Karaman, U.; Koloren, Z.; Seferoglu, O.; Ayaz, E.; Demirel, E. Presence of Parasites in Environmental Waters in Samsun and Its Districts. *Turk. Parazitol. Derg.* **2017**, *41*, 19–21. [CrossRef] [PubMed]

272. Hatam-Nahavandi, K.; Mahvi, A.H.; Mohebali, M.; Keshavarz, H.; Mobedi, I.; Rezaeian, M. Detection of parasitic particles in domestic and urban wastewaters and assessment of removal efficiency of treatment plants in Tehran, Iran. *J. Environ. Health Sci. Eng.* **2015**, *13*, 4. [CrossRef] [PubMed]

273. Mora, L.; Martinez, I.; Figuera, L.; Segura, M.; Del Valle, G. Protozoans in superficial waters and faecal samples of individuals of rural populations of the Montes municipality, Sucre state, Venezuela. *Investig. Clin.* **2010**, *51*, 457–466.

274. Elshazly, A.M.; Elsheikha, H.M.; Soltan, D.M.; Mohammad, K.A.; Morsy, T.A. Protozoal pollution of surface water sources in Dakahlia Governorate, Egypt. *J. Egypt. Soc. Parasitol.* **2007**, *37*, 51–64. [PubMed]

275. Squier, C.; Yu, V.L.; Stout, J.E. Waterborne Nosocomial Infections. *Curr. Infect. Dis. Rep.* **2000**, *2*, 490–496. [CrossRef] [PubMed]

276. Wang, J.L.; Chen, M.L.; Lin, Y.E.; Chang, S.C.; Chen, Y.C. Association between contaminated faucets and colonization or infection by nonfermenting gram-negative bacteria in intensive care units in Taiwan. *J. Clin. Microbiol.* **2009**, *47*, 3226–3230. [CrossRef] [PubMed]

277. Moore, L.S.; Owens, D.S.; Jepson, A.; Turton, J.F.; Ashworth, S.; Donaldson, H.; Holmes, A.H. Waterborne Elizabethkingia meningoseptica in Adult Critical Care. *Emerg. Infect. Dis.* **2016**, *22*, 9–17. [CrossRef] [PubMed]

278. Wang, X.W.; Li, J.; Guo, T.; Zhen, B.; Kong, Q.; Yi, B.; Li, Z.; Song, N.; Jin, M.; Xiao, W.; et al. Concentration and detection of SARS coronavirus in sewage from Xiao Tang Shan Hospital and the 309th Hospital of the Chinese People's Liberation Army. *Water Sci. Technol.* **2005**, *52*, 213–221. [CrossRef] [PubMed]

279. Nichols, G.L.C.; Hadfield, S.J. Molecular epidemiology of human cryptosporidiosis. In *Cryptosporidium: Parasite and Disease*; Cacciò, S.M., Widmer, G., SpringerLink (Online Service), Eds.; Springer: London, UK, 2014; pp. 81–133.

280. Polgreen, P.M.; Sparks, J.D.; Polgreen, L.A.; Yang, M.; Harris, M.L.; Pentella, M.A.; Cavanaugh, J.E. A statewide outbreak of *Cryptosporidium* and its association with the distribution of public swimming pools. *Epidemiol. Infect.* **2012**, *140*, 1439–1445. [CrossRef] [PubMed]

281. Moreira, N.A.; Bondelind, M. Safe drinking water and waterborne outbreaks. *J. Water Health* **2017**, *15*, 83–96. [CrossRef] [PubMed]

282. Mahon, M.; Doyle, S. Waterborne outbreak of cryptosporidiosis in the South East of Ireland: Weighing up the evidence. *Ir. J. Med. Sci.* **2017**, *186*, 989–994. [CrossRef] [PubMed]

283. Efstratiou, A.; Ongerth, J.E.; Karanis, P. Waterborne transmission of protozoan parasites: Review of worldwide outbreaks—An update 2011–2016. *Water Res.* **2017**, *114*, 14–22. [CrossRef] [PubMed]

284. Serdarevic, F.; Jones, R.C.; Weaver, K.N.; Black, S.R.; Ritger, K.A.; Guichard, F.; Dombroski, P.; Emanuel, B.P.; Miller, L.; Gerber, S.I. Multi-pathogen waterborne disease outbreak associated with a dinner cruise on Lake Michigan. *Epidemiol. Infect.* **2012**, *140*, 621–625. [CrossRef] [PubMed]

285. Neira-Munoz, E.; Okoro, C.; McCarthy, N.D. Outbreak of waterborne cryptosporidiosis associated with low oocyst concentrations. *Epidemiol. Infect.* **2007**, *135*, 1159–1164. [CrossRef] [PubMed]

286. Samadder, S.R.; Ziegler, P.; Murphy, T.M.; Holden, N.M. Spatial distribution of risk factors for Cryptosporidium spp. transport in an Irish catchment. *Water Environ. Res.* **2010**, *82*, 750–758. [CrossRef] [PubMed]

287. Jagai, J.S.; Castronovo, D.A.; Monchak, J.; Naumova, E.N. Seasonality of cryptosporidiosis: A meta-analysis approach. *Environ. Res.* **2009**, *109*, 465–478. [CrossRef] [PubMed]

288. Atherton, F.; Newman, C.P.; Casemore, D.P. An outbreak of waterborne cryptosporidiosis associated with a public water supply in the UK. *Epidemiol. Infect.* **1995**, *115*, 123–131. [CrossRef] [PubMed]

289. Bridgman, S.A.; Robertson, R.M.; Syed, Q.; Speed, N.; Andrews, N.; Hunter, P.R. Outbreak of cryptosporidiosis associated with a disinfected groundwater supply. *Epidemiol. Infect.* **1995**, *115*, 555–566. [CrossRef] [PubMed]

290. Harrison, S.L.; Nelder, R.; Hayek, L.; Mackenzie, I.F.; Casemore, D.P.; Dance, D. Managing a large outbreak of cryptosporidiosis: How to investigate and when to decide to lift a 'boil water' notice. *Commun. Dis. Public Health PHLS* **2002**, *5*, 230–239.

291. Patel, S.; Pedraza-Diaz, S.; McLauchlin, J.; Casemore, D.P. Molecular characterisation of Cryptosporidium parvum from two large suspected waterborne outbreaks. Outbreak Control Team South and West Devon 1995, Incident Management Team and Further Epidemiological and Microbiological Studies Subgroup North Thames 1997. *Commun. Dis. Public Health PHLS* **1998**, *1*, 231–233.

292. Smith, H.V.; Patterson, W.J.; Hardie, R.; Greene, L.A.; Benton, C.; Tulloch, W.; Gilmour, R.A.; Girdwood, R.W.; Sharp, J.C.; Forbes, G.I. An outbreak of waterborne cryptosporidiosis caused by post-treatment contamination. *Epidemiol. Infect.* **1989**, *103*, 703–715. [CrossRef] [PubMed]

293. Risebro, H.L.; Doria, M.F.; Andersson, Y.; Medema, G.; Osborn, K.; Schlosser, O.; Hunter, P.R. Fault tree analysis of the causes of waterborne outbreaks. *J. Water Health* **2007**, *5* (Suppl. 1), 1–18. [CrossRef] [PubMed]

294. Wuhib, T.; Silva, T.M.; Newman, R.D.; Garcia, L.S.; Pereira, M.L.; Chaves, C.S.; Wahlquist, S.P.; Bryan, R.T.; Guerrant, R.L.; Sousa Ade, Q.; et al. Cryptosporidial and microsporidial infections in human immunodeficiency virus-infected patients in northeastern Brazil. *J. Infect. Dis.* **1994**, *170*, 494–497. [CrossRef] [PubMed]

295. Scheepstra, D.R.; Lipman, L.J. Swimming in summer. *Tijdschr. Diergeneeskd.* **2006**, *131*, 742–744. [PubMed]

296. Herwaldt, B.L.; Craun, G.F.; Stokes, S.L.; Juranek, D.D. Waterborne-disease outbreaks, 1989–1990. *MMWR CDC Surveill. Summ.* **1991**, *40*, 1–21. [PubMed]

297. Griffiths, D.J.; Saker, M.L. The Palm Island mystery disease 20 years on: A review of research on the cyanotoxin cylindrospermopsin. *Environ. Toxicol.* **2003**, *18*, 78–93. [CrossRef] [PubMed]

298. Hawkins, P.R.; Runnegar, M.T.; Jackson, A.R.; Falconer, I.R. Severe hepatotoxicity caused by the tropical cyanobacterium (blue-green alga) Cylindrospermopsis raciborskii (Woloszynska) Seenaya and Subba Raju isolated from a domestic water supply reservoir. *Appl. Environ. Microbiol.* **1985**, *50*, 1292–1295. [PubMed]

299. Osborne, N.J.; Shaw, G.R.; Webb, P.M. Health effects of recreational exposure to Moreton Bay, Australia waters during a Lyngbya majuscula bloom. *Environ. Int.* **2007**, *33*, 309–314. [CrossRef] [PubMed]

300. Soares, R.M.; Yuan, M.; Servaites, J.C.; Delgado, A.; Magalhaes, V.F.; Hilborn, E.D.; Carmichael, W.W.; Azevedo, S.M. Sublethal exposure from microcystins to renal insufficiency patients in Rio de Janeiro, Brazil. *Environ. Toxicol.* **2006**, *21*, 95–103. [CrossRef] [PubMed]

301. Azevedo, S.M.; Carmichael, W.W.; Jochimsen, E.M.; Rinehart, K.L.; Lau, S.; Shaw, G.R.; Eaglesham, G.K. Human intoxication by microcystins during renal dialysis treatment in Caruaru-Brazil. *Toxicology* **2002**, *181–182*, 441–446. [CrossRef]

302. Jochimsen, E.M.; Carmichael, W.W.; An, J.S.; Cardo, D.M.; Cookson, S.T.; Holmes, C.E.; Antunes, M.B.; de Melo Filho, D.A.; Lyra, T.M.; Barreto, V.S.; et al. Liver failure and death after exposure to microcystins at a hemodialysis center in Brazil. *N. Engl. J. Med.* **1998**, *338*, 873–878. [CrossRef] [PubMed]

303. El Saadi, O.E.; Esterman, A.J.; Cameron, S.; Roder, D.M. Murray River water, raised cyanobacterial cell counts, and gastrointestinal and dermatological symptoms. *Med. J. Aust.* **1995**, *162*, 122–125. [PubMed]

304. Ortega, Y.R.; Sanchez, R. Update on Cyclospora cayetanensis, a food-borne and waterborne parasite. *Clin. Microbiol. Rev.* **2010**, *23*, 218–234. [CrossRef] [PubMed]

305. Wurtz, R. Cyclospora: A newly identified intestinal pathogen of humans. *Clin. Infect. Dis.* **1994**, *18*, 620–623. [CrossRef] [PubMed]

306. Tichadou, L.; Glaizal, M.; Armengaud, A.; Grossel, H.; Lemee, R.; Kantin, R.; Lasalle, J.L.; Drouet, G.; Rambaud, L.; Malfait, P.; et al. Health impact of unicellular algae of the Ostreopsis genus blooms in the Mediterranean Sea: Experience of the French Mediterranean coast surveillance network from 2006 to 2009. *Clin. Toxicol.* **2010**, *48*, 839–844. [CrossRef] [PubMed]

307. Hossen, V.; Jourdan-da Silva, N.; Guillois-Becel, Y.; Marchal, J.; Krys, S. Food poisoning outbreaks linked to mussels contaminated with okadaic acid and ester dinophysistoxin-3 in France, June 2009. *Euro Surveill.* **2011**, *16*. [CrossRef]

308. Barroso Garcia, P.; Rueda de la Puerta, P.; Parron Carreno, T.; Marin Martinez, P.; Guillen Enriquez, J. An epidemic outbreak with respiratory symptoms in the province of Almeria [Spain] due to toxic microalgae exposure. *Gac. Sanit.* **2008**, *22*, 578–584. [PubMed]

309. Watkins, S.M.; Reich, A.; Fleming, L.E.; Hammond, R. Neurotoxic shellfish poisoning. *Mar. Drugs* **2008**, *6*, 431–455. [CrossRef] [PubMed]

310. La Barbera-Sanchez, A.; Franco Soler, J.; Rojas de Astudillo, L.; Chang-Yen, I. Paralytic shellfish poisoning (PSP) in Margarita Island, Venezuela. *Rev. Biol. Trop.* **2004**, *52* (Suppl. 1), 89–98. [PubMed]

311. Anderson, D.M.; Sullivan, J.J.; Reguera, B. Paralytic shellfish poisoning in northwest Spain: The toxicity of the dinoflagellate Gymnodinium catenatum. *Toxicon* **1989**, *27*, 665–674. [CrossRef]

312. Tan, C.T.; Lee, E.J. Paralytic shellfish poisoning in Singapore. *Ann. Acad. Med. Singap.* **1986**, *15*, 77–79. [PubMed]

313. Kat, M. Diarrhetic mussel poisoning in the Netherlands related to the dinoflagellate *Dinophysis acuminata*. *Antonie Van Leeuwenhoek* **1983**, *49*, 417–427. [PubMed]

314. Popkiss, M.E.; Horstman, D.A.; Harpur, D. Paralytic shellfish poisoning. A report of 17 cases in Cape Town. *S. Afr. Med. J.* **1979**, *55*, 1017–1023. [PubMed]

315. Morse, E.V. Paralytic shellfish poisoning: A review. *J. Am. Vet. Med. Assoc.* **1977**, *171*, 1178–1180. [PubMed]

316. Pottier, I.; Vernoux, J.P.; Lewis, R.J. Ciguatera fish poisoning in the Caribbean islands and Western Atlantic. *Rev. Environ. Contam. Toxicol.* **2001**, *168*, 99–141. [PubMed]

317. Juranovic, L.R.; Park, D.L. Foodborne toxins of marine origin: Ciguatera. *Rev. Environ. Contam. Toxicol.* **1991**, *117*, 51–94. [PubMed]

318. Taylor, M.; McIntyre, L.; Ritson, M.; Stone, J.; Bronson, R.; Bitzikos, O.; Rourke, W.; Galanis, E.; Team, O.I. Outbreak of diarrhetic shellfish poisoning associated with mussels, British Columbia, Canada. *Mar. Drugs* **2013**, *11*, 1669–1676. [CrossRef] [PubMed]

319. Chen, T.; Xu, X.; Wei, J.; Chen, J.; Miu, R.; Huang, L.; Zhou, X.; Fu, Y.; Yan, R.; Wang, Z.; et al. Food-borne disease outbreak of diarrhetic shellfish poisoning due to toxic mussel consumption: The first recorded outbreak in China. *PLoS ONE* **2013**, *8*, e65049. [CrossRef] [PubMed]

320. Economou, V.; Papadopoulou, C.; Brett, M.; Kansouzidou, A.; Charalabopoulos, K.; Filioussis, G.; Seferiadis, K. Diarrheic shellfish poisoning due to toxic mussel consumption: The first recorded outbreak in Greece. *Food Addit. Contam.* **2007**, *24*, 297–305. [CrossRef] [PubMed]

321. Chung, P.H.; Chuang, S.K.; Tsang, T. Consumption of viscera as the most important risk factor in the largest outbreak of shellfish poisoning in Hong Kong, 2005. *Southeast Asian J. Trop. Med. Public Health* **2006**, *37*, 120–125. [PubMed]

322. De Schrijver, K.; Maes, I.; De Man, L.; Michelet, J. An outbreak of diarrhoeic shellfish poisoning in Antwerp, Belgium. *Euro Surveill.* **2002**, *7*, 138–141. [CrossRef] [PubMed]

323. James, K.J.; Gillman, M.; Amandi, M.F.; Lopez-Rivera, A.; Puente, P.F.; Lehane, M.; Mitrovic, S.; Furey, A. Amnesic shellfish poisoning toxins in bivalve molluscs in Ireland. *Toxicon* **2005**, *46*, 852–858. [CrossRef] [PubMed]

324. James, K.J.; Furey, A.; Lehane, M.; Ramstad, H.; Aune, T.; Hovgaard, P.; Morris, S.; Higman, W.; Satake, M.; Yasumoto, T. First evidence of an extensive northern European distribution of azaspiracid poisoning (AZP) toxins in shellfish. *Toxicon* **2002**, *40*, 909–915. [CrossRef]

325. Pfannkuchen, M.; Godrijan, J.; Pfannkuchen, D.M.; Ivesa, L.; Kruzic, P.; Ciminiello, P.; Dell'Aversano, C.; Dello Iacovo, E.; Fattorusso, E.; Forino, M.; et al. Toxin-producing Ostreopsis cf. ovata are likely to bloom undetected along coastal areas. *Environ. Sci. Technol.* **2012**, *46*, 5574–5582. [CrossRef] [PubMed]

326. Jeffery, B.; Barlow, T.; Moizer, K.; Paul, S.; Boyle, C. Amnesic shellfish poison. *Food Chem. Toxicol.* **2004**, *42*, 545–557. [CrossRef] [PubMed]

327. Perl, T.M.; Bedard, L.; Kosatsky, T.; Hockin, J.C.; Todd, E.C.; McNutt, L.A.; Remis, R.S. Amnesic shellfish poisoning: A new clinical syndrome due to domoic acid. *Can. Dis. Wkly. Rep.* **1990**, *16* (Suppl. 1E), 7–8. [PubMed]

328. Pulido, O.M. Domoic acid toxicologic pathology: A review. *Mar. Drugs* **2008**, *6*, 180–219. [CrossRef] [PubMed]

329. Vale, P.; de M Sampayo, M.A. First confirmation of human diarrhoeic poisonings by okadaic acid esters after ingestion of razor clams (Solen marginatus) and green crabs (Carcinus maenas) in Aveiro lagoon, Portugal and detection of okadaic acid esters in phytoplankton. *Toxicon* **2002**, *40*, 989–996. [CrossRef]

330. Garcia, C.; Schonstedt, V.; Santelices, J.P.; Lagos, N. High amount of dinophysistoxin-3 in Mytilus chilensis collected in Seno de Reloncavi, chile, during massive human intoxication associated with outbreak of Vibrio parahaemolyticus. *J. Toxicol. Sci.* **2006**, *31*, 305–314. [CrossRef] [PubMed]

331. McIntyre, L.; Cassis, D.; Haigh, N. Formation of a volunteer harmful algal bloom network in British Columbia, Canada, following an outbreak of diarrhetic shellfish poisoning. *Mar. Drugs* **2013**, *11*, 4144–4157. [CrossRef] [PubMed]

332. Chevallier, O.P.; Graham, S.F.; Alonso, E.; Duffy, C.; Silke, J.; Campbell, K.; Botana, L.M.; Elliott, C.T. New insights into the causes of human illness due to consumption of azaspiracid contaminated shellfish. *Sci. Rep.* **2015**, *5*, 9818. [CrossRef] [PubMed]

333. Farrell, H.; Zammit, A.; Manning, J.; Shadbolt, C.; Szabo, L.; Harwood, D.T.; McNabb, P.; Turahui, J.A.; van den Berg, D.J. Clinical diagnosis and chemical confirmation of ciguatera fish poisoning in New South Wales, Australia. *Commun. Dis. Intell. Q. Rep.* **2016**, *40*, E1–E6. [PubMed]

334. Schlaich, C.; Hagelstein, J.G.; Burchard, G.D.; Schmiedel, S. Outbreak of ciguatera fish poisoning on a cargo ship in the port of hamburg. *J. Travel Med.* **2012**, *19*, 238–242. [CrossRef] [PubMed]

335. Nunez, D.; Matute, P.; Garcia, A.; Garcia, P.; Abadia, N. Outbreak of ciguatera food poisoning by consumption of amberjack (*Seriola* spp.) in the Canary Islands, May 2012. *Euro Surveill.* **2012**, *17*, 20188. [PubMed]

336. Yoshikawa Ebesu, J.S.; Hokama, Y. Study of an outbreak of ciguatera fish poisoning in Hong Kong. *Toxicon* **2006**, *48*, 467–469. [CrossRef] [PubMed]

337. Wong, C.K.; Hung, P.; Lee, K.L.; Kam, K.M. Study of an outbreak of ciguatera fish poisoning in Hong Kong. *Toxicon* **2005**, *46*, 563–571. [CrossRef] [PubMed]

338. Ng, S.; Gregory, J. An outbreak of ciguatera fish poisoning in Victoria. *Commun. Dis. Intell.* **2000**, *24*, 344–346. [PubMed]

339. Poli, M.A.; Lewis, R.J.; Dickey, R.W.; Musser, S.M.; Buckner, C.A.; Carpenter, L.G. Identification of Caribbean ciguatoxins as the cause of an outbreak of fish poisoning among U.S. soldiers in Haiti. *Toxicon* **1997**, *35*, 733–741. [CrossRef]

340. Lucas, R.E.; Lewis, R.J.; Taylor, J.M. Pacific ciguatoxin-1 associated with a large common-source outbreak of ciguatera in east Arnhem Land, Australia. *Nat. Toxins* **1997**, *5*, 136–140. [CrossRef] [PubMed]

341. Morris, P.D.; Campbell, D.S.; Freeman, J.I. Ciguatera fish poisoning: An outbreak associated with fish caught from North Carolina coastal waters. *South. Med. J.* **1990**, *83*, 379–382. [CrossRef] [PubMed]

342. Frenette, C.; MacLean, J.D.; Gyorkos, T.W. A large common-source outbreak of ciguatera fish poisoning. *J. Infect. Dis.* **1988**, *158*, 1128–1131. [CrossRef] [PubMed]

343. Engleberg, N.C.; Morris, J.G., Jr.; Lewis, J.; McMillan, J.P.; Pollard, R.A.; Blake, P.A. Ciguatera fish poisoning: A major common-source outbreak in the U.S. Virgin Islands. *Ann. Intern. Med.* **1983**, *98*, 336–337. [CrossRef] [PubMed]

344. Tonge, J.I.; Battey, Y.; Forbes, J.J.; Grant, E.M. Ciguatera poisoning: A report of two outbreaks and a probable fatal case in Queensland. *Med. J. Aust.* **1967**, *2*, 1088–1090. [PubMed]

345. Durando, P.; Ansaldi, F.; Oreste, P.; Moscatelli, P.; Marensi, L.; Grillo, C.; Gasparini, R.; Icardi, G. The Collaborative Group for the Ligurian Syndromic Algal Surveillance. Ostreopsis ovata and human health: Epidemiological and clinical features of respiratory syndrome outbreaks from a two-year syndromic surveillance, 2005–2006, in north-west Italy. *Euro Surveill.* **2007**, *12*, 3212. [PubMed]

346. Centers for Disease Control and Prevention. Renewed transmission of dracunculiasis—Chad, 2010. *MMWR Morb. Mortal. Wkly. Rep.* **2011**, *60*, 744–748.

347. Centers for Disease Control and Prevention. Progress toward global eradication of dracunculiasis, January 2010–June 2011. *MMWR Morb. Mortal. Wkly. Rep.* **2011**, *60*, 1450–1453.

348. Miri, E.S.; Hopkins, D.R.; Ruiz-Tiben, E.; Keana, A.S.; Withers, P.C., Jr.; Anagbogu, I.N.; Sadiq, L.K.; Kale, O.O.; Edungbola, L.D.; Braide, E.I.; et al. Nigeria's triumph: Dracunculiasis eradicated. *Am. J. Trop. Med. Hyg.* **2010**, *83*, 215–225. [CrossRef] [PubMed]

349. Hunter, J.M. Bore holes and the vanishing of guinea worm disease in Ghana's upper region. *Soc. Sci. Med.* **1997**, *45*, 71–89. [CrossRef]

350. Hopkins, D.R.; Azam, M.; Ruiz-Tiben, E.; Kappus, K.D. Eradication of dracunculiasis from Pakistan. *Lancet* **1995**, *346*, 621–624. [CrossRef]

351. Edungbola, L.D.; Watts, S. An outbreak of dracunculiasis in a peri-urban community of Ilorin, Kwara State, Nigeria. *Acta Trop.* **1984**, *41*, 155–163. [PubMed]

352. Steib, K.; Mayer, P. Epidemiology and vectors of Dracunculus medinensis in northwest Burkina Faso, West Africa. *Ann. Trop. Med. Parasitol.* **1988**, *82*, 189–199. [CrossRef] [PubMed]

353. Ilegbodu, V.A.; Christensen, B.L.; Wise, R.A.; Ilegbodu, A.E.; Kale, O.O. Source of drinking water supply and transmission of guinea worm disease in Nigeria. *Ann. Trop. Med. Parasitol.* **1987**, *81*, 713–718. [CrossRef] [PubMed]

354. Barwick, R.S.; Uzicanin, A.; Lareau, S.; Malakmadze, N.; Imnadze, P.; Iosava, M.; Ninashvili, N.; Wilson, M.; Hightower, A.W.; Johnston, S.; et al. Outbreak of amebiasis in Tbilisi, Republic of Georgia, 1998. *Am. J. Trop. Med. Hyg.* **2002**, *67*, 623–631. [CrossRef] [PubMed]

355. Chen, K.T.; Chen, C.J.; Chiu, J.P. A school waterborne outbreak involving both Shigella sonnei and Entamoeba histolytica. *J. Environ. Health* **2001**, *64*, 9–13. [PubMed]

356. Carmona, C.; Perdomo, R.; Carbo, A.; Alvarez, C.; Monti, J.; Grauert, R.; Stern, D.; Perera, G.; Lloyd, S.; Bazini, R.; et al. Risk factors associated with human cystic echinococcosis in Florida, Uruguay: Results of a mass screening study using ultrasound and serology. *Am. J. Trop. Med. Hyg.* **1998**, *58*, 599–605. [CrossRef] [PubMed]

357. Kramer, M.H. Association between asymptomatic cystic echinococcosis and some types of rural water supplies. *Am. J. Trop. Med. Hyg.* **1999**, *60*, 520. [CrossRef] [PubMed]

358. Hernandez, A.; Cardozo, G.; Dematteis, S.; Baz, A.; Trias, N.; Nunez, H.; Barrague, A.; Lopez, L.; Fuentes, J.; Lopez, O.; et al. Cystic echinococcosis: Analysis of the serological profile related to the risk factors in individuals without ultrasound liver changes living in an endemic area of Tacuarembo, Uruguay. *Parasitology* **2005**, *130*, 455–460. [CrossRef] [PubMed]

359. Larrieu, E.J.; Costa, M.T.; del Carpio, M.; Moguillansky, S.; Bianchi, G.; Yadon, Z.E. A case-control study of the risk factors for cystic echinococcosis among the children of Rio Negro province, Argentina. *Ann. Trop. Med. Parasitol.* **2002**, *96*, 43–52. [CrossRef] [PubMed]

360. Schantz, P.M.; Wang, H.; Qiu, J.; Liu, F.J.; Saito, E.; Emshoff, A.; Ito, A.; Roberts, J.M.; Delker, C. Echinococcosis on the Tibetan Plateau: Prevalence and risk factors for cystic and alveolar echinococcosis in Tibetan populations in Qinghai Province, China. *Parasitology* **2003**, *127*, S109–S120. [CrossRef] [PubMed]

361. Yamamoto, N.; Kishi, R.; Katakura, Y.; Miyake, H. Risk factors for human alveolar echinococcosis: A case-control study in Hokkaido, Japan. *Ann. Trop. Med. Parasitol.* **2001**, *95*, 689–696. [CrossRef] [PubMed]

362. Yamamoto, N.; Ohmiya, Y.; Kishi, R.; Miyake, H. Risk factors in serologically diagnosed echinococcosis—A case control study in the Nemuro region of Hokkaido Prefecture. *Nihon Koshu Eisei Zasshi* **1996**, *43*, 446–456. [PubMed]

363. Yang, Y.R.; Sun, T.; Li, Z.; Zhang, J.; Teng, J.; Liu, X.; Liu, R.; Zhao, R.; Jones, M.K.; Wang, Y.; et al. Community surveys and risk factor analysis of human alveolar and cystic echinococcosis in Ningxia Hui Autonomous Region, China. *Bull. World Health Organ.* **2006**, *84*, 714–721. [CrossRef] [PubMed]

364. Matthys, B.; Bobieva, M.; Karimova, G.; Mengliboeva, Z.; Jean-Richard, V.; Hoimnazarova, M.; Kurbonova, M.; Lohourignon, L.K.; Utzinger, J.; Wyss, K. Prevalence and risk factors of helminths and intestinal protozoa infections among children from primary schools in western Tajikistan. *Parasites Vectors* **2011**, *4*, 195. [CrossRef] [PubMed]

365. Agi, P.I. Comparative helminth infections of man in two rural communities of the Niger Delta, Nigeria. *West Afr. J. Med.* **1997**, *16*, 232–236. [PubMed]

366. Agi, P.I. Pattern of infection of intestinal parasites in Sagbama community of the Niger Delta, Nigeria. *West Afr. J. Med.* **1995**, *14*, 39–42. [PubMed]

367. Amvrosieva, T.V.; Paklonskaya, N.V.; Biazruchka, A.A.; Kazinetz, O.N.; Bohush, Z.F.; Fisenko, E.G. Enteroviral infection outbreak in the Republic of Belarus: Principal characteristics and phylogenetic analysis of etiological agents. *Cent. Eur. J. Public Health* **2006**, *14*, 67–73. [CrossRef] [PubMed]

368. El-Esnawy, N.A.; Ali, M.A.; Bayoumi, F.S.; Abo-El-Khir, A.; Abdel-Wahab, K.S. Waterborne viruses associated with repeated abortion. *J. Egypt. Public Health Assoc.* **2001**, *76*, 487–503. [PubMed]

369. Lu, X.D.; Cui, L.L.; Ma, Y.; Zu, R.Q.; Shen, T.; Li, J.Q.; Yao, J.X.; Shan, J.; Xie, Q.; Shi, C.; et al. A viral meningitis outbreak associated with Echo30 in drinking water. *Zhonghua Liu Xing Bing Xue Za Zhi* **2012**, *33*, 1067–1071. [PubMed]

370. Sirbu, A.; Pistol, A.; Popovici, F.; Zaharia, A. Considerations in an outbreak of viral acute gastroenteritis, Cisnadie 2009. *Bacteriol. Virusol. Parazitol. Epidemiol.* **2010**, *55*, 41–44. [PubMed]

371. Park, S.K.; Park, B.; Ki, M.; Kim, H.; Lee, K.; Jung, C.; Sohn, Y.M.; Choi, S.M.; Kim, D.K.; Lee, D.S.; et al. Transmission of seasonal outbreak of childhood enteroviral aseptic meningitis and hand-foot-mouth disease. *J. Korean Med. Sci.* **2010**, *25*, 677–683. [CrossRef] [PubMed]

372. Scarcella, C.; Carasi, S.; Cadoria, F.; Macchi, L.; Pavan, A.; Salamana, M.; Alborali, G.L.; Losio, M.M.; Boni, P.; Lavazza, A.; et al. An outbreak of viral gastroenteritis linked to municipal water supply, Lombardy, Italy, June 2009. *Euro Surveill.* **2009**, *14*, 19274. [CrossRef] [PubMed]

373. Jean, J.S.; Guo, H.R.; Chen, S.H.; Liu, C.C.; Chang, W.T.; Yang, Y.J.; Huang, M.C. The association between rainfall rate and occurrence of an enterovirus epidemic due to a contaminated well. *J. Appl. Microbiol.* **2006**, *101*, 1224–1231. [CrossRef] [PubMed]

374. McBride, G.B.; Stott, R.; Miller, W.; Bambic, D.; Wuertz, S. Discharge-based QMRA for estimation of public health risks from exposure to stormwater-borne pathogens in recreational waters in the United States. *Water Res.* **2013**, *47*, 5282–5297. [CrossRef] [PubMed]

375. Hauri, A.M.; Schimmelpfennig, M.; Walter-Domes, M.; Letz, A.; Diedrich, S.; Lopez-Pila, J.; Schreier, E. An outbreak of viral meningitis associated with a public swimming pond. *Epidemiol. Infect.* **2005**, *133*, 291–298. [CrossRef] [PubMed]

376. McLaughlin, J.B.; Gessner, B.D.; Lynn, T.V.; Funk, E.A.; Middaugh, J.P. Association of regulatory issues with an echovirus 18 meningitis outbreak at a children's summer camp in Alaska. *Pediatr. Infect. Dis. J.* **2004**, *23*, 875–877. [CrossRef] [PubMed]

377. Amvrosieva, T.V.; Titov, L.P.; Mulders, M.; Hovi, T.; Dyakonova, O.V.; Votyakov, V.I.; Kvacheva, Z.B.; Eremin, V.F.; Sharko, R.M.; Orlova, S.V.; et al. Viral water contamination as the cause of aseptic meningitis outbreak in Belarus. *Cent. Eur. J. Public Health* **2001**, *9*, 154–157. [PubMed]

378. Yatsuyanagi, J.; Saito, S.; Miyajima, Y.; Amano, K.; Enomoto, K. Characterization of atypical enteropathogenic *Escherichia coli* strains harboring the astA gene that were associated with a waterborne outbreak of diarrhea in Japan. *J. Clin. Microbiol.* **2003**, *41*, 2033–2039. [CrossRef] [PubMed]

379. Huerta, M.; Grotto, I.; Gdalevich, M.; Mimouni, D.; Gavrieli, B.; Yavzori, M.; Cohen, D.; Shpilberg, O. A waterborne outbreak of gastroenteritis in the Golan Heights due to enterotoxigenic *Escherichia coli*. *Infection* **2000**, *28*, 267–271. [CrossRef] [PubMed]

380. Daniels, N.A.; Neimann, J.; Karpati, A.; Parashar, U.D.; Greene, K.D.; Wells, J.G.; Srivastava, A.; Tauxe, R.V.; Mintz, E.D.; Quick, R. Traveler's diarrhea at sea: Three outbreaks of waterborne enterotoxigenic *Escherichia coli* on cruise ships. *J. Infect. Dis.* **2000**, *181*, 1491–1495. [CrossRef] [PubMed]

381. O'Mahony, M.; Noah, N.D.; Evans, B.; Harper, D.; Rowe, B.; Lowes, J.A.; Pearson, A.; Goode, B. An outbreak of gastroenteritis on a passenger cruise ship. *J. Hyg. (Lond.)* **1986**, *97*, 229–236. [CrossRef] [PubMed]

382. Muramatsu, K.; Wada, M.; Kobayashi, M. An outbreak of well water-associated food-poisoning caused by enterotoxigenic *Escherichia coli* O159:H20. *Kansenshogaku Zasshi* **1986**, *60*, 1–6. [CrossRef] [PubMed]

383. Rosenberg, M.L.; Koplan, J.P.; Wachsmuth, I.K.; Wells, J.G.; Gangarosa, E.J.; Guerrant, R.L.; Sack, D.A. Epidemic diarrhea at Crater Lake from enterotoxigenic *Escherichia coli*. A large waterborne outbreak. *Ann. Intern. Med.* **1977**, *86*, 714–718. [CrossRef] [PubMed]

384. Cortes-Ortiz, I.A.; Rodriguez-Angeles, G.; Moreno-Escobar, E.A.; Tenorio-Lara, J.M.; Torres-Mazadiego, B.P.; Montiel-Vazquez, E. Outbreak caused by *Escherichia coli* in Chalco, Mexico. *Salud Publica Mex.* **2002**, *44*, 297–302. [CrossRef] [PubMed]

385. Rangel, J.M.; Sparling, P.H.; Crowe, C.; Griffin, P.M.; Swerdlow, D.L. Epidemiology of *Escherichia coli* O157:H7 outbreaks, United States, 1982–2002. *Emerg. Infect. Dis.* **2005**, *11*, 603–609. [CrossRef] [PubMed]

386. Hrudey, S.E.; Payment, P.; Huck, P.M.; Gillham, R.W.; Hrudey, E.J. A fatal waterborne disease epidemic in Walkerton, Ontario: Comparison with other waterborne outbreaks in the developed world. *Water Sci. Technol.* **2003**, *47*, 7–14. [CrossRef] [PubMed]

387. Samadpour, M.; Stewart, J.; Steingart, K.; Addy, C.; Louderback, J.; McGinn, M.; Ellington, J.; Newman, T. Laboratory investigation of an *E. coli* O157:H7 outbreak associated with swimming in Battle Ground Lake, Vancouver, Washington. *J. Environ. Health* **2002**, *64*, 16–20. [PubMed]

388. Yarze, J.C.; Chase, M.P.E. *E. coli* O157:H7—Another waterborne outbreak! *Am. J. Gastroenterol.* **2000**, *95*, 1096. [CrossRef] [PubMed]

389. Swerdlow, D.L.; Woodruff, B.A.; Brady, R.C.; Griffin, P.M.; Tippen, S.; Donnell, H.D., Jr.; Geldreich, E.; Payne, B.J.; Meyer, A., Jr.; Wells, J.G.; et al. A waterborne outbreak in Missouri of *Escherichia coli* O157:H7 associated with bloody diarrhea and death. *Ann. Intern. Med.* **1992**, *117*, 812–819. [CrossRef] [PubMed]

390. Dev, V.J.; Main, M.; Gould, I. Waterborne outbreak of *Escherichia coli* O157. *Lancet* **1991**, *337*, 1412. [CrossRef]

391. Corso, P.S.; Kramer, M.H.; Blair, K.A.; Addiss, D.G.; Davis, J.P.; Haddix, A.C. Cost of illness in the 1993 waterborne Cryptosporidium outbreak, Milwaukee, Wisconsin. *Emerg. Infect. Dis.* **2003**, *9*, 426–431. [CrossRef] [PubMed]

392. Yatsuyanagi, J.; Saito, S.; Kinouchi, Y.; Sato, H.; Morita, M.; Itoh, K. Characteristics of enterotoxigenic *Escherichia coli* and *E. coli* harboring enteroaggregative *E. coli* heat-stable enterotoxin-1 (EAST-1) gene isolated from a water-borne outbreak. *Kansenshogaku Zasshi* **1996**, *70*, 215–223. [CrossRef] [PubMed]

393. Chen, J.X.; Chen, M.X.; Ai, L.; Xu, X.N.; Jiao, J.M.; Zhu, T.J.; Su, H.Y.; Zang, W.; Luo, J.J.; Guo, Y.H.; et al. An Outbreak of Human Fascioliasis gigantica in Southwest China. *PLoS ONE* **2013**, *8*, e71520. [CrossRef] [PubMed]

394. Ashrafi, K.; Bargues, M.D.; O'Neill, S.; Mas-Coma, S. Fascioliasis: A worldwide parasitic disease of importance in travel medicine. *Travel Med. Infect. Dis.* **2014**, *12*, 636–649. [CrossRef] [PubMed]

395. Fox, N.J.; White, P.C.; McClean, C.J.; Marion, G.; Evans, A.; Hutchings, M.R. Predicting impacts of climate change on Fasciola hepatica risk. *PLoS ONE* **2011**, *6*, e16126. [CrossRef] [PubMed]

396. Mochankana, M.E.; Robertson, I.D. A retrospective study of the prevalence of bovine fasciolosis at major abattoirs in Botswana. *Onderstepoort J. Vet. Res.* **2016**, *83*, a1015. [CrossRef] [PubMed]

397. Selemetas, N.; Ducheyne, E.; Phelan, P.; O'Kiely, P.; Hendrickx, G.; de Waal, T. Spatial analysis and risk mapping of Fasciola hepatica infection in dairy herds in Ireland. *Geospat. Health* **2015**, *9*, 281–291. [CrossRef] [PubMed]

398. Selemetas, N.; de Waal, T. Detection of major climatic and environmental predictors of liver fluke exposure risk in Ireland using spatial cluster analysis. *Vet. Parasitol.* **2015**, *209*, 242–253. [CrossRef] [PubMed]

399. Ducheyne, E.; Charlier, J.; Vercruysse, J.; Rinaldi, L.; Biggeri, A.; Demeler, J.; Brandt, C.; De Waal, T.; Selemetas, N.; Hoglund, J.; et al. Modelling the spatial distribution of Fasciola hepatica in dairy cattle in Europe. *Geospat. Health* **2015**, *9*, 261–270. [CrossRef] [PubMed]

400. Graczyk, T.K.; Gilman, R.H.; Fried, B. Fasciolopsiasis: Is it a controllable food-borne disease? *Parasitol. Res.* **2001**, *87*, 80–83. [CrossRef] [PubMed]

401. Bello, J.; Nunez, F.A.; Gonzalez, O.M.; Fernandez, R.; Almirall, P.; Escobedo, A.A. Risk factors for Giardia infection among hospitalized children in Cuba. *Ann. Trop. Med. Parasitol.* **2011**, *105*, 57–64. [CrossRef] [PubMed]

402. Teixeira, J.C.; Heller, L.; Barreto, M.L. Giardia duodenalis infection: Risk factors for children living in sub-standard settlements in Brazil. *Cad. Saude Publica* **2007**, *23*, 1489–1493. [CrossRef] [PubMed]

403. Cifuentes, E.; Suarez, L.; Espinosa, M.; Juarez-Figueroa, L.; Martinez-Palomo, A. Risk of Giardia intestinalis infection in children from an artificially recharged groundwater area in Mexico City. *Am. J. Trop. Med. Hyg.* **2004**, *71*, 65–70. [CrossRef] [PubMed]

404. Melius, E.J.; Davis, S.I.; Redd, J.T.; Lewin, M.; Herlihy, R.; Henderson, A.; Sobel, J.; Gold, B.; Cheek, J.E. Estimating the prevalence of active Helicobacter pylori infection in a rural community with global positioning system technology-assisted sampling. *Epidemiol. Infect.* **2013**, *141*, 472–480. [CrossRef] [PubMed]

405. Hanafi, M.I.; Mohamed, A.M. Helicobacter pylori infection: Seroprevalence and predictors among healthy individuals in Al Madinah, Saudi Arabia. *J. Egypt. Public Health Assoc.* **2013**, *88*, 40–45. [CrossRef] [PubMed]

406. Bahrami, A.R.; Rahimi, E.; Ghasemian Safaei, H. Detection of Helicobacter pylori in city water, dental units' water, and bottled mineral water in Isfahan, Iran. *Sci. World J.* **2013**, *2013*, 280510. [CrossRef] [PubMed]

407. Queiroz, D.M.; Carneiro, J.G.; Braga-Neto, M.B.; Fialho, A.B.; Fialho, A.M.; Goncalves, M.H.; Rocha, G.A.; Rocha, A.M.; Braga, L.L. Natural history of Helicobacter pylori infection in childhood: Eight-year follow-up cohort study in an urban community in northeast of Brazil. *Helicobacter* **2012**, *17*, 23–29. [CrossRef] [PubMed]

408. Khan, A.; Farooqui, A.; Kazmi, S.U. Presence of Helicobacter pylori in drinking water of Karachi, Pakistan. *J. Infect. Dev. Ctries.* **2012**, *6*, 251–255. [CrossRef] [PubMed]

409. Travis, P.B.; Goodman, K.J.; O'Rourke, K.M.; Groves, F.D.; Sinha, D.; Nicholas, J.S.; VanDerslice, J.; Lackland, D.; Mena, K.D. The association of drinking water quality and sewage disposal with Helicobacter pylori incidence in infants: The potential role of water-borne transmission. *J. Water Health* **2010**, *8*, 192–203. [CrossRef] [PubMed]

410. Strebel, K.; Rolle-Kampczyk, U.; Richter, M.; Kindler, A.; Richter, T.; Schlink, U. A rigorous small area modelling-study for the Helicobacter pylori epidemiology. *Sci. Total Environ.* **2010**, *408*, 3931–3942. [CrossRef] [PubMed]

411. Mansour-Ghanaei, F.; Taefeh, N.; Joukar, F.; Besharati, S.; Naghipour, M.; Nassiri, R. Recurrence of Helicobacter pylori infection 1 year after successful eradication: A prospective study in Northern Iran. *Med. Sci. Monit.* **2010**, *16*, CR144–CR148. [PubMed]

412. Hestvik, E.; Tylleskar, T.; Kaddu-Mulindwa, D.H.; Ndeezi, G.; Grahnquist, L.; Olafsdottir, E.; Tumwine, J.K. Helicobacter pylori in apparently healthy children aged 0–12 years in urban Kampala, Uganda: A community-based cross sectional survey. *BMC Gastroenterol.* **2010**, *10*, 62. [CrossRef] [PubMed]

413. Zhang, L.H.; Zhou, Y.N.; Zhang, Z.Y.; Zhang, F.H.; Li, G.Z.; Li, Q.; Wu, Z.Q.; Ren, B.L.; Zou, S.J.; Wang, J.X. Epidemiological study on status of Helicobacter pylori in children and teenagers in Wuwei city, Gansu province. *Zhonghua Yi Xue Za Zhi* **2009**, *89*, 2682–2685. [PubMed]

414. Zaterka, S.; Eisig, J.N.; Chinzon, D.; Rothstein, W. Factors related to Helicobacter pylori prevalence in an adult population in Brazil. *Helicobacter* **2007**, *12*, 82–88. [CrossRef] [PubMed]

415. Karita, M.; Teramukai, S.; Matsumoto, S. Risk of Helicobacter pylori transmission from drinking well water is higher than that from infected intrafamilial members in Japan. *Dig. Dis. Sci.* **2003**, *48*, 1062–1067. [CrossRef] [PubMed]

416. Bellou, M.; Kokkinos, P.; Vantarakis, A. Shellfish-borne viral outbreaks: A systematic review. *Food Environ. Virol.* **2013**, *5*, 13–23. [CrossRef] [PubMed]

417. Tauil Mde, C.; Ferreira, P.M.; Abreu, M.C.; Lima, H.C.; Nobrega, A.A. Hepatite A outbreak in an urban area of Luziania, State of Goias, Brazil, 2009. *Rev. Soc. Bras. Med. Trop.* **2010**, *43*, 740–742. [PubMed]

418. Zhang, L.J.; Wang, X.J.; Bai, J.M.; Fang, G.; Liu, L.G.; Zhang, Y.; Fontaine, R.E. An outbreak of hepatitis A in recently vaccinated students from ice snacks made from contaminated well water. *Epidemiol. Infect.* **2009**, *137*, 428–433. [CrossRef] [PubMed]

419. Chobe, L.P.; Arankalle, V.A. Investigation of a hepatitis A outbreak from Shimla Himachal Pradesh. *Indian J. Med. Res.* **2009**, *130*, 179–184. [PubMed]

420. Sowmyanarayanan, T.V.; Mukhopadhya, A.; Gladstone, B.P.; Sarkar, R.; Kang, G. Investigation of a hepatitis A outbreak in children in an urban slum in Vellore, Tamil Nadu, using geographic information systems. *Indian J. Med. Res.* **2008**, *128*, 32–37. [PubMed]

421. Bosch, A.; Sanchez, G.; Le Guyader, F.; Vanaclocha, H.; Haugarreau, L.; Pinto, R.M. Human enteric viruses in Coquina clams associated with a large hepatitis A outbreak. *Water Sci. Technol.* **2001**, *43*, 61–65. [CrossRef] [PubMed]

422. Barrimah, E.; Salem, K.A.; Gabal, M.S. An outbreak of hepatitis A associated with treated waste water used for irrigation. *J. Egypt. Public Health Assoc.* **1999**, *74*, 227–239. [PubMed]

423. Poonawagul, U.; Warintrawat, S.; Snitbhan, R.; Kitisriwarapoj, S.; Chaiyakunt, V.; Foy, H.M. Outbreak of hepatitis A in a college traced to contaminated water reservoir in cafeteria. *Southeast Asian J. Trop. Med. Public Health* **1995**, *26*, 705–708. [PubMed]

424. Divizia, M.; Gnesivo, C.; Amore Bonapasta, R.; Morace, G.; Pisani, G.; Pana, A. Hepatitis A virus identification in an outbreak by enzymatic amplification. *Eur. J. Epidemiol.* **1993**, *9*, 203–208. [CrossRef] [PubMed]

425. Lee, C.S.; Lee, J.H.; Kwon, K.S. Outbreak of hepatitis A in Korean military personnel. *Jpn. J. Infect. Dis.* **2008**, *61*, 239–241. [CrossRef] [PubMed]

426. Gharbi-Khelifi, H.; Sdiri, K.; Ferre, V.; Harrath, R.; Berthome, M.; Billaudel, S.; Aouni, M. A 1-year study of the epidemiology of hepatitis A virus in Tunisia. *Clin. Microbiol. Infect.* **2007**, *13*, 25–32. [CrossRef] [PubMed]

427. Villar, L.M.; De Paula, V.S.; Gaspar, A.M. Seasonal variation of hepatitis A virus infection in the city of Rio de Janeiro, Brazil. *Rev. Inst. Med. Trop. Sao Paulo* **2002**, *44*, 289–292. [CrossRef] [PubMed]

428. Bile, K.; Isse, A.; Mohamud, O.; Allebeck, P.; Nilsson, L.; Norder, H.; Mushahwar, I.K.; Magnius, L.O. Contrasting roles of rivers and wells as sources of drinking water on attack and fatality rates in a hepatitis E epidemic in Somalia. *Am. J. Trop. Med. Hyg.* **1994**, *51*, 466–474. [CrossRef] [PubMed]

429. Abravanel, F.; Lhomme, S.; Dubois, M.; Peron, J.M.; Alric, L.; Kamar, N.; Izopet, J. Hepatitis E virus. *Med. Mal. Infect.* **2013**, *43*, 263–270. [CrossRef] [PubMed]

430. Sailaja, B.; Murhekar, M.V.; Hutin, Y.J.; Kuruva, S.; Murthy, S.P.; Reddy, K.S.; Rao, G.M.; Gupte, M.D. Outbreak of waterborne hepatitis E in Hyderabad, India, 2005. *Epidemiol. Infect.* **2009**, *137*, 234–240. [CrossRef] [PubMed]

431. Sarguna, P.; Rao, A.; Sudha Ramana, K.N. Outbreak of acute viral hepatitis due to hepatitis E virus in Hyderabad. *Indian J. Med. Microbiol.* **2007**, *25*, 378–382. [CrossRef] [PubMed]

432. Shrestha, S.M. Hepatitis E in Nepal. *Kathmandu Univ. Med. J. (KUMJ)* **2006**, *4*, 530–544.

433. Leclerc, H.; Schwartzbrod, L.; Dei-Cas, E. Microbial agents associated with waterborne diseases. *Crit. Rev. Microbiol.* **2002**, *28*, 371–409. [CrossRef] [PubMed]

434. Corwin, A.L.; Khiem, H.B.; Clayson, E.T.; Pham, K.S.; Vo, T.T.; Vu, T.Y.; Cao, T.T.; Vaughn, D.; Merven, J.; Richie, T.L.; et al. A waterborne outbreak of hepatitis E virus transmission in southwestern Vietnam. *Am. J. Trop. Med. Hyg.* **1996**, *54*, 559–562. [CrossRef] [PubMed]

435. Naik, S.R.; Aggarwal, R.; Salunke, P.N.; Mehrotra, N.N. A large waterborne viral hepatitis E epidemic in Kanpur, India. *Bull. World Health Organ.* **1992**, *70*, 597–604. [PubMed]

436. Tsega, E.; Krawczynski, K.; Hansson, B.G.; Nordenfelt, E.; Negusse, Y.; Alemu, W.; Bahru, Y. Outbreak of acute hepatitis E virus infection among military personnel in northern Ethiopia. *J. Med. Virol.* **1991**, *34*, 232–236. [CrossRef] [PubMed]

437. Struelens, M.J.; Maes, N.; Rost, F.; Deplano, A.; Jacobs, F.; Liesnard, C.; Bornstein, N.; Grimont, F.; Lauwers, S.; McIntyre, M.P.; et al. Genotypic and phenotypic methods for the investigation of a nosocomial Legionella pneumophila outbreak and efficacy of control measures. *J. Infect. Dis.* **1992**, *166*, 22–30. [CrossRef] [PubMed]

438. Yiallouros, P.K.; Papadouri, T.; Karaoli, C.; Papamichael, E.; Zeniou, M.; Pieridou-Bagatzouni, D.; Papageorgiou, G.T.; Pissarides, N.; Harrison, T.G.; Hadjidemetriou, A. First outbreak of nosocomial Legionella infection in term neonates caused by a cold mist ultrasonic humidifier. *Clin. Infect. Dis.* **2013**, *57*, 48–56. [CrossRef] [PubMed]

439. Ferre, M.R.; Arias, C.; Oliva, J.M.; Pedrol, A.; Garcia, M.; Pellicer, T.; Roura, P.; Dominguez, A. A community outbreak of Legionnaires' disease associated with a cooling tower in Vic and Gurb, Catalonia (Spain) in 2005. *Eur. J. Clin. Microbiol. Infect. Dis.* **2009**, *28*, 153–159. [CrossRef] [PubMed]

440. Burnsed, L.J.; Hicks, L.A.; Smithee, L.M.; Fields, B.S.; Bradley, K.K.; Pascoe, N.; Richards, S.M.; Mallonee, S.; Littrell, L.; Benson, R.F.; et al. A large, travel-associated outbreak of legionellosis among hotel guests: Utility of the urine antigen assay in confirming Pontiac fever. *Clin. Infect. Dis.* **2007**, *44*, 222–228. [CrossRef] [PubMed]

441. Okada, M.; Kawano, K.; Kura, F.; Amemura-Maekawa, J.; Watanabe, H.; Yagita, K.; Endo, T.; Suzuki, S. The largest outbreak of legionellosis in Japan associated with spa baths: Epidemic curve and environmental investigation. *Kansenshogaku Zasshi* **2005**, *79*, 365–374. [CrossRef] [PubMed]

442. Garcia-Fulgueiras, A.; Navarro, C.; Fenoll, D.; Garcia, J.; Gonzalez-Diego, P.; Jimenez-Bunuales, T.; Rodriguez, M.; Lopez, R.; Pacheco, F.; Ruiz, J.; et al. Legionnaires' disease outbreak in Murcia, Spain. *Emerg. Infect. Dis.* **2003**, *9*, 915–921. [CrossRef] [PubMed]

443. Den Boer, J.W.; Yzerman, E.P.; Schellekens, J.; Lettinga, K.D.; Boshuizen, H.C.; Van Steenbergen, J.E.; Bosman, A.; Van den Hof, S.; Van Vliet, H.A.; Peeters, M.F.; et al. A large outbreak of Legionnaires' disease at a flower show, the Netherlands, 1999. *Emerg. Infect. Dis.* **2002**, *8*, 37–43. [CrossRef] [PubMed]

444. Fisman, D.N.; Lim, S.; Wellenius, G.A.; Johnson, C.; Britz, P.; Gaskins, M.; Maher, J.; Mittleman, M.A.; Spain, C.V.; Haas, C.N.; et al. It's not the heat, it's the humidity: Wet weather increases legionellosis risk in the greater Philadelphia metropolitan area. *J. Infect. Dis.* **2005**, *192*, 2066–2073. [CrossRef] [PubMed]

445. Herrera-Lara, S.; Fernandez-Fabrellas, E.; Cervera-Juan, A.; Blanquer-Olivas, R. Do seasonal changes and climate influence the etiology of community acquired pneumonia? *Arch. Bronconeumol.* **2013**, *49*, 140–145. [CrossRef] [PubMed]

446. Garcia-Vidal, C.; Labori, M.; Viasus, D.; Simonetti, A.; Garcia-Somoza, D.; Dorca, J.; Gudiol, F.; Carratala, J. Rainfall is a risk factor for sporadic cases of Legionella pneumophila pneumonia. *PLoS ONE* **2013**, *8*, e61036. [CrossRef] [PubMed]

447. Kool, J.L.; Warwick, M.C.; Pruckler, J.M.; Brown, E.W.; Butler, J.C. Outbreak of Legionnaires' disease at a bar after basement flooding. *Lancet* **1998**, *351*, 1030. [CrossRef]

448. Cacciapuoti, B.; Ciceroni, L.; Maffei, C.; Di Stanislao, F.; Strusi, P.; Calegari, L.; Lupidi, R.; Scalise, G.; Cagnoni, G.; Renga, G. A waterborne outbreak of leptospirosis. *Am. J. Epidemiol.* **1987**, *126*, 535–545. [CrossRef] [PubMed]

449. Stern, E.J.; Galloway, R.; Shadomy, S.V.; Wannemuehler, K.; Atrubin, D.; Blackmore, C.; Wofford, T.; Wilkins, P.P.; Ari, M.D.; Harris, L.; et al. Outbreak of leptospirosis among Adventure Race participants in Florida, 2005. *Clin. Infect. Dis.* **2010**, *50*, 843–849. [CrossRef] [PubMed]

450. Chumakov, M.E. Leptospirosis in the Republic of Mordovia. *Med. Parazitol. (Mosk)* **2004**, 45–50.

451. Ramakrishnan, R.; Patel, M.S.; Gupte, M.D.; Manickam, P.; Venkataraghavan, S. An institutional outbreak of leptospirosis in Chennai, South India. *J. Commun. Dis.* **2003**, *35*, 1–8. [PubMed]

452. Perra, A.; Servas, V.; Terrier, G.; Postic, D.; Baranton, G.; Andre-Fontaine, G.; Vaillant, V.; Capek, I. Clustered cases of leptospirosis in Rochefort, France, June 2001. *Euro Surveill.* **2002**, *7*, 131–136. [CrossRef] [PubMed]

453. Ashford, D.A.; Kaiser, R.M.; Spiegel, R.A.; Perkins, B.A.; Weyant, R.S.; Bragg, S.L.; Plikaytis, B.; Jarquin, C.; De Lose Reyes, J.O.; Amador, J.J. Asymptomatic infection and risk factors for leptospirosis in Nicaragua. *Am. J. Trop. Med. Hyg.* **2000**, *63*, 249–254. [CrossRef] [PubMed]

454. Ciceroni, L.; Stepan, E.; Pinto, A.; Pizzocaro, P.; Dettori, G.; Franzin, L.; Lupidi, R.; Mansueto, S.; Manera, A.; Ioli, A.; et al. Epidemiological trend of human leptospirosis in Italy between 1994 and 1996. *Eur. J. Epidemiol.* **2000**, *16*, 79–86. [CrossRef] [PubMed]

455. Vanasco, N.B.; Sequeira, G.; Dalla Fontana, M.L.; Fusco, S.; Sequeira, M.D.; Enria, D. Report on a leptospirosis outbreak in the city of Santa Fe, Argentina, March–April 1998. *Rev. Panam. Salud Publ.* **2000**, *7*, 35–40. [CrossRef]

456. Corwin, A.; Ryan, A.; Bloys, W.; Thomas, R.; Deniega, B.; Watts, D. A waterborne outbreak of leptospirosis among United States military personnel in Okinawa, Japan. *Int. J. Epidemiol.* **1990**, *19*, 743–748. [CrossRef] [PubMed]

457. Radl, C.; Muller, M.; Revilla-Fernandez, S.; Karner-Zuser, S.; de Martin, A.; Schauer, U.; Karner, F.; Stanek, G.; Balcke, P.; Hallas, A.; et al. Outbreak of leptospirosis among triathlon participants in Langau, Austria, 2010. *Wien. Klin. Wochenschr.* **2011**, *123*, 751–755. [CrossRef] [PubMed]

458. Sohan, L.; Shyamal, B.; Kumar, T.S.; Malini, M.; Ravi, K.; Venkatesh, V.; Veena, M.; Lal, S. Studies on leptospirosis outbreaks in Peddamandem Mandal of Chittoor district, Andhra Pradesh. *J. Commun. Dis.* **2008**, *40*, 127–132. [PubMed]

459. Gaynor, K.; Katz, A.R.; Park, S.Y.; Nakata, M.; Clark, T.A.; Effler, P.V. Leptospirosis on Oahu: An outbreak associated with flooding of a university campus. *Am. J. Trop. Med. Hyg.* **2007**, *76*, 882–885. [CrossRef] [PubMed]

460. Narita, M.; Fujitani, S.; Haake, D.A.; Paterson, D.L. Leptospirosis after recreational exposure to water in the Yaeyama islands, Japan. *Am. J. Trop. Med. Hyg.* **2005**, *73*, 652–656. [CrossRef] [PubMed]

461. Boland, M.; Sayers, G.; Coleman, T.; Bergin, C.; Sheehan, N.; Creamer, E.; O'Connell, M.; Jones, L.; Zochowski, W. A cluster of leptospirosis cases in canoeists following a competition on the River Liffey. *Epidemiol. Infect.* **2004**, *132*, 195–200. [CrossRef] [PubMed]

462. Karande, S.; Bhatt, M.; Kelkar, A.; Kulkarni, M.; De, A.; Varaiya, A. An observational study to detect leptospirosis in Mumbai, India, 2000. *Arch. Dis. Child.* **2003**, *88*, 1070–1075. [CrossRef] [PubMed]

463. Gol'denshtein, Z.A.; Kalashnikov, I.A.; Mkrtchan, M.O.; Popov, V.A.; Pshenichnyi, V.N. Causes of leptospirosis morbidity in Krasnodar region. *Zh Mikrobiol. Epidemiol. Immunobiol.* **2001**, 77–79.

464. Karande, S.; Kulkarni, H.; Kulkarni, M.; De, A.; Varaiya, A. Leptospirosis in children in Mumbai slums. *Indian J. Pediatr.* **2002**, *69*, 855–858. [CrossRef] [PubMed]

465. Morgan, J.; Bornstein, S.L.; Karpati, A.M.; Bruce, M.; Bolin, C.A.; Austin, C.C.; Woods, C.W.; Lingappa, J.; Langkop, C.; Davis, B.; et al. Outbreak of leptospirosis among triathlon participants and community residents in Springfield, Illinois, 1998. *Clin. Infect. Dis.* **2002**, *34*, 1593–1599. [CrossRef] [PubMed]

466. Trevejo, R.T.; Rigau-Perez, J.G.; Ashford, D.A.; McClure, E.M.; Jarquin-Gonzalez, C.; Amador, J.J.; de los Reyes, J.O.; Gonzalez, A.; Zaki, S.R.; Shieh, W.J.; et al. Epidemic leptospirosis associated with pulmonary hemorrhage-Nicaragua, 1995. *J. Infect. Dis.* **1998**, *178*, 1457–1463. [CrossRef] [PubMed]

467. Centers for Disease Control and Prevention. Outbreak of leptospirosis among white-water rafters—Costa Rica, 1996. *MMWR Morb. Mortal. Wkly. Rep.* **1997**, *46*, 577–579.

468. Anderson, D.C.; Folland, D.S.; Fox, M.D.; Patton, C.M.; Kaufmann, A.F. Leptospirosis: A common-source outbreak due to leptospires of the grippotyphosa serogroup. *Am. J. Epidemiol.* **1978**, *107*, 538–544. [CrossRef] [PubMed]

469. Perez, J.; Brescia, F.; Becam, J.; Mauron, C.; Goarant, C. Rodent abundance dynamics and leptospirosis carriage in an area of hyper-endemicity in New Caledonia. *PLoS Negl. Trop. Dis.* **2011**, *5*, e1361. [CrossRef] [PubMed]

470. Robertson, C.; Nelson, T.A.; Stephen, C. Spatial epidemiology of suspected clinical leptospirosis in Sri Lanka. *Epidemiol. Infect.* **2012**, *140*, 731–743. [CrossRef] [PubMed]

471. Desai, S.; van Treeck, U.; Lierz, M.; Espelage, W.; Zota, L.; Sarbu, A.; Czerwinski, M.; Sadkowska-Todys, M.; Avdicova, M.; Reetz, J.; et al. Resurgence of field fever in a temperate country: An epidemic of leptospirosis among seasonal strawberry harvesters in Germany in 2007. *Clin. Infect. Dis.* **2009**, *48*, 691–697. [CrossRef] [PubMed]

472. Hicks, L.A.; Rose, C.E., Jr.; Fields, B.S.; Drees, M.L.; Engel, J.P.; Jenkins, P.R.; Rouse, B.S.; Blythe, D.; Khalifah, A.P.; Feikin, D.R.; et al. Increased rainfall is associated with increased risk for legionellosis. *Epidemiol. Infect.* **2007**, *135*, 811–817. [CrossRef] [PubMed]

473. Lau, C.L.; Watson, C.H.; Lowry, J.H.; David, M.C.; Craig, S.B.; Wynwood, S.J.; Kama, M.; Nilles, E.J. Human Leptospirosis Infection in Fiji: An Eco-epidemiological Approach to Identifying Risk Factors and Environmental Drivers for Transmission. *PLoS Negl. Trop. Dis.* **2016**, *10*, e0004405. [CrossRef] [PubMed]

474. Decraene, V.; Lebbad, M.; Botero-Kleiven, S.; Gustavsson, A.M.; Lofdahl, M. First reported foodborne outbreak associated with microsporidia, Sweden, October 2009. *Epidemiol. Infect.* **2012**, *140*, 519–527. [CrossRef] [PubMed]

475. Cotte, L.; Rabodonirina, M.; Chapuis, F.; Bailly, F.; Bissuel, F.; Raynal, C.; Gelas, P.; Persat, F.; Piens, M.A.; Trepo, C. Waterborne outbreak of intestinal microsporidiosis in persons with and without human immunodeficiency virus infection. *J. Infect. Dis.* **1999**, *180*, 2003–2008. [CrossRef] [PubMed]

476. Hunter, P.R. Waterborne outbreak of microsporidiosis. *J. Infect. Dis.* **2000**, *182*, 380–381. [CrossRef] [PubMed]

477. Sengupta, J.; Saha, S.; Khetan, A.; Pal, D.; Gangopadhyay, N.; Banerjee, D. Characteristics of microsporidial keratoconjunctivitis in an eastern Indian cohort: A case series. *Indian J. Pathol. Microbiol.* **2011**, *54*, 565–568. [CrossRef] [PubMed]

478. Sharma, S.; Das, S.; Joseph, J.; Vemuganti, G.K.; Murthy, S. Microsporidial keratitis: Need for increased awareness. *Surv. Ophthalmol.* **2011**, *56*, 1–22. [CrossRef] [PubMed]

479. Das, S.; Sharma, S.; Sahu, S.K.; Nayak, S.S.; Kar, S. New microbial spectrum of epidemic keratoconjunctivitis: Clinical and laboratory aspects of an outbreak. *Br. J. Ophthalmol.* **2008**, *92*, 861–862. [PubMed]

480. Didier, E.S.; Weiss, L.M. Microsporidiosis: Not just in AIDS patients. *Curr. Opin. Infect. Dis.* **2011**, *24*, 490–495. [CrossRef] [PubMed]

481. Tu, E.Y.; Joslin, C.E. Microsporidia and Acanthamoeba: The role of emerging corneal pathogens. *Eye* **2012**, *26*, 222–227. [CrossRef] [PubMed]

482. Merlani, G.M.; Francioli, P. Established and emerging waterborne nosocomial infections. *Curr. Opin. Infect. Dis.* **2003**, *16*, 343–347. [CrossRef] [PubMed]

483. Moraga-McHaley, S.A.; Landen, M.; Krapfl, H.; Sewell, C.M. Hypersensitivity pneumonitis with Mycobacterium avium complex among spa workers. *Int. J. Occup. Environ. Health* **2013**, *19*, 55–61. [CrossRef] [PubMed]

484. Brown-Elliott, B.A.; Wallace, R.J., Jr.; Tichindelean, C.; Sarria, J.C.; McNulty, S.; Vasireddy, R.; Bridge, L.; Mayhall, C.G.; Turenne, C.; Loeffelholz, M. Five-year outbreak of community- and hospital-acquired Mycobacterium porcinum infections related to public water supplies. *J. Clin. Microbiol.* **2011**, *49*, 4231–4238. [CrossRef] [PubMed]

485. Yuan, J.; Liu, Y.; Yang, Z.; Cai, Y.; Deng, Z.; Qin, P.; Li, T.; Dong, Z.; Yan, Z.; Zhou, D.; et al. Mycobacterium abscessus post-injection abscesses from extrinsic contamination of multiple-dose bottles of normal saline in a rural clinic. *Int. J. Infect. Dis.* **2009**, *13*, 537–542. [CrossRef] [PubMed]

486. Livni, G.; Yaniv, I.; Samra, Z.; Kaufman, L.; Solter, E.; Ashkenazi, S.; Levy, I. Outbreak of Mycobacterium mucogenicum bacteraemia due to contaminated water supply in a paediatric haematology-oncology department. *J. Hosp. Infect.* **2008**, *70*, 253–258. [CrossRef] [PubMed]

487. Kline, S.; Cameron, S.; Streifel, A.; Yakrus, M.A.; Kairis, F.; Peacock, K.; Besser, J.; Cooksey, R.C. An outbreak of bacteremias associated with Mycobacterium mucogenicum in a hospital water supply. *Infect. Control Hosp. Epidemiol.* **2004**, *25*, 1042–1049. [CrossRef] [PubMed]

488. Falkinham, J.O., 3rd. Mycobacterial aerosols and respiratory disease. *Emerg. Infect. Dis.* **2003**, *9*, 763–767. [CrossRef] [PubMed]

489. Winthrop, K.L.; Abrams, M.; Yakrus, M.; Schwartz, I.; Ely, J.; Gillies, D.; Vugia, D.J. An outbreak of mycobacterial furunculosis associated with footbaths at a nail salon. *N. Engl. J. Med.* **2002**, *346*, 1366–1371. [CrossRef] [PubMed]

490. Astagneau, P.; Desplaces, N.; Vincent, V.; Chicheportiche, V.; Botherel, A.; Maugat, S.; Lebascle, K.; Leonard, P.; Desenclos, J.; Grosset, J.; et al. Mycobacterium xenopi spinal infections after discovertebral surgery: Investigation and screening of a large outbreak. *Lancet* **2001**, *358*, 747–751. [CrossRef]

491. Wenger, J.D.; Spika, J.S.; Smithwick, R.W.; Pryor, V.; Dodson, D.W.; Carden, G.A.; Klontz, K.C. Outbreak of Mycobacterium chelonae infection associated with use of jet injectors. *JAMA* **1990**, *264*, 373–376. [CrossRef] [PubMed]

492. Bolan, G.; Reingold, A.L.; Carson, L.A.; Silcox, V.A.; Woodley, C.L.; Hayes, P.S.; Hightower, A.W.; McFarland, L.; Brown, J.W., 3rd; Petersen, N.J.; et al. Infections with Mycobacterium chelonei in patients receiving dialysis and using processed hemodialyzers. *J. Infect. Dis.* **1985**, *152*, 1013–1019. [CrossRef] [PubMed]

493. Apisarnthanarak, A.; Warren, D.K.; Glen Mayhall, C. Healthcare-associated infections and their prevention after extensive flooding. *Curr. Opin. Infect. Dis.* **2013**, *26*, 359–365. [CrossRef] [PubMed]

494. Falkinham, J.O., 3rd; Iseman, M.D.; de Haas, P.; van Soolingen, D. Mycobacterium avium in a shower linked to pulmonary disease. *J. Water Health* **2008**, *6*, 209–213. [CrossRef] [PubMed]

495. Yagupsky, P.; Menegus, M.A. Waterborne Mycobacterium avium infection. *JAMA* **1989**, *261*, 994. [CrossRef] [PubMed]

496. Marciano-Cabral, F. Biology of *Naegleria* spp. *Microbiol. Rev.* **1988**, *52*, 114–133. [PubMed]

497. Levy, D.A.; Bens, M.S.; Craun, G.F.; Calderon, R.L.; Herwaldt, B.L. Surveillance for waterborne-disease outbreaks—United States, 1995–1996. *MMWR CDC Surveill. Summ.* **1998**, *47*, 1–34. [PubMed]

498. Diaz, J. Seasonal primary amebic meningoencephalitis (PAM) in the south: Summertime is PAM time. *J. La. State Med. Soc.* **2012**, *164*, 148–150. [PubMed]

499. Weeks, J.; Mooney, P.; Lipscomb, G.; Pearson, J.M.; Ong, A.; Singh, S. An unexpected finding on gastroscopy: Gastro-gastric fistula with Helicobacter pylori and Giardia lamblia. *Endoscopy* **2013**, *45* (Suppl. 2). [CrossRef] [PubMed]

500. Zhou, X.; Li, H.; Sun, L.; Mo, Y.; Chen, S.; Wu, X.; Liang, J.; Zheng, H.; Ke, C.; Varma, J.K.; et al. Epidemiological and molecular analysis of a waterborne outbreak of norovirus GII.4. *Epidemiol. Infect.* **2012**, *140*, 2282–2289. [CrossRef] [PubMed]

501. Nenonen, N.P.; Hannoun, C.; Larsson, C.U.; Bergstrom, T. Marked genomic diversity of norovirus genogroup I strains in a waterborne outbreak. *Appl. Environ. Microbiol.* **2012**, *78*, 1846–1852. [CrossRef] [PubMed]

502. Matthews, J.E.; Dickey, B.W.; Miller, R.D.; Felzer, J.R.; Dawson, B.P.; Lee, A.S.; Rocks, J.J.; Kiel, J.; Montes, J.S.; Moe, C.L.; et al. The epidemiology of published norovirus outbreaks: A review of risk factors associated with attack rate and genogroup. *Epidemiol. Infect.* **2012**, *140*, 1161–1172. [CrossRef] [PubMed]

503. Arvelo, W.; Sosa, S.M.; Juliao, P.; Lopez, M.R.; Estevez, A.; Lopez, B.; Morales-Betoulle, M.E.; Gonzalez, M.; Gregoricus, N.A.; Hall, A.J.; et al. Norovirus outbreak of probable waterborne transmission with high attack rate in a Guatemalan resort. *J. Clin. Virol.* **2012**, *55*, 8–11. [CrossRef] [PubMed]

504. Riera-Montes, M.; Brus Sjolander, K.; Allestam, G.; Hallin, E.; Hedlund, K.O.; Lofdahl, M. Waterborne norovirus outbreak in a municipal drinking-water supply in Sweden. *Epidemiol. Infect.* **2011**, *139*, 1928–1935. [CrossRef] [PubMed]

505. Koh, S.J.; Cho, H.G.; Kim, B.H.; Choi, B.Y. An outbreak of gastroenteritis caused by norovirus-contaminated groundwater at a waterpark in Korea. *J. Korean Med. Sci.* **2011**, *26*, 28–32. [CrossRef] [PubMed]

506. Breitenmoser, A.; Fretz, R.; Schmid, J.; Besl, A.; Etter, R. Outbreak of acute gastroenteritis due to a washwater-contaminated water supply, Switzerland, 2008. *J. Water Health* **2011**, *9*, 569–576. [CrossRef] [PubMed]

507. Werber, D.; Lausevic, D.; Mugosa, B.; Vratnica, Z.; Ivanovic-Nikolic, L.; Zizic, L.; Alexandre-Bird, A.; Fiore, L.; Ruggeri, F.M.; Di Bartolo, I.; et al. Massive outbreak of viral gastroenteritis associated with consumption of municipal drinking water in a European capital city. *Epidemiol. Infect.* **2009**, *137*, 1713–1720. [CrossRef] [PubMed]

508. Podewils, L.J.; Zanardi Blevins, L.; Hagenbuch, M.; Itani, D.; Burns, A.; Otto, C.; Blanton, L.; Adams, S.; Monroe, S.S.; Beach, M.J.; et al. Outbreak of norovirus illness associated with a swimming pool. *Epidemiol. Infect.* **2007**, *135*, 827–833. [CrossRef] [PubMed]

509. Doyle, A.; Barataud, D.; Gallay, A.; Thiolet, J.M.; Le Guyaguer, S.; Kohli, E.; Vaillant, V. Norovirus foodborne outbreaks associated with the consumption of oysters from the Etang de Thau, France, December 2002. *Euro Surveill.* **2004**, *9*, 24–26. [CrossRef] [PubMed]

510. Kramer, M.H.; Herwaldt, B.L.; Craun, G.F.; Calderon, R.L.; Juranek, D.D. Surveillance for waterborne-disease outbreaks—United States, 1993–1994. *MMWR CDC Surveill. Summ.* **1996**, *45*, 1–33. [PubMed]

511. Moore, A.C.; Herwaldt, B.L.; Craun, G.F.; Calderon, R.L.; Highsmith, A.K.; Juranek, D.D. Surveillance for waterborne disease outbreaks—United States, 1991–1992. *MMWR CDC Surveill. Summ.* **1993**, *42*, 1–22. [PubMed]

512. Levine, W.C.; Stephenson, W.T.; Craun, G.F. Waterborne disease outbreaks, 1986-1988. *MMWR CDC Surveill. Summ.* **1990**, *39*, 1–13. [CrossRef] [PubMed]

513. Tsukamoto, T.; Kinoshita, Y.; Shimada, T.; Sakazaki, R. Two epidemics of diarrhoeal disease possibly caused by Plesiomonas shigelloides. *J. Hyg. (Lond.)* **1978**, *80*, 275–280. [CrossRef] [PubMed]

514. Van Asperen, I.A.; de Rover, C.M.; Schijven, J.F.; Oetomo, S.B.; Schellekens, J.F.; van Leeuwen, N.J.; Colle, C.; Havelaar, A.H.; Kromhout, D.; Sprenger, M.W. Risk of otitis externa after swimming in recreational fresh water lakes containing Pseudomonas aeruginosa. *BMJ* **1995**, *311*, 1407–1410. [CrossRef] [PubMed]

515. Arseculeratne, S.N.; Sumathipala, S.; Eriyagama, N.B. Patterns of rhinosporidiosis in Sri Lanka: Comparison with international data. *Southeast Asian J. Trop. Med. Public Health* **2010**, *41*, 175–191. [PubMed]

516. Radovanovic, Z.; Vukovic, Z.; Jankovic, S. Attitude of involved epidemiologists toward the first European outbreak of rhinosporidiosis. *Eur. J. Epidemiol.* **1997**, *13*, 157–160. [CrossRef] [PubMed]

517. Vukovic, Z.; Bobic-Radovanovic, A.; Latkovic, Z.; Radovanovic, Z. An epidemiological investigation of the first outbreak of rhinosporidiosis in Europe. *J. Trop. Med. Hyg.* **1995**, *98*, 333–337. [PubMed]

518. Kozlica, J.; Claudet, A.L.; Solomon, D.; Dunn, J.R.; Carpenter, L.R. Waterborne outbreak of Salmonella I 4,[5],12:i. *Foodborne Pathog. Dis.* **2010**, *7*, 1431–1433. [CrossRef] [PubMed]

519. Gallay, A.; De Valk, H.; Cournot, M.; Ladeuil, B.; Hemery, C.; Castor, C.; Bon, F.; Megraud, F.; Le Cann, P.; Desenclos, J.C.; et al. A large multi-pathogen waterborne community outbreak linked to faecal contamination of a groundwater system, France, 2000. *Clin. Microbiol. Infect.* **2006**, *12*, 561–570. [CrossRef] [PubMed]

520. Koroglu, M.; Yakupogullari, Y.; Otlu, B.; Ozturk, S.; Ozden, M.; Ozer, A.; Sener, K.; Durmaz, R. A waterborne outbreak of epidemic diarrhea due to group A rotavirus in Malatya, Turkey. *New Microbiol.* **2011**, *34*, 17–24. [PubMed]

521. Siqueira, A.A.; Santelli, A.C.; Alencar, L.R., Jr.; Dantas, M.P.; Dimech, C.P.; Carmo, G.M.; Santos, D.A.; Alves, R.M.; Lucena, M.B.; Morais, M.; et al. Outbreak of acute gastroenteritis in young children with death due to rotavirus genotype G9 in Rio Branco, Brazilian Amazon region, 2005. *Int. J. Infect. Dis.* **2010**, *14*, e898–e903. [CrossRef] [PubMed]

522. Timenetsky, M.C.; Gouvea, V.; Santos, N.; Alge, M.E.; Kisiellius, J.J.; Carmona, R.C. Outbreak of severe gastroenteritis in adults and children associated with type G2 rotavirus. Study Group on Diarrhea of the Instituto Adolfo Lutz. *J. Diarrhoeal Dis. Res.* **1996**, *14*, 71–74. [PubMed]

523. Hopkins, R.S.; Gaspard, G.B.; Williams, F.P., Jr.; Karlin, R.J.; Cukor, G.; Blacklow, N.R. A community waterborne gastroenteritis outbreak: Evidence for rotavirus as the agent. *Am. J. Public Health* **1984**, *74*, 263–265. [CrossRef] [PubMed]

524. Sutmoller, F.; Azeredo, R.S.; Lacerda, M.D.; Barth, O.M.; Pereira, H.G.; Hoffer, E.; Schatzmayr, H.G. An outbreak of gastroenteritis caused by both rotavirus and *Shigella sonnei* in a private school in Rio de Janeiro. *J. Hyg. (Lond.)* **1982**, *88*, 285–293. [CrossRef] [PubMed]

525. Zamotin, B.A.; Libiiainen, L.T.; Bortnik, F.L.; Chernitskaia, E.P.; Enina, Z.I. Water-borne outbreaks of rotavirus infections. *Zh Mikrobiol. Epidemiol. Immunobiol.* **1981**, 100–102.

526. Gouvea, V.S.; Dias, G.S.; Aguiar, E.A.; Pedro, A.R.; Fichman, E.R.; Chinem, E.S.; Gomes, S.P.; Domingues, A.L. Acute gastroenteritis in a pediatric hospital in rio de janeiro in pre- and post-rotavirus vaccination settings. *Open Virol. J.* **2009**, *3*, 26–30. [CrossRef] [PubMed]

527. Cervia, J.S.; Farber, B.; Armellino, D.; Klocke, J.; Bayer, R.L.; McAlister, M.; Stanchfield, I.; Canonica, F.P.; Ortolano, G.A. Point-of-use water filtration reduces healthcare-associated infections in bone marrow transplant recipients. *Transpl. Infect. Dis.* **2010**, *12*, 238–241. [CrossRef] [PubMed]

528. Dignani, M.C.; Grazziutti, M.; Anaissie, E. Stenotrophomonas maltophilia infections. *Semin. Respir. Crit. Care Med.* **2003**, *24*, 89–98. [CrossRef] [PubMed]

529. Sakhnini, E.; Weissmann, A.; Oren, I. Fulminant Stenotrophomonas maltophilia soft tissue infection in immunocompromised patients: An outbreak transmitted via tap water. *Am. J. Med. Sci.* **2002**, *323*, 269–272. [CrossRef] [PubMed]

530. Verweij, P.E.; Meis, J.F.; Christmann, V.; Van der Bor, M.; Melchers, W.J.; Hilderink, B.G.; Voss, A. Nosocomial outbreak of colonization and infection with Stenotrophomonas maltophilia in preterm infants associated with contaminated tap water. *Epidemiol. Infect.* **1998**, *120*, 251–256. [CrossRef] [PubMed]

531. Weber, D.J.; Rutala, W.A.; Blanchet, C.N.; Jordan, M.; Gergen, M.F. Faucet aerators: A source of patient colonization with *Stenotrophomonas maltophilia*. *Am. J. Infect. Control* **1999**, *27*, 59–63. [CrossRef]

532. Ailes, E.; Budge, P.; Shankar, M.; Collier, S.; Brinton, W.; Cronquist, A.; Chen, M.; Thornton, A.; Beach, M.J.; Brunkard, J.M. Economic and health impacts associated with a Salmonella Typhimurium drinking water outbreak-Alamosa, CO, 2008. *PLoS ONE* **2013**, *8*, e57439. [CrossRef] [PubMed]

533. Anita, S.; Amir, K.M.; Fadzilah, K.; Ahamad, J.; Noorhaida, U.; Marina, K.; Paid, M.Y.; Hanif, Z. Risk factors for typhoid outbreak in Sungai Congkak Recreational Park, Selangor 2009. *Med. J. Malays.* **2012**, *67*, 12–16.

534. Bayram, Y.; Guducuoglu, H.; Otlu, B.; Aypak, C.; Gursoy, N.C.; Uluc, H.; Berktas, M. Epidemiological characteristics and molecular typing of *Salmonella enterica* serovar Typhi during a waterborne outbreak in Eastern Anatolia. *Ann. Trop. Med. Parasitol.* **2011**, *105*, 359–365. [CrossRef] [PubMed]

535. Bhunia, R.; Hutin, Y.; Ramakrishnan, R.; Pal, N.; Sen, T.; Murhekar, M. A typhoid fever outbreak in a slum of South Dumdum municipality, West Bengal, India, 2007: Evidence for foodborne and waterborne transmission. *BMC Public Health* **2009**, *9*, 115. [CrossRef] [PubMed]

536. Farooqui, A.; Khan, A.; Kazmi, S.U. Investigation of a community outbreak of typhoid fever associated with drinking water. *BMC Public Health* **2009**, *9*, 476. [CrossRef] [PubMed]

537. Franklin, L.J.; Fielding, J.E.; Gregory, J.; Gullan, L.; Lightfoot, D.; Poznanski, S.Y.; Vally, H. An outbreak of Salmonella Typhimurium 9 at a school camp linked to contamination of rainwater tanks. *Epidemiol. Infect.* **2009**, *137*, 434–440. [CrossRef] [PubMed]

538. King, C.C.; Chen, C.J.; You, S.L.; Chuang, Y.C.; Huang, H.H.; Tsai, W.C. Community-wide epidemiological investigation of a typhoid outbreak in a rural township in Taiwan, Republic of China. *Int. J. Epidemiol.* **1989**, *18*, 254–260. [CrossRef] [PubMed]

539. Molinero, M.E.; Fernandez, I.; Garcia-Calabuig, M.A.; Peiro, E. Investigation of a water-borne Salmonella ohio outbreak. *Enferm. Infecc. Microbiol. Clin.* **1998**, *16*, 230–232. [PubMed]

540. Muramatsu, K.; Nishizawa, S. An outbreak of municipal water-associated food poisoning caused by Salmonella enteritidis. *Kansenshogaku Zasshi* **1992**, *66*, 754–760. [CrossRef] [PubMed]

541. Nishina, T.; Shiozawa, K.; Hayashi, M.; Akiyama, M.; Sahara, K.; Miwa, N.; Nakatsugawa, S.; Murakami, M.; Nakamura, A. A waterborne outbreak of typhoid fever associated with a small drinking water supply system in Fuji city. *Kansenshogaku Zasshi* **1989**, *63*, 240–247. [CrossRef] [PubMed]

542. Swaddiwudhipong, W.; Kanlayanaphotporn, J. A common-source water-borne outbreak of multidrug-resistant typhoid fever in a rural Thai community. *J. Med. Assoc. Thai.* **2001**, *84*, 1513–1517. [PubMed]

543. Taylor, R.; Sloan, D.; Cooper, T.; Morton, B.; Hunter, I. A waterborne outbreak of Salmonella Saintpaul. *Commun. Dis. Intell.* **2000**, *24*, 336–340. [PubMed]

544. Usera, M.A.; Echeita, A.; Aladuena, A.; Alvarez, J.; Carreno, C.; Orcau, A.; Planas, C. Investigation of an outbreak of water-borne typhoid fever in Catalonia in 1994. *Enferm. Infecc. Microbiol. Clin.* **1995**, *13*, 450–454. [PubMed]

545. Yang, H.H.; Gong, J.; Zhang, J.; Wang, M.L.; Yang, J.; Wu, G.Z.; Quan, W.L.; Gong, H.M.; Szu, S.C. An outbreak of Salmonella Paratyphi A in a boarding school: A community-acquired enteric fever and carriage investigation. *Epidemiol. Infect.* **2010**, *138*, 1765–1774. [CrossRef] [PubMed]

546. Fleury, M.; Charron, D.F.; Holt, J.D.; Allen, O.B.; Maarouf, A.R. A time series analysis of the relationship of ambient temperature and common bacterial enteric infections in two Canadian provinces. *Int. J. Biometeorol.* **2006**, *50*, 385–391. [CrossRef] [PubMed]

547. Karkey, A.; Jombart, T.; Walker, A.W.; Thompson, C.N.; Torres, A.; Dongol, S.; Tran Vu Thieu, N.; Pham Thanh, D.; Tran Thi Ngoc, D.; Voong Vinh, P.; et al. The Ecological Dynamics of Fecal Contamination and Salmonella Typhi and Salmonella Paratyphi A in Municipal Kathmandu Drinking Water. *PLoS Negl. Trop. Dis.* **2016**, *10*, e0004346. [CrossRef] [PubMed]

548. Kim, S. Salmonella serovars from foodborne and waterborne diseases in Korea, 1998–2007: Total isolates decreasing versus rare serovars emerging. *J. Korean Med. Sci.* **2010**, *25*, 1693–1699. [CrossRef] [PubMed]

549. Iseri, L.; Bayraktar, M.R.; Aktas, E.; Durmaz, R. Investigation of an outbreak of Salmonella Typhi in Battalgazi district, Malatya-Turkey. *Braz. J. Microbiol.* **2009**, *40*, 170–173. [CrossRef] [PubMed]

550. Karkey, A.; Arjyal, A.; Anders, K.L.; Boni, M.F.; Dongol, S.; Koirala, S.; My, P.V.; Nga, T.V.; Clements, A.C.; Holt, K.E.; et al. The burden and characteristics of enteric fever at a healthcare facility in a densely populated area of Kathmandu. *PLoS ONE* **2010**, *5*, e13988. [CrossRef] [PubMed]

551. Krol-Turminska, K.; Olender, A. Human infections caused by free-living amoebae. *Ann. Agric. Environ. Med.* **2017**, *24*, 254–260. [CrossRef] [PubMed]

552. Gelman, B.B.; Popov, V.; Chaljub, G.; Nader, R.; Rauf, S.J.; Nauta, H.W.; Visvesvara, G.S. Neuropathological and ultrastructural features of amebic encephalitis caused by *Sappinia diploidea*. *J. Neuropathol. Exp. Neurol.* **2003**, *62*, 990–998. [CrossRef] [PubMed]

553. Makhija, M. Histological identification of muscular sarcocystis: A report of two cases. *Indian J. Pathol. Microbiol.* **2012**, *55*, 552–554. [CrossRef] [PubMed]

554. Nichols, G.L. Food-borne protozoa. *Br. Med. Bull.* **2000**, *56*, 209–235. [CrossRef] [PubMed]

555. Remais, J. Modelling environmentally-mediated infectious diseases of humans: Transmission dynamics of schistosomiasis in China. *Adv. Exp. Med. Boil.* **2010**, *673*, 79–98.

556. Barbosa, C.S.; Domingues, A.L.; Abath, F.; Montenegro, S.M.; Guida, U.; Carneiro, J.; Tabosa, B.; Moraes, C.N.; Spinelli, V. An outbreak of acute schistosomiasis at Porto de Galinhas beach, Pernambuco, Brazil. *Cad. Saude Publica* **2001**, *17*, 725–728. [CrossRef] [PubMed]

557. Bavia, M.E.; Hale, L.F.; Malone, J.B.; Braud, D.H.; Shane, S.M. Geographic information systems and the environmental risk of schistosomiasis in Bahia, Brazil. *Am. J. Trop. Med. Hyg.* **1999**, *60*, 566–572. [CrossRef] [PubMed]

558. Ekpo, U.F.; Mafiana, C.F.; Adeofun, C.O.; Solarin, A.R.; Idowu, A.B. Geographical information system and predictive risk maps of urinary schistosomiasis in Ogun State, Nigeria. *BMC Infect. Dis.* **2008**, *8*, 74. [CrossRef] [PubMed]

559. Kabatereine, N.B.; Brooker, S.; Tukahebwa, E.M.; Kazibwe, F.; Onapa, A.W. Epidemiology and geography of Schistosoma mansoni in Uganda: Implications for planning control. *Trop. Med. Int. Health* **2004**, *9*, 372–380. [CrossRef] [PubMed]

560. Raso, G.; Matthys, B.; N'Goran, E.K.; Tanner, M.; Vounatsou, P.; Utzinger, J. Spatial risk prediction and mapping of Schistosoma mansoni infections among schoolchildren living in western Cote d'Ivoire. *Parasitology* **2005**, *131*, 97–108. [CrossRef] [PubMed]

561. Remais, J.; Liang, S.; Spear, R.C. Coupling hydrologic and infectious disease models to explain regional differences in schistosomiasis transmission in southwestern China. *Environ. Sci. Technol.* **2008**, *42*, 2643–2649. [CrossRef] [PubMed]

562. Remais, J.; Zhong, B.; Carlton, E.J.; Spear, R.C. Model approaches for estimating the influence of time-varying socio-environmental factors on macroparasite transmission in two endemic regions. *Epidemics* **2009**, *1*, 213–220. [CrossRef] [PubMed]

563. Yang, J.; Zhao, Z.; Li, Y.; Krewski, D.; Wen, S.W. A multi-level analysis of risk factors for Schistosoma japonicum infection in China. *Int. J. Infect. Dis.* **2009**, *13*, e407–e412. [CrossRef] [PubMed]

564. Visser, L.G.; Polderman, A.M.; Stuiver, P.C. Outbreak of schistosomiasis among travelers returning from Mali, West Africa. *Clin. Infect. Dis.* **1995**, *20*, 280–285. [CrossRef] [PubMed]

565. Blach, O.; Rai, B.; Oates, K.; Franklin, G.; Bramwell, S. An outbreak of schistosomiasis in travellers returning from endemic areas: The importance of rigorous tracing in peer groups exposed to risk of infection. *J. Public Health* **2012**, *34*, 32–36. [CrossRef] [PubMed]

566. Lin, C.Y.; Fan, P.C.; Wu, C.C. An outbreak of schistosomiasis japonica in Taiwan. *Zhonghua Minguo Wei Sheng Wu Ji Mian Yi Xue Za Zhi* **1988**, *21*, 243–246. [PubMed]

567. Chunge, C.N.; Chunge, R.N.; Masinde, M.S.; Atinga, J.N. An outbreak of acute schistosomiasis following a church retreat to Mwanza, Tanzania, 2008. *J. Travel Med.* **2011**, *18*, 408–410. [CrossRef] [PubMed]

568. Enk, M.J.; Amorim, A.; Schall, V.T. Acute schistosomiasis outbreak in the metropolitan area of Belo Horizonte, Minas Gerais: Alert about the risk of unnoticed transmission increased by growing rural tourism. *Mem. Inst. Oswaldo Cruz* **2003**, *98*, 745–750. [CrossRef] [PubMed]

569. Talla, I.; Kongs, A.; Verle, P.; Belot, J.; Sarr, S.; Coll, A.M. Outbreak of intestinal schistosomiasis in the Senegal River Basin. *Ann. Soc. Belg. Med. Trop.* **1990**, *70*, 173–180. [PubMed]

570. Saeed, A.A.; Magzoub, M. An outbreak of Schistosoma mansoni infection in a European community in Darfur, Western Sudan. *Ann. Trop. Med. Parasitol.* **1974**, *68*, 405–413. [CrossRef] [PubMed]

571. Sayasone, S.; Rasphone, O.; Vanmany, M.; Vounatsou, P.; Utzinger, J.; Tanner, M.; Akkhavong, K.; Hatz, C.; Odermatt, P. Severe morbidity due to Opisthorchis viverrini and Schistosoma mekongi infection in Lao People's Democratic Republic. *Clin. Infect. Dis.* **2012**, *55*, e54–e57. [CrossRef] [PubMed]

572. Sinuon, M.; Tsuyuoka, R.; Socheat, D.; Odermatt, P.; Ohmae, H.; Matsuda, H.; Montresor, A.; Palmer, K. Control of Schistosoma mekongi in Cambodia: Results of eight years of control activities in the two endemic provinces. *Trans. R. Soc. Trop. Med. Hyg.* **2007**, *101*, 34–39. [CrossRef] [PubMed]

573. Urbani, C.; Sinoun, M.; Socheat, D.; Pholsena, K.; Strandgaard, H.; Odermatt, P.; Hatz, C. Epidemiology and control of mekongi schistosomiasis. *Acta Trop.* **2002**, *82*, 157–168. [CrossRef]

574. Attwood, S.W. Schistosomiasis in the Mekong region: Epidemiology and phylogeography. *Adv. Parasitol.* **2001**, *50*, 87–152. [PubMed]

575. Kullavanijaya, P.; Wongwaisayawan, H. Outbreak of cercarial dermatitis in Thailand. *Int. J. Dermatol.* **1993**, *32*, 113–115. [CrossRef] [PubMed]

576. Bracaglia, G.; Ranno, S.; Mancinelli, L.; Santoro, M.; Cerroni, L.; Massone, C.; Sangueza, O.; Bravo, F.G.; Diociaiuti, A.; Nicastri, E.; et al. A waterborn zoonotic helminthiase in an Italian diver: A case report of a cutaneous Sparganum infection and a review of European cases. *Pathog. Glob. Health* **2015**, *109*, 383–386. [CrossRef] [PubMed]

577. Wiwanitkit, V. A review of human sparganosis in Thailand. *Int. J. Infect. Dis.* **2005**, *9*, 312–316. [CrossRef] [PubMed]

578. Pampiglione, S.; Fioravanti, M.L.; Rivasi, F. Human sparganosis in Italy. Case report and review of the European cases. *APMIS* **2003**, *111*, 349–354. [CrossRef] [PubMed]

579. Chang, J.H.; Lin, O.S.; Yeh, K.T. Subcutaneous sparganosis—A case report and a review of human sparganosis in Taiwan. *Kaohsiung J. Med. Sci.* **1999**, *15*, 567–571. [PubMed]

580. Ma, Q.; Xu, X.; Luo, M.; Wang, J.; Yang, C.; Hu, X.; Liang, B.; Wu, F.; Yang, X.; Wang, J.; et al. A Waterborne Outbreak of Shigella sonnei with Resistance to Azithromycin and Third-Generation Cephalosporins in China in 2015. *Antimicrob. Agents Chemother.* **2017**, *61*. [CrossRef] [PubMed]

581. Godoy, P.; Bartolome, R.; Torres, J.; Espinet, L.; Escobar, A.; Nuin, C.; Dominguez, A. Waterborne outbreak of Shigella sonnei caused by consumption of public supply water. *Gac. Sanit.* **2011**, *25*, 363–367. [CrossRef] [PubMed]

582. Arias, C.; Sala, M.R.; Dominguez, A.; Bartolome, R.; Benavente, A.; Veciana, P.; Pedrol, A.; Hoyo, G.; Outbreak Working, G. Waterborne epidemic outbreak of *Shigella sonnei* gastroenteritis in Santa Maria de Palautordera, Catalonia, Spain. *Epidemiol. Infect.* **2006**, *134*, 598–604. [CrossRef] [PubMed]

583. Alamanos, Y.; Maipa, V.; Levidiotou, S.; Gessouli, E. A community waterborne outbreak of gastro-enteritis attributed to *Shigella sonnei*. *Epidemiol. Infect.* **2000**, *125*, 499–503. [CrossRef] [PubMed]

584. Sergevnin, V.I.; Sarmometov, E.V.; Skovorodin, A.N. Waterborne outbreaks of Sonne dysentery are not myth but reality. *Zh Mikrobiol. Epidemiol. Immunobiol.* **1999**, 112–113.

585. Vaudaux, J.D.; Muccioli, C.; James, E.R.; Silveira, C.; Magargal, S.L.; Jung, C.; Dubey, J.P.; Jones, J.L.; Doymaz, M.Z.; Bruckner, D.A.; et al. Identification of an atypical strain of toxoplasma gondii as the cause of a waterborne outbreak of toxoplasmosis in Santa Isabel do Ivai, Brazil. *J. Infect. Dis.* **2010**, *202*, 1226–1233. [CrossRef] [PubMed]

586. Balasundaram, M.B.; Andavar, R.; Palaniswamy, M.; Venkatapathy, N. Outbreak of acquired ocular toxoplasmosis involving 248 patients. *Arch. Ophthalmol.* **2010**, *128*, 28–32. [CrossRef] [PubMed]

587. De Moura, L.; Bahia-Oliveira, L.M.; Wada, M.Y.; Jones, J.L.; Tuboi, S.H.; Carmo, E.H.; Ramalho, W.M.; Camargo, N.J.; Trevisan, R.; Graca, R.M.; et al. Waterborne toxoplasmosis, Brazil, from field to gene. *Emerg. Infect. Dis.* **2006**, *12*, 326–329. [CrossRef] [PubMed]

588. Heukelbach, J.; Meyer-Cirkel, V.; Moura, R.C.; Gomide, M.; Queiroz, J.A.; Saweljew, P.; Liesenfeld, O. Waterborne toxoplasmosis, northeastern Brazil. *Emerg. Infect. Dis.* **2007**, *13*, 287–289. [CrossRef] [PubMed]

589. Palanisamy, M.; Madhavan, B.; Balasundaram, M.B.; Andavar, R.; Venkatapathy, N. Outbreak of ocular toxoplasmosis in Coimbatore, India. *Indian J. Ophthalmol.* **2006**, *54*, 129–131. [CrossRef] [PubMed]

590. Bowie, W.R.; King, A.S.; Werker, D.H.; Isaac-Renton, J.L.; Bell, A.; Eng, S.B.; Marion, S.A. Outbreak of toxoplasmosis associated with municipal drinking water. The BC Toxoplasma Investigation Team. *Lancet* **1997**, *350*, 173–177. [CrossRef]

591. Eng, S.B.; Werker, D.H.; King, A.S.; Marion, S.A.; Bell, A.; Issac-Renton, J.L.; Irwin, G.S.; Bowie, W.R. Computer-generated dot maps as an epidemiologic tool: Investigating an outbreak of toxoplasmosis. *Emerg. Infect. Dis.* **1999**, *5*, 815–819. [CrossRef] [PubMed]

592. Caron, Y.; Cabaraux, A.; Marechal, F.; Losson, B. Swimmer's Itch in Belgium: First Recorded Outbreaks, Molecular Identification of the Parasite Species and Intermediate Hosts. *Vector-Borne Zoonotic Dis.* **2017**, *17*, 190–194. [CrossRef] [PubMed]

593. Lawton, S.P.; Lim, R.M.; Dukes, J.P.; Cook, R.T.; Walker, A.J.; Kirk, R.S. Identification of a major causative agent of human cercarial dermatitis, Trichobilharzia franki (Muller and Kimmig 1994), in southern England and its evolutionary relationships with other European populations. *Parasit Vectors* **2014**, *7*, 277. [CrossRef] [PubMed]

594. Kolarova, L.; Horak, P.; Skirnisson, K.; Mareckova, H.; Doenhoff, M. Cercarial dermatitis, a neglected allergic disease. *Clin. Rev. Allergy Immunol.* **2013**, *45*, 63–74. [CrossRef] [PubMed]

595. Brant, S.V.; Loker, E.S. Schistosomes in the southwest United States and their potential for causing cercarial dermatitis or 'swimmer's itch'. *J. Helminthol.* **2009**, *83*, 191–198. [CrossRef] [PubMed]

596. Valdovinos, C.; Balboa, C. Cercarial dermatitis and lake eutrophication in south-central Chile. *Epidemiol. Infect.* **2008**, *136*, 391–394. [CrossRef] [PubMed]

597. Schets, F.M.; Lodder, W.J.; van Duynhoven, Y.T.; de Roda Husman, A.M. Cercarial dermatitis in the Netherlands caused by *Trichobilharzia* spp. *J. Water Health* **2008**, *6*, 187–195. [CrossRef] [PubMed]

598. Kolarova, L. Schistosomes causing cercarial dermatitis: A mini-review of current trends in systematics and of host specificity and pathogenicity. *Folia Parasitol.* **2007**, *54*, 81–87. [CrossRef] [PubMed]

599. Kolarova, L.; Gottwaldova, V.; Cechova, D.; Sevcova, M. The occurrence of cercarial dermatitis in Central Bohemia. *Zentralbl. Hyg. Umweltmed.* **1989**, *189*, 1–13. [PubMed]

600. Ranjbar, R.; Rahbar, M.; Naghoni, A.; Farshad, S.; Davari, A.; Shahcheraghi, F. A cholera outbreak associated with drinking contaminated well water. *Arch. Iran. Med.* **2011**, *14*, 339–340. [PubMed]

601. Mridha, P.; Biswas, A.K.; Ramakrishnan, R.; Murhekar, M.V. The 2010 outbreak of cholera among workers of a jute mill in Kolkata, West Bengal, India. *J. Health Popul. Nutr.* **2011**, *29*, 9–13. [CrossRef] [PubMed]

602. Bhunia, R.; Ramakrishnan, R.; Hutin, Y.; Gupte, M.D. Cholera outbreak secondary to contaminated pipe water in an urban area, West Bengal, India, 2006. *Indian J. Gastroenterol.* **2009**, *28*, 62–64. [CrossRef] [PubMed]

603. De Magny, G.C.; Thiaw, W.; Kumar, V.; Manga, N.M.; Diop, B.M.; Gueye, L.; Kamara, M.; Roche, B.; Murtugudde, R.; Colwell, R.R. Cholera outbreak in Senegal in 2005: Was climate a factor? *PLoS ONE* **2012**, *7*, e44577. [CrossRef]

604. Goel, A.K.; Jiang, S.C. Association of Heavy Rainfall on Genotypic Diversity in V. cholerae Isolates from an Outbreak in India. *Int. J. Microbiol.* **2011**, *2011*, 230597. [CrossRef] [PubMed]

605. Martinez-Urtaza, J.; Huapaya, B.; Gavilan, R.G.; Blanco-Abad, V.; Ansede-Bermejo, J.; Cadarso-Suarez, C.; Figueiras, A.; Trinanes, J. Emergence of Asiatic Vibrio diseases in South America in phase with El Nino. *Epidemiology* **2008**, *19*, 829–837. [CrossRef] [PubMed]

606. Oberbeckmann, S.; Fuchs, B.M.; Meiners, M.; Wichels, A.; Wiltshire, K.H.; Gerdts, G. Seasonal dynamics and modeling of a Vibrio community in coastal waters of the North Sea. *Microb. Ecol.* **2012**, *63*, 543–551. [CrossRef] [PubMed]

607. Constantin de Magny, G.; Murtugudde, R.; Sapiano, M.R.; Nizam, A.; Brown, C.W.; Busalacchi, A.J.; Yunus, M.; Nair, G.B.; Gil, A.I.; Lanata, C.F.; et al. Environmental signatures associated with cholera epidemics. *Proc. Natl. Acad. Sci. USA* **2008**, *105*, 17676–17681. [CrossRef] [PubMed]

608. Huq, A.; Sack, R.B.; Nizam, A.; Longini, I.M.; Nair, G.B.; Ali, A.; Morris, J.G., Jr.; Khan, M.N.; Siddique, A.K.; Yunus, M.; et al. Critical factors influencing the occurrence of Vibrio cholerae in the environment of Bangladesh. *Appl. Environ. Microbiol.* **2005**, *71*, 4645–4654. [CrossRef] [PubMed]

609. Dziuban, E.J.; Liang, J.L.; Craun, G.F.; Hill, V.; Yu, P.A.; Painter, J.; Moore, M.R.; Calderon, R.L.; Roy, S.L.; Beach, M.J.; et al. Surveillance for waterborne disease and outbreaks associated with recreational water—United States, 2003–2004. *Morb. Mortal. Wkly. Rep. Surveill. Summ.* **2006**, *55*, 1–30.

610. Shaw, K.S.; Sapkota, A.R.; Jacobs, J.M.; He, X.; Crump, B.C. Recreational swimmers' exposure to Vibrio vulnificus and Vibrio parahaemolyticus in the Chesapeake Bay, Maryland, USA. *Environ. Int.* **2015**, *74*, 99–105. [CrossRef] [PubMed]

611. Paz, S.; Bisharat, N.; Paz, E.; Kidar, O.; Cohen, D. Climate change and the emergence of Vibrio vulnificus disease in Israel. *Environ. Res.* **2007**, *103*, 390–396. [CrossRef] [PubMed]

612. Wallace, B.J.; Guzewich, J.J.; Cambridge, M.; Altekruse, S.; Morse, D.L. Seafood-associated disease outbreaks in New York, 1980–1994. *Am. J. Prev. Med.* **1999**, *17*, 48–54. [CrossRef]

613. Esteves, K.; Hervio-Heath, D.; Mosser, T.; Rodier, C.; Tournoud, M.G.; Jumas-Bilak, E.; Colwell, R.R.; Monfort, P. Rapid proliferation of Vibrio parahaemolyticus, Vibrio vulnificus, and Vibrio cholerae during freshwater flash floods in French Mediterranean coastal lagoons. *Appl. Environ. Microbiol.* **2015**, *81*, 7600–7609. [CrossRef] [PubMed]

614. Banakar, V.; Constantin de Magny, G.; Jacobs, J.; Murtugudde, R.; Huq, A.; Wood, R.J.; Colwell, R.R. Temporal and spatial variability in the distribution of Vibrio vulnificus in the Chesapeake Bay: A hindcast study. *Ecohealth* **2011**, *8*, 456–467. [CrossRef] [PubMed]

615. Inoue, Y.; Miyasaka, J.; Ono, T.; Ihn, H. The growth of Vibrio vulnificus and the habitat of infected patients in Kumamoto. *Biosci. Trends* **2007**, *1*, 134–139. [PubMed]

616. Fuenzalida, L.; Armijo, L.; Zabala, B.; Hernandez, C.; Rioseco, M.L.; Riquelme, C.; Espejo, R.T. Vibrio parahaemolyticus strains isolated during investigation of the summer 2006 seafood related diarrhea outbreaks in two regions of Chile. *Int. J. Food Microbiol.* **2007**, *117*, 270–275. [CrossRef] [PubMed]

617. McLaughlin, J.B.; DePaola, A.; Bopp, C.A.; Martinek, K.A.; Napolilli, N.P.; Allison, C.G.; Murray, S.L.; Thompson, E.C.; Bird, M.M.; Middaugh, J.P. Outbreak of *Vibrio parahaemolyticus* gastroenteritis associated with Alaskan oysters. *N. Engl. J. Med.* **2005**, *353*, 1463–1470. [CrossRef] [PubMed]

618. Meusburger, S.; Reichart, S.; Kapfer, S.; Schableger, K.; Fretz, R.; Allerberger, F. Outbreak of acute gastroenteritis of unknown etiology caused by contaminated drinking water in a rural village in Austria, August 2006. *Wien. Klin. Wochenschr.* **2007**, *119*, 717–721. [CrossRef] [PubMed]

619. Inoue, M.; Nakashima, H.; Ishida, T.; Tsubokura, M. Three outbreaks of yersinia pseudotuberculosis infection. *Zentralbl. Bakteriol. Mikrobiol. Hyg. B* **1988**, *186*, 504–511. [PubMed]

620. Thompson, J.S.; Gravel, M.J. Family outbreak of gastroenteritis due to Yersinia enterocolitica serotype 0:3 from well water. *Can. J. Microbiol.* **1986**, *32*, 700–701. [CrossRef] [PubMed]

621. Kozisek, F.; Jeligova, H.; Dvorakova, A. Waterborne diseases outbreaks in the Czech Republic, 1995–2005. *Epidemiol. Mikrobiol. Imunol.* **2009**, *58*, 124–131. [PubMed]

622. Schuster, C.J.; Ellis, A.G.; Robertson, W.J.; Charron, D.F.; Aramini, J.J.; Marshall, B.J.; Medeiros, D.T. Infectious disease outbreaks related to drinking water in Canada, 1974–2001. *Can. J. Public Health* **2005**, *96*, 254–258. [PubMed]

623. Eregno, F.E.; Tryland, I.; Tjomsland, T.; Myrmel, M.; Robertson, L.; Heistad, A. Quantitative microbial risk assessment combined with hydrodynamic modelling to estimate the public health risk associated with bathing after rainfall events. *Sci. Total Environ.* **2016**, *548–549*, 270–279. [CrossRef] [PubMed]

Review

El Niño Southern Oscillation (ENSO) and Health: An Overview for Climate and Health Researchers

Glenn R. McGregor [1,*] and Kristie Ebi [2]

[1] Department of Geography, Durham University, Durham DH1 3LE, UK
[2] Department of Global Health, Centre for Global Health and Environment, University of Washington, Seattle, WA 98105, USA; krisebi@uw.edu
* Correspondence: glenn.mcgregor@durham.ac.uk

Received: 29 June 2018; Accepted: 16 July 2018; Published: 19 July 2018

Abstract: The El Niño Southern Oscillation (ENSO) is an important mode of climatic variability that exerts a discernible impact on ecosystems and society through alterations in climate patterns. For this reason, ENSO has attracted much interest in the climate and health science community, with many analysts investigating ENSO health links through considering the degree of dependency of the incidence of a range of climate diseases on the occurrence of El Niño events. Because of the mounting interest in the relationship between ENSO as a major mode of climatic variability and health, this paper presents an overview of the basic characteristics of the ENSO phenomenon and its climate impacts, discusses the use of ENSO indices in climate and health research, and outlines the present understanding of ENSO health associations. Also touched upon are ENSO-based seasonal health forecasting and the possible impacts of climate change on ENSO and the implications this holds for future assessments of ENSO health associations. The review concludes that there is still some way to go before a thorough understanding of the association between ENSO and health is achieved, with a need to move beyond analyses undertaken through a purely statistical lens, with due acknowledgement that ENSO is a complex non-canonical phenomenon, and that simple ENSO health associations should not be expected.

Keywords: El Niño Southern Oscillation; ENSO; health; climatic variability; climate-sensitive disease

1. Introduction

Through a cascade of processes that link variability in the ocean-atmosphere system and the surface environment, weather and climate can have a discernible impact on health. Such impacts may be direct, indirect, or diffuse [1], and occur over a range of temporal and spatial scales [2]. There is a burgeoning literature on the assessment of the impacts of climate on health, generally focusing on the health risks of climate variability and climate change.

The climate variability and health literature is generally concerned with establishing the impact on health of variations in weather conditions at intra-seasonal, inter-annual, and inter-decadal time periods. In general, climatic variability is connected with variations in the state of the atmospheric and ocean circulation and land surface properties (e.g., soil moisture) at the intra-seasonal to inter-decadal timescales. Therefore, climate variability and health studies explore relationships between historical climate and health data at a variety of temporal and spatial scales. Climate change-related health studies generally focus on the risks of a systematic change in the statistical properties of climate (e.g., mean and variance) over a prolonged period (e.g., several decades and beyond). Projections of the health risks of climate change use established associations, often derived from quantifications of associations between health outcomes and climate variability, and "force" these associations with projected changes in climate variables to make projections about the possible outcomes arising from anthropogenic climate change.

Assessments of the health risks of climate change rely on both known associations between health and climate variability, and on projections of how the magnitude and pattern of risks could change with additional climate change. Therefore, it is important to understand the range of modes of climate variability, generally defined as quasi-periodic variations in ocean and atmospheric circulation patterns that possess an oscillatory behaviour, which might influence health. A large number of modes of climate variability, not all independent of each other, have been identified [3–5], all of which could be considered as potential moderators of intra-seasonal to inter-annual to inter-decadal variability in health outcomes (Table 1).

Table 1. A selection of teleconnection patterns and indices (Source: McGregor [6]).

Arctic Oscillation (AO)/Northern Annual Mode (NAM)	El Niño Southern Oscillation (ENSO)
North Atlantic Oscillation (NAO)	East Atlantic Pattern (EA)
West Pacific Pattern (WP)	East Pacific/North Pacific Pattern (EP/NP)
Pacific/North American Pattern (PNA)	East Atlantic/West Russia Pattern (EA/WR)
Scandinavia Pattern (SCA)	Northern Hemisphere Pattern (TNH)
Polar/Eurasia Pattern (POL)	Pacific Transition Pattern (PT)
Pacific South American Pattern (PSA)	Southern Annular Mode (SAM)/Antarctic Oscillation (AO)
Indian Ocean Dipole (IOD)	South Pacific Wave Pattern (SPW)
Quasi Biennial Oscillation (QBO)	Madden Julian Oscillation (MJO)
Pacific Decadal Oscillation (PDO)	Atlantic Meridional Oscillation (AMO)

A major driver of inter-annual climate variability associated with adverse health outcomes is the El Niño Southern Oscillation (ENSO) phenomenon [7]. ENSO can account for a considerable proportion of climate variance across a range of geographical scales [8,9] and thus impact health sensitive environmental conditions, including land- and ocean-based temperature and precipitation extremes, ecosystem health, drought and riverine and coastal flooding. Because strong ENSO-related climate anomalies have discernible impacts on health in some regions and because ENSO generally accounts for the largest proportion of the inter-annual variation in climate [8], especially in regions where health systems are less resilient to "climate shocks", the purpose of this paper is to present an overview of the basic characteristics of the ENSO phenomenon and its climate impacts, discuss the use of ENSO indices in climate and health research and outline our present understanding of ENSO health associations. ENSO-based seasonal health forecasting, and the possible impacts of climate change on ENSO, and the implications this holds for future assessments of ENSO health associations, will also be briefly discussed.

2. The ENSO Phenomenon

The ENSO phenomenon refers to the variations in atmospheric and ocean conditions, or in the climate conditions, arising from variations in sea surface temperatures and atmospheric pressure across the tropical Pacific Ocean. ENSO is comprised of two major components that reflect its complex coupled nature, the El Niño or ocean, and the Southern Oscillation or atmospheric component. Human society has chronicled the impacts of El Niño for far longer than its atmospheric counterpart. Peruvian fishermen in the 1500s understood well the impact on fisheries of unusually warm waters that occasionally occurred off the coast of Peru around Christmas time. Because El Niño events are associated with anomalously high sea surface temperatures, they are also referred to as "warm events". The "cool event" counterpart carries the name of La Niña.

H. Hidebrandsson, working in the late 1890s, is often credited with unearthing the rudiments of what we now know as the Southern Oscillation or the atmospheric pressure "seesaw" (barometric seesaw) between the eastern and western Pacific [10]. Building on this work and that of Norman and Lockyer in 1902, and extensive research of his own and that of his collaborator E W Bliss, Gilbert Walker in 1928 named and presented the first coherent ideas about the Southern Oscillation (SO) and extensively described the implications of the SO for inter-annual climate variability across the tropics, including how a Southern Oscillation Index (SOI) could be applied to climate forecasting a season ahead [10]. While Troup [11] reconfirmed and refined many of the earlier long distance associations (now referred to as teleconnections) discovered by Walker, Bjerknes [12] conceptualised the association between El Niño and the Southern Oscillation as an outcome of air-sea or ocean-atmosphere interaction that led to the coining of the term ENSO.

El Niño and La Niña are part of the ENSO cycle that lasts from 12–18 months, over periods of 2–7 years, associated with alterations of the SO; although some El Niño and La Niña events can last beyond 24 months. The ENSO cycle refers to the alteration of climate fields associated with the development, peak, and decay of sea surface temperature anomalies in the eastern and central Pacific along with alterations to the atmospheric circulation and weather patterns across vast areas. El Niño (La Niña) or warm (cool) event conditions first begin to manifest as positive (negative) sea surface temperature (SST) anomalies in the central and eastern Pacific around July. These continue to develop as the ENSO cycle progresses, reaching a peak in the Northern Hemisphere around January to February of the following year, trailed by a decay or lessening of SST anomalies in the subsequent months of March to July/August, and cessation of the El Niño (La Niña) event by the end of summer. The swing between El Niño and La Niña phases is not immediate and successive. Rather, El Niño and La Niña events can be punctuated by "neutral" conditions when SST conditions in the eastern and central Pacific are in and around "normal".

ENSO events can be viewed as a self-sustained and naturally oscillatory mode of the coupled ocean-atmosphere system, or a stable mode triggered by stochastic forcing [13], and positive ocean-atmosphere feedback processes, with negative feedbacks required to terminate events [12]. No two ENSO events are alike. From exhaustive analyses of ocean and atmosphere climate fields, two broad types of ENSO have emerged: Eastern Pacific (EP) and Central Pacific (CP). The two types were identified in relation to where maximum SST anomalies tend to occur, with the CP type also referred to by a variety of other names particularly the widely used "El Niño Modoki" [13]. While in the context of ENSO and health the different types or flavours of ENSO events might appear inconsequential, the nuanced differences in their climate impacts may hold implications for the temporal and spatial dynamics of ENSO-related health responses.

A further characteristic of ENSO that holds possible implications for climate and health associations is the multi-decadal changes observed for ENSO's amplitude, period, propagation characteristics, asymmetry, onset, and predictability [13,14]. For example, the variance of the 2–7 year periodicity of ENSO was relatively high during the periods 1875–1920 and 1960–1990, but relatively low from 1920 to 1960 [14]. A clear shift in the amplitude of SST anomalies in the EP occurred in and around the mid-1970s. Such a shift and the variation in the variance of the 2–7 year periodicity appears to be related to the background climate state of the Pacific Ocean, or the phase of the Pacific Decadal Oscillation (PDO). As the Pacific Ocean transitions from a cool (warm) phase with lower (higher) than normal SSTs to a warm (cool) phase with positive (negative) SST anomalies over a period of 3–4 decades, ENSO characteristics and their link to climate impacts are affected [9,15–19]. The implication is that ENSO-related health impacts may be non-stationary at the multi-decadal scale. That there are non-symmetric relationships between ENSO and the Indian Ocean Dipole (IOD), another form of ocean climate variability [20], also raises the question as to whether the strength and direction of ENSO health links in the broad region of IOD influence might be IOD phase dependent.

3. ENSO (Teleconnection) Indices

A teleconnection index is used to describe the temporal behavior of a particular mode of climate variability such as ENSO. Essentially statistical constructs, teleconnection indices are presented in the form of a single number for the temporal scale of interest (e.g., monthly, annual) with the assumption that a specific index captures the range of often complex, ocean and/or atmospheric process interactions that give rise to what is a multifaceted mode of climatic variability. While each teleconnection index has a commonly accepted acronym (Table 1), there may be various versions of a particular index because different methods, data sets, atmosphere and ocean variables, criteria, and sampling periods might have been used in their derivation [6].

A plethora of ENSO indices have been developed to measure, monitor, and summarise ENSO status. These can be broadly divided into atmospheric, oceanic and blended indices. Some of the more common ENSO atmospheric and oceanic indices, reported monthly by NOAA in its Climate Diagnostics Bulletin, are presented in Table 2.

Table 2. El Niño Southern Oscillation (ENSO) indices commonly reported by NOAA on a monthly basis. Modified from the Climate Diagnostics Bulletin http://www.cpc.ncep.noaa.gov/products/CDB.

Index	Index Name	Variable	Form of Index Value	Region (in Degrees)
Type				
Atmospheric				
	Southern Oscillation Index (SOI)	Pressure	Standardised pressure difference between Tahiti and Darwin	Uses the single location of Darwin and Tahiti
	850-hPa Trade Wind Index [1]	Wind direction and speed, Southwest Pacific	Standardised by mean annual standard deviation for reference period [2]	5 N–5 S 135 E–180
	850-hPa Trade Wind Index [1]	Wind direction and speed, South Central Pacific	Standardised by mean annual standard deviation for reference period	5 N–5 S 175 W–140 W
	850-hPa Trade Wind Index [1]	Wind direction and speed, Southeast Pacific	Standardised by mean annual standard deviation for reference period	5 N–5 S 135 W–120 W
	200-hPa Zonal Wind Index [3]	Wind direction and speed, Central to Eastern Equatorial Pacific	Standardised by mean annual standard deviation for reference period	5 N–5 S 165 W–110 W
	Outgoing Longwave Radiation Index (OLR) [4]	Longwave radiation in W/m^2	Standardised by mean annual standard deviation for reference period	5 N–5 S 160 E–160 W
Oceanic				
	NIÑO 1 + 2	Pacific sea surface temperature (SST) in °C	Departures from 1981–2010 mean	0–10 S 90 W–80 W
	NIÑO 3	Pacific SST in °C	Departures from 1981–2010 mean	5 N–5 S 150 W–90 W
	NIÑO 3.4	Pacific SST in °C	Departures from 1981–2010 mean	5 N–5 S 170 W–120 W
	NIÑO 4	Pacific SST in °C	Departures from 1981–2010 mean	5 N–5 S 160 E–150 W
	North Atlantic	Atlantic SST in °C	Departures from 1981–2010 mean	5 N–20 N 60 W–30 W
	South Atlantic	Atlantic SST in °C	Departures from 1981–2010 mean	0–20 S 30 W–10 E
	Global tropics	Global Tropics SST in °C	Departures from 1981–2010 mean	10 N–10 S 0 W–360 W

[1] Positive (negative) values of 850-hPa zonal wind indices imply easterly (westerly) anomalies. [2] Currently NOAA uses 1981–2010 as the base period. [3] Positive (negative) values of 200-hPa zonal wind index imply westerly (easterly) anomalies. [4] Positive (negative) values indicate large amounts of outgoing longwave radiation.

3.1. Atmospheric Indices

Of the indices presented in Table 2, the SOI has the longest history [21,22]. It is composed of the standardised pressure difference between Tahiti and Darwin. These locations are sometimes referred as "centres of action" because they are in the general region at either end of the barometric seesaw that straddles the Pacific and so demonstrate the maximum climate station-based variance in pressure during an ENSO event. The SOI swings between positive and negative values with a phase shift from La Niña to El Niño such that when the pressure is above (below) average in Darwin and below (above) average in Tahiti, as found during an El Niño (La Niña) event, the SOI is negative (positive) (Figure 1).

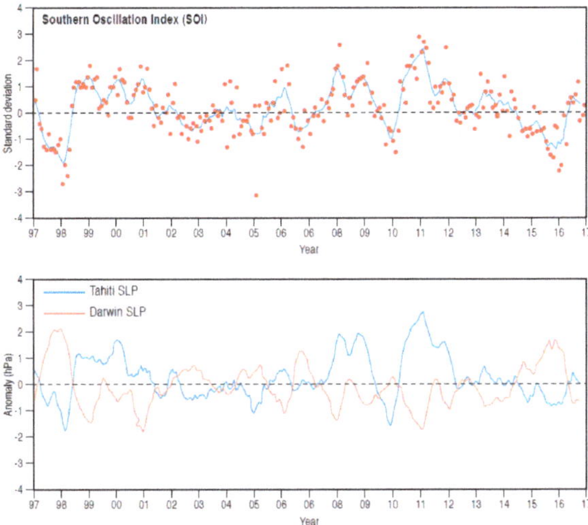

Figure 1. Time series of Southern Oscillation Index (SOI) and Tahiti and Darwin pressure, 1997–2016. Sourced and redrawn from the Climate Diagnostics Bulletin (http://www.cpc.ncep.noaa.gov/products/CDB).

The SOI is calculated in two stages. First, sea level pressure is standardised in relation to a set reference period, separately for Darwin and Tahiti. The differences in the standardised values between the two locations are then standardised. The resulting SOI values vary between −2.5 and 2.5, with roughly 66 percent of the values occurring between −1.0 and 1.0 [21]. Although this range of values implies symmetry around a mean of zero, there is a slight asymmetric distribution of SOI values because the strongest El Niño events tend to produce greater negative departures from zero compared to the positive departures for strong La Niña events. The strongest El Niño events are more intense than the strongest La Niña events. Although SOI values can be calculated for daily and weekly timescales, it is best if monthly to seasonal values are used in health impact analyses. This is because short term fluctuations in pressure at the two reference stations can occur due to weather and climate phenomena other than ENSO. The method of averaging over longer time scales therefore facilitates identification of continued periods of positive or negative departure of the SOI that is most likely due to ENSO.

While there are good historic reasons as to why Darwin and Tahiti were selected as the reference locations for the development of the SOI, their location is slightly south of the main equatorial region where ENSO manifests itself. Accordingly, an alternative form of the SOI was developed: the Equatorial SOI (EQ SOI) [21]. This is calculated using re-analysis as opposed to observed atmospheric pressure data, as the standardised anomaly of the difference between the area-average monthly sea level pressure between largely oceanic equatorial regions in the eastern Pacific (80° W–130° W, 5° N–5° S) and Indonesia (90° E–140° E, 5° N–5° S) (Figure 2). Although the EQ SOI may have advantages over the SOI in that it is derived for equatorial slices that more closely map onto ENSO centres of action, the record only extends back to 1949 (historical extent of the re-analysis data); the SOI is available from the late 19th century. Further in relation to the EQ SOI, it is worth mentioning that prior to the satellite era (pre-1979), the re-analyses on which the EQ SOI is based, possess some uncertainties, as in situ observations were sparse, thus compromising the quality of the re-analysis data.

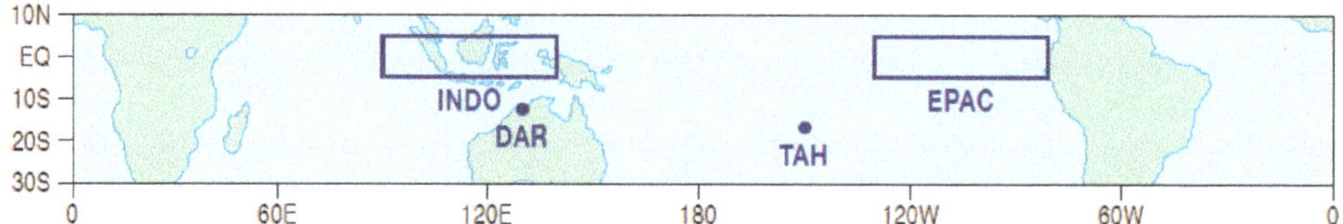

Figure 2. Locations for calculation of SOI and Equatorial SOI. EQ is equator, DAR is Darwin, TAH is Tahiti, EPAC is eastern Pacific and INDO is Indonesia. Sourced and redrawn from Barnston [21] https://www.climate.gov/news-features/blogs/enso/why-are-there-so-many-enso-indexes-instead-just-one.

During ENSO events, the major zones of deep convection that produce thunderstorm-related rainfall move eastward away from their "normal" regions of predominance in the western Pacific. This shift is evident from rainfall observations and from space as changes in cloud patterns captured by satellite images of the global tropics. Because clouds, like all other objects, emit longwave radiation, satellite-based measurements since the late 1970s have been used to construct an Outgoing Longwave Radiation (OLR) index that has proven to be a good indicator of ENSO events (Figure 3). As described by the Stefan-Boltzman Law, the cooler an object, the lower the amount of longwave radiation emitted. Therefore, deep convective storms that reach high into the troposphere and produce substantial rainfall will have very low cloud top temperatures. Accordingly, such clouds will emit less OLR than their warmer and shallower counterparts, such that low (high) values of outgoing longwave are taken to mean enhanced (suppressed) thunderstorm activity and anomalously high (low) rainfall. Although the OLR index has yet to be used to explore ENSO health links, it may offer some potential for exploring rainfall-sensitive health outcomes, especially for the geographic region for which the index is derived (Table 2), because it is a proxy of thunderstorm/rainfall activity.

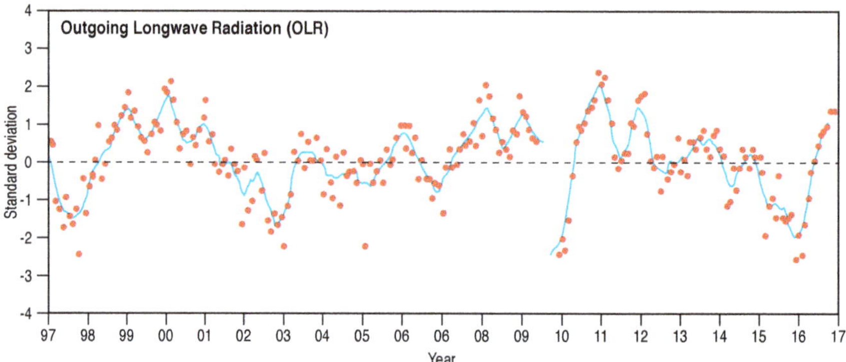

Figure 3. Time series of outgoing longwave radiation (see Table 2), 1997–2016, showing clearly the below average OLR for the 1997–1998 and 2015–2016 El Niño events. Sourced and redrawn from the Climate Diagnostics Bulletin http://www.cpc.ncep.noaa.gov/products/CDB.

Multiple lower and upper atmosphere wind indices have been developed for monitoring the flow of air in the lower and upper branches of the Pacific Walker Circulation [21]. The three 850 hPa indices (Table 2) represent the strength of the easterly trade winds in ENSO critical regions along the equator (Figure 4). The trade winds form the lower east to west branch of the Walker Circulation. The 200 hPa zonal wind index (Table 2) provides a measure of wind strength in the upper west to east branch of the Walker Circulation (Figure 4). At the western and eastern extremities of the Walker Circulation, air ascends and descends, respectively, thus forming the ascending and descending branches of the along the equator circulation. Positive (negative) values of the 850 hPa wind indices indicate strong (weak) trade winds. The weakened trade winds of the 1997–1998 and 2015–2016 ENSO events are clearly visible in time series of this index for the three reference regions, especially for the 1997–1998 event (Figure 4). How inter-annual variations in the trade wind strength might play out in terms of health impacts, especially in countries directly affected by these anomalies across the wider Pacific Basin, remains to be explored.

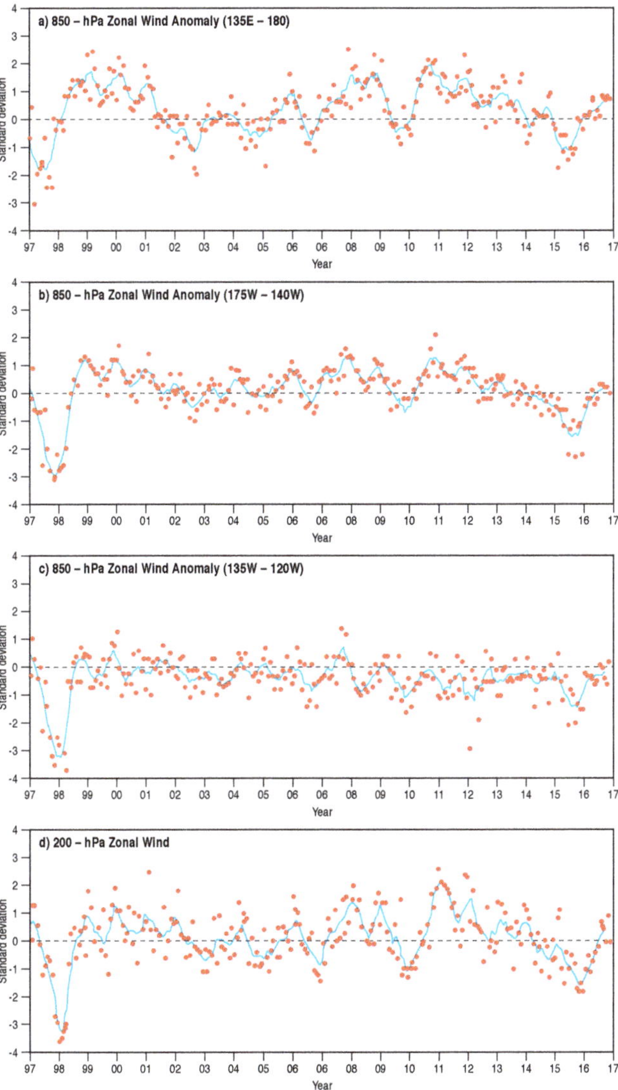

Figure 4. Time series of ENSO wind indices for various regions (see Table 2) 1997–2016. The weakening of the zonal winds is especially apparent for the 1997–1998 El Niño event. Sourced and redrawn from the Climate Diagnostics Bulletin http://www.cpc.ncep.noaa.gov/products/CDB.

3.2. Oceanic Indices

Because ENSO is very much a phenomenon associated with ocean-atmosphere interaction, a logical parameter for monitoring its behaviour is sea surface temperature (SST) as initially recognised by Bjerknes [12] and later fully explored by Rasmussen and Carpenter [23]. The three main oceanic indices used are based on SST anomalies for a number of ocean regions distributed along the equator. These are named after numbered shipping routes as Niño 1, 2, 3, and 4, because a vast array of ships crossing the Pacific for operational reasons recorded SST over a number of decades (Figure 5). Based on studies of SST variability in relation to ENSO events, Niño 1 and 2 were combined into region Niño 1 + 2; a new region, Niño 3.4 that straddles Niño 3 and 4, is now used (Figure 5).

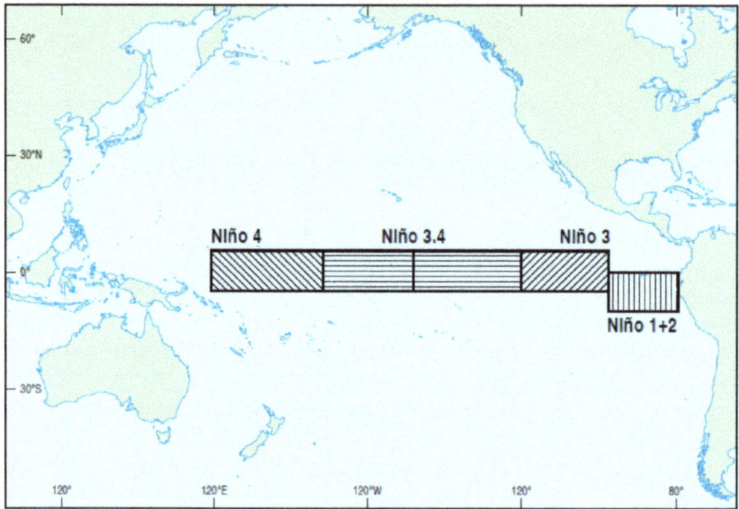

Figure 5. Niño sea surface temperature regions (see Table 2). Sourced and redrawn from Barnston [21] https://www.climate.gov/news-features/blogs/enso/why-are-there-so-many-enso-indexes-instead-just-one.

Niño 1 + 2 is the smallest Niño region. It sits directly off the coast of South America and tends to have the largest variation in SST when compared with the other Niño SST regions (Figure 6). Initially, Niño 3 was favoured as the key region for observing and forecasting El Niño; however, it was realised that in terms of critical ENSO related ocean-atmosphere interactions, an area further to the west, Niño 3.4, had greater diagnostic power [24]. Niño 3.4 anomalies capture the average equatorial SSTs across the Pacific from around the dateline to the South American coast (Figure 6). It is one of two official NOAA ENSO indices used for classifying ENSO events: when the 5-month running mean of Niño 3.4 SST anomalies exceeds +0.4 °C (−0.4 °C) for six months or more, an El Niño (La Niña) is defined to have occurred. Complementing the Niño 3.4 index is the Oceanic Niño Index (ONI), the other official index, and the one used for operational definitions of ENSO events by NOAA. While the ONI uses the same SST region as the Niño 3.4 index, it classifies ENSO events differently. A 3-month running mean is used, with "fully-fledged" El Niño (La Niña) events defined when SST anomalies exceed +0.5 °C (−0.5 °C) for at least five consecutive months. The ONI is also used for defining El Niño (La Niña) onset. When the Niño 3.4 anomaly exceeds +0.5 °C (−0.5 °C) for a 3-month period El Niño (La Niña) onset is declared. Niño 4 covers the central equatorial Pacific. It displays the least SST variance of all the Niño regions (Figure 6) and is infrequently used in ENSO analyses.

In addition to the Niño regionally-based oceanic indices, the Trans-Niño Index (TNI) was developed by Trenberth and Stepaniak [24], who suggest that the TNI be used in tandem with the

Niño 3.4 index. The TNI is defined as the difference in the standardised SST anomalies between Niño 1 + 2 and Niño 4 regions. The physical justification is that it captures the SST gradient between the central and eastern Pacific, and thus may be useful for identifying El Niño Modoki events, as they arise when the central to eastern Pacific SST gradient is steep, for example with sizeable positive (negative) SST anomalies in Niño 4 (Niño 1 + 2) regions. However, as noted by Hanley et al. [25], the TNI has non-consistent lag correlations with the Niño 3.4 index related to the transition of the Pacific Ocean from a cool to warm PDO phases in the mid-1970s. Accordingly, Hanley et al. [25] do not include the TNI as an index for identifying individual events and comparison of ENSO years.

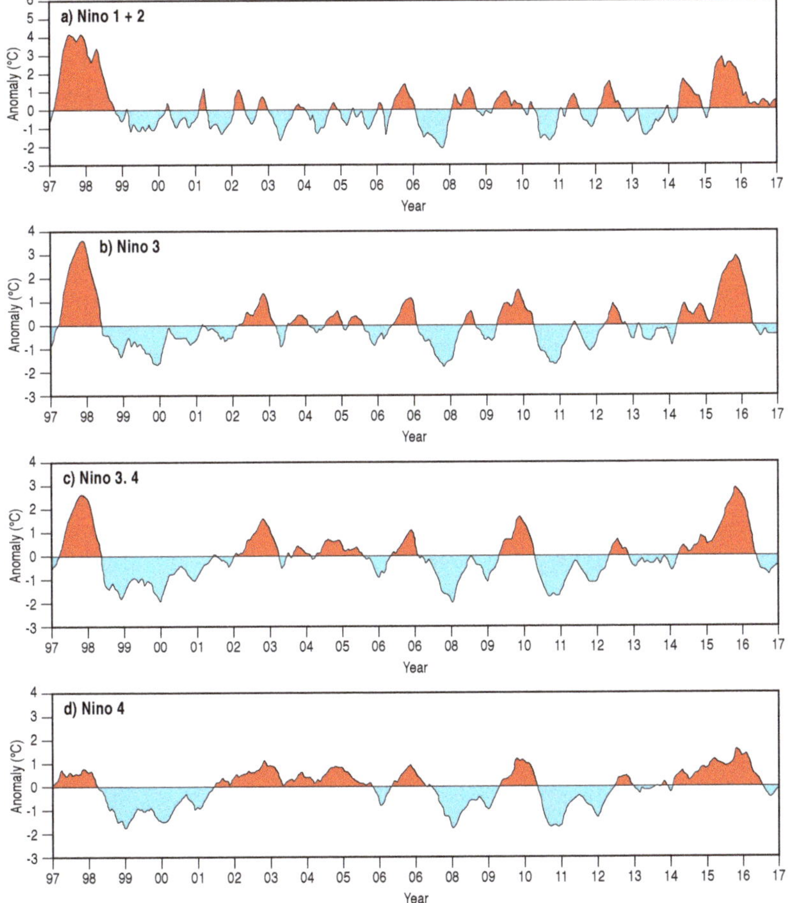

Figure 6. Time series of SST anomalies (°C) for NIÑO regions (see Table 2 and Figure 5) 1997–2016. The strong positive SST anomalies are especially apparent for the 1997–1998 and 2015–2016 El Niño events. Notice also how the SST anomalies generally decrease in magnitude to the west. Sourced and redrawn from the Climate Diagnostics Bulletin http://www.cpc.ncep.noaa.gov/products/CDB.

3.3. Blended Indices

Blended indices are a combination of a number of single variables with the blending technique achieved through a variety of methods. The justification for blended indices is that ENSO is a multivariate phenomenon [26], thus any index should be comprised of more than one variable.

Most commonly, blended indices are used to diagnose ENSO events. While researchers recognised the physical complexity of ENSO events and the challenges associated with diagnosing them [26, 27] it wasn't until the early 21st century that attempts to produce blended indices first appeared in the literature. For example, a Bivariate ENSO Timeseries (BEST) was produced by Smith and Sardeshmukh [28] by combining the SOI and Niño 3.4 SST, while Allan [29] blended SST and SLP fields using empirical orthogonal function (EOF) analysis. More recent attempts to produce new blended indices are largely based on different SST variables and apply various forms of EOF analysis [30–37]. While such new indices proved useful for diagnosing various aspects of ENSO, they have not gained particularly strong traction for ENSO monitoring.

Perhaps the most widely used blended index for ENSO monitoring and impact studies is the Multivariate ENSO Index (MEI) of Wolter and Timlin [38]. Unlike other blended indices, the MEI amalgamates more than one ocean and atmospheric variable, being based on six observed variables recorded over the tropical Pacific: sea level pressure, zonal and meridional surface wind components, SST, surface air temperature and total cloudiness fraction of the sky. The MEI is actually the first unrotated Principal Component (PC) of the aforementioned variables. It is calculated separately for 12 sliding bi-monthly seasons (December/January, January/February, … , November/December) with all bi-monthly values standardised for each season based on a 1950–1993 reference period. More recently Wolter and Timlin [39] extended the MEI back to 1871 (MEI.etx) using a reduced set of variables because of possible errors associated with wind variables prior to 1950.

The extended Multivariate ENSO Index (MEI.ext) is based on reconstructed values of SST and SLP, and is calculated as the first principal component of SST and SLP fields (similar to the MEI). As noted by Wolter and Timlin [39], the MEI.ext confirms many of the postulated ENSO characteristics evident from analyses of the original but shorter MEI time series, including ENSO activity subsided in the early to mid-20th century, and ENSO was about as predominant a century ago as it is currently. Further, Wolter and Timlin [39] were able to detect strong associations between ENSO amplitude and duration plus amplitude and periodicity using the MEI.ext.

Although not strictly blended in nature, a number of alternative ENSO indices based on variables other than the more traditional ones (e.g., SLP, SST, wind, and OLR) have emerged recently, including an ozone-based ENSO index [40], an atmospheric electrical index [41], and an ENSO salinity index [42].

4. ENSO Indices for Climate and Health Analyses

Faced with a range of ENSO indices, questions that are likely to arise when planning analyses of ENSO health associations are "which index will be optimal for exploring ENSO-related health impacts?" and "are there different ways of defining ENSO events?"

Currently there is no common consensus in the climate science community as to which ENSO index best describes ENSO phases. This appears to contrast with the climate and health science community in which there appears to be a substantial amount of blind faith applied to the use of ENSO indices for exploring possible ENSO driven health variations at a range of temporal and spatial scales. In many ways, the choice of an ENSO index for health analyses may depend on geographical location, its relation to a health-sensitive climate variable or even disease outcome. For example, for regions in close proximity to the Niño oceanic regions, the SST-based indices may be appropriate, especially for exploring rainfall, air and sea temperature-related health outcomes. Similarly, the OLR index may be useful for exploring rainfall-related health issues in the central Pacific. ENSO-related wind indices, because they represent the surface trade wind strength, are more than likely to be only useful in the trade wind regions of the Pacific Basin. For locations distant from the Pacific Basin, such as southern and eastern Asia, one of the pressure-based indices might be more suitable, as these represent variations in atmospheric circulation conditions over a larger geographical range.

While some researchers might be tempted to use one of the more recently developed blended or multivariate ENSO indices, it is worth bearing in mind that these are complex indices made up of multiple interacting variables. Accordingly, they may not be appropriate if the purpose of an analysis was to uncover a direct link between a specific climate attribute such as temperature and a disease outcome. Further, blended indices have not yet been widely adopted by ENSO forecasting centres. This is most likely because of the prediction error associated with the individual input variables such that the cumulative error for a predicted value of a blended index could be large when compared to a single variable based index. Notwithstanding this, blended indices such as that of Wolter and Timlin [38] have been applied on a number of occasions in ENSO-health studies (see Section 6).

In selecting ENSO indices, researchers also need to be aware of the producing agency because of the way ENSO phases and events are identified can vary between agencies. As noted above, NOAA uses the Niño 3.4 region (5° N–5° S, 170° W–120° W) based on ONI and the persistence of SST anomalies in excess of $+/-0.5$ °C for five months for identifying ENSO phases. In contrast, the Japan Meteorological Agency (JMA) uses a slightly different formulation for the calculation of their ENSO index, which is a 5-month running mean of spatially averaged SST anomalies over the geographical range of 4° S–4° N, 150° W–90° W; this is essentially a latitude-restricted Niño 3 region. Further, the JMA define an ENSO year as October through to the following September. If the JMA index values exceed +0.5 °C or −0.5 °C for six consecutive 5-month periods, including October to December, the ENSO year is declared as either El Niño or La Niña. The Australian Bureau of Meteorology (BoM), like NOAA and the JMA, uses SST anomalies as a basis for the ENSO phase definition but sets a higher threshold. For an El Niño (La Niña) event to be called, SST anomalies in the Niño 3 and Niño 3.4 regions must exceed +0.8 °C (−0.8 °C). Further to the SST criteria, BoM specifies much weaker (stronger) trade winds over the western or central equatorial Pacific for the previous 3–4 months, as well as a SOI value of −7 (+7) to be necessary for an El Niño (La Niña) event to be specified. Note that in the case of BoM, the SOI values are quite different from those associated with the NOAA SOI index because the BoM uses the Troup SOI, the standardised anomaly of the mean sea level pressure difference between Tahiti and Darwin. The BoM calculation uses the period 1933 to 1992 as the climatology; this contrasts with the NOAA and JMA. Once the Tromp SOI is calculated, it is multiplied by 10. Using this convention, the BoM SOI takes on values in the range of −35 (strong El Niño) to about +35 (strong La Niña).

NOAA, JMA, and BoM use SST anomalies in their definitions of the El Niño and La Niña phases of ENSO. In addition to the slight differences in the criteria used for defining events (e.g., SST anomaly, anomaly period, and region), a further source of difference between the oceanic indices are the SST data sets employed for constructing the SST anomalies. For example, the JMA uses the Centennial In Situ Observation-Based Estimates (COBE)-SST data set for ENSO monitoring with sliding climatological values based on the most recent 30-year period as described by Ishii et al. [43] and JMA [44]. In contrast, NOAA and BoM use the Extended Reconstructed Sea Surface Temperature, Version 5 (ERSSTv5) data set with anomalies based on centred 30-year periods updated every five years as described by Huang et al. [45]. Although not described here, the United Kingdom's Met Office's Hadley Centre applies yet another SST data set for deriving historical SST-based ENSO measures, the HadISST data set [46]. Important to note in the consideration of possible pre-1950 ENSO health associations is that for this period, because of observational uncertainties, the ERSSTv5 and HadISST, from which ENSO indices are derived, demonstrate significant differences [47]. Given this, researchers need to be aware of the differences in SST data sets for deriving ENSO indices in terms of the SST observations drawn upon, statistical and data assimilation methods applied, and spatial resolution of the final SST products [48] because these data set properties may affect the degree to which meaningful associations between ENSO and health outcomes can be quantified.

Hanley et al. [25] provided a useful comparison of ENSO indices in terms of their ability to describe ENSO events. They found that El Niño only engenders a weak SST response in the Niño 4 region, whereas La Niña produces quite a strong SST signal in the Niño 1 + 2 region (see Figure 6). They also concluded, based on an analysis of the sensitivity of a range of ENSO indices relative to each other, that the choice of an index for analysing ENSO related risks is somewhat dependent on the ENSO phase. For example, in the case of La Niña events, the JMA ENSO Index was more sensitive than other atmospheric and oceanic indices. In contrast for El Niño events, the SOI, Niño 3.4, and Niño 4 indices are almost equally sensitive but more sensitive than the JMA, Niño 1 + 2, and Niño 3 indices.

ENSO indices have also been used to classify the strength of ENSO phases. Such classifications may be of interest to climate and health researchers because El Niño or La Niña strength may be an indicator of the potential scale of climate-related health risks, all other variables, such as socio-economic conditions or vulnerability, being equal. A recent classification of ENSO phase strength for the period 1950–2016 was produced by Santoso et al. [47] using Niño 3.4 SST anomalies averaged across four SST reanalysis products (ERSSTv4, ERSSTv5, HadISST, and COBE) over the months of November-December-January (NDJ) and December–January–February (DJF). Because the classification is based on SST anomalies for NDJ and DJF, it identifies the year of the development phase of an ENSO event. For strong (weak) events, the averaged Niño 3.4 anomaly must exceed 1 (be between 0.5 and 1) standard deviation. A neutral phase is deemed to be associated with a standard deviation of less than 0.5. Strong and weak El Niño and La Niña years are listed in Table 3. The classification, while identifying the often cited extreme 1972/1973, 1982/1983, 1997/1998, and 2015/2016 El Niño events as extremes, also highlights other strong El Niño events that have received little attention in the literature. Recalling that two broad types of El Niño events occur, Santoso et al. [47] used the Niño4 index DJF average to identify Central Pacific El Niño events; the Niño 4 average must be greater than 0.5 °C and greater than Niño3 to be classified as a CP event (Table 3). Usefully, Santoso et al. [47] also analyse the sensitivity of El Niño and La Niña strength classification to varying criteria; this has implications for general qualitative statements about ENSO strength and health associations.

Table 3. Years in which strong and weak ENSO phases developed (Source: Santoso et al. [48]).

Strong El Niño	1957, 1965, 1972, 1982, 1991, 1997, 2009, 2015
Weak El Niño	1963, 1968, 1976, 1977, 1987, 1994, 2002, 2006
Strong La Niña	1973, 1975, 1988, 1998, 1999, 2007, 2010
Weak La Niña	1950, 1954, 1955, 1964, 1970, 1971, 1984, 1995, 2000, 2005, 2008, 2011
Central Pacific (CP) El Niño	1958, 1968, 1977, 1979, 1987, 1990, 1994, 2002, 2004, 2006, 2009, 2014

Perhaps even more apposite when considering the application of ENSO indices to the analysis of ENSO-health associations would be the reflection on the conceptual links between ENSO and health-sensitive climate fields given that the ENSO signature may vary considerably with season and location.

5. ENSO and Health-Sensitive Climate Impacts

In most conceptualisations of climate and health links, the variables of rainfall and temperature dominate as climate drivers of hypothesised and actual health outcomes. Further, an often unstated assumption in many analyses of the relationship between ENSO and disease is that ENSO "signals" will be found in disease incidence time series. While this might be self-evident, the way in which this axiom is applied is often naïve. This is because ENSO forcing of adverse health outcomes is usually explored without prior investigation of the extent to which a disease-relevant climate variable, such as rainfall or temperature, is ENSO sensitive for a specific location, region, or time period.

There is no doubt that ENSO has a marked impact on climate fields, with this impact being geographically and seasonally dependent. Given this, graphics of ENSO climate-related impacts (Figures 7 and 8) should be useful indicators of where direct ENSO/climate-driven (rainfall/temperature) variations in disease might occur. In effect, such canonical patterns, as appear in El Niño and La Niña composites (Figures 7 and 8), should assist with identifying potential ENSO -health "hotspots". However, having said that, an essential in the approach to any ENSO-health study informed by canonical patterns of health-sensitive climate impacts is the recognition that ENSO composites of rainfall and temperature patterns are only averages (the climatology) and as such mask much El Niño/La Niña inter-event variability in climate impacts. For example, and beyond the broad central Pacific (CP–"Modoki") and eastern Pacific (EP) ENSO types, Johnson [48] identified nine different "flavours" of ENSO; a distinct climate outcome is associated with each one. Paek et al. [49] also highlight that no two ENSO events are the same, and provide a useful analysis of the differences of the strong 1997–1998 and 2015–2016 El Niño events. It is perhaps no surprise that some ENSO-health studies do not find consistent temporal or spatial ENSO-health associations, as the "strength" of climate forcing may vary from ENSO event to event, both temporally and geographically. Although there have been no systematic studies of the way in which different flavours of ENSO might manifest in variable regional or local ENSO-health associations, the contrasting rainfall fields for El Niño and El Niño-Modoki events hint at potential temporal and spatial inconsistencies of ENSO-health associations (Figure 9). For example, in El Niño-Modoki events not only is the degree of departure of rainfall from the long-term mean weak, but the spatial pattern of both positive and negative rainfall anomalies is fragmented and somewhat different, and for some regions opposite (e.g., equatorial South America, equatorial eastern Pacific) when compared to El Niño events (Figure 9). The implications for ENSO-health studies of such contrasting climate responses for different ENSO types is clear, especially if an ENSO index that does not discriminate between ENSO types is applied bluntly to the analysis of disease incidence time series.

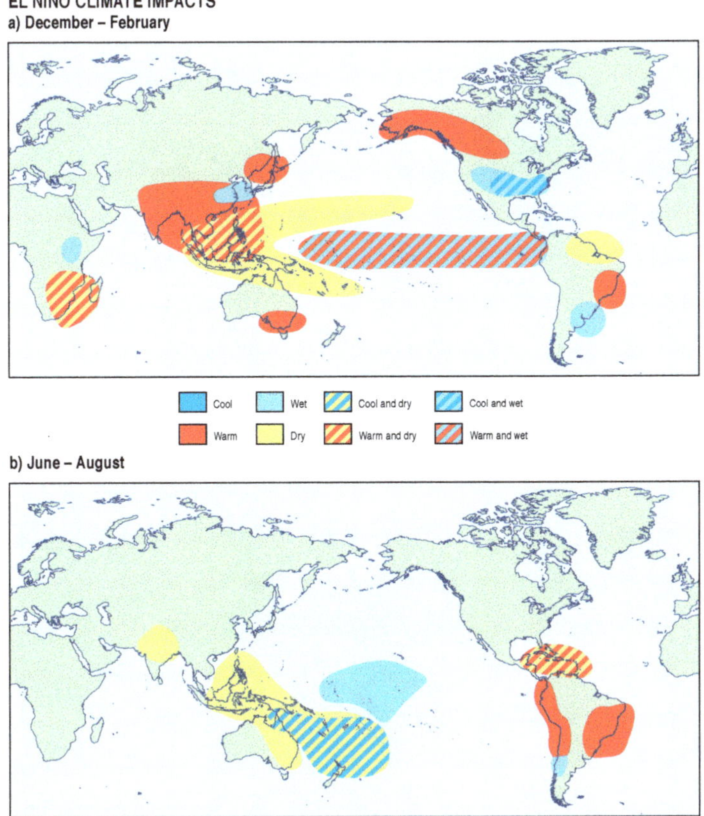

Figure 7. Canonical climate impact patterns of El Niño for (**a**) December–February and (**b**) June–August. Sourced and redrawn from the Climate Predicition Centre http://www.cpc.ncep.noaa.gov/products/ analysis_monitoring/ensostuff/ensofaq.shtml#GLOBALimpacts.

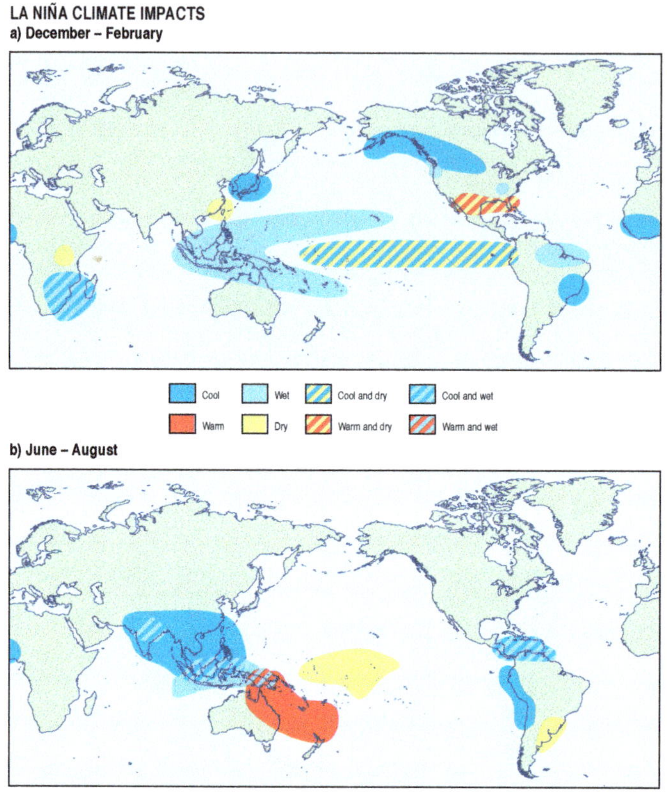

Figure 8. Canonical climate impact patterns of La Niña for (**a**) December–February and (**b**) June–August. Sourced and redrawn from the Climate Predicition Centre http://www.cpc.ncep.noaa.gov/products/ analysis_monitoring/ensostuff/ensofaq.shtml#GLOBALimpacts.

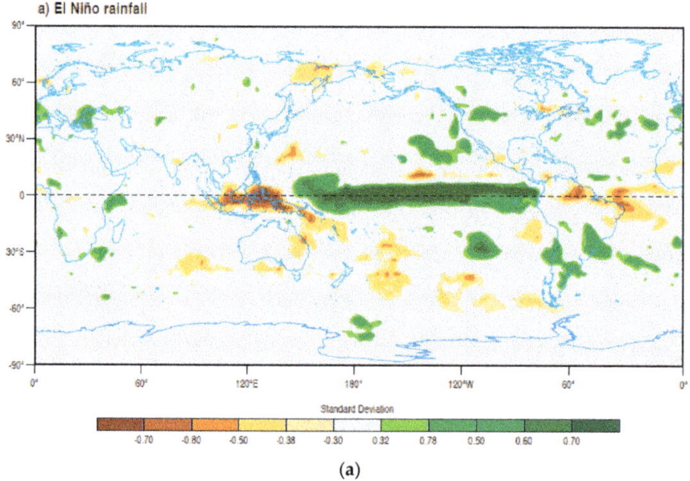

(a)

Figure 9. *Cont.*

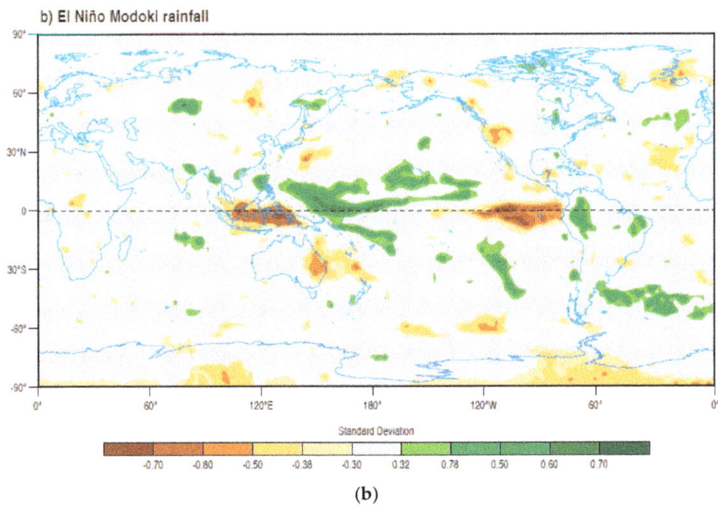

(b)

Figure 9. Rainfall anomalies for (**a**) El Niño and (**b**) El Niño Modoki events. Sourced and redrawn from Japan Agency for Marine, Earth Science and Technology. http://www.jamstec.go.jp/frcgc/research/d1/iod/enmodoki_home_s.html.en.

6. ENSO and Health Impacts

The World Health Organisation posited a number of potential ENSO, or more specifically, El Niño-related health impacts (Figure 10), based either on known health outcomes arising from past ENSO events or conceptual relationships of climate and health, given that ENSO produces discernible variations in health sensitive climate variables. Kovats et al. [7] provided a useful overview of the impact of El Niño on infectious diseases and recommendations related to the assessment and reporting of interactions between ENSO and health. Amongst the potentially ENSO-sensitive infectious diseases reviewed were malaria, dengue, and diarrhoea. These make significant contributions to the burden of climate-sensitive disease. For example in the case of malaria and dengue, the per capita mortality rate is almost 300 times greater in developing nations than in developed regions [50], with many of the affected nations lying in regions impacted by ENSO events. Accordingly, there is a growing interest in establishing the veracity of ENSO-malaria and -dengue associations based on well-known climate vector relationships such as the broad dependence of the distribution of insect vectors on temperature, humidity and rainfall patterns, and at the insect scale, the modulating effect that climate variables have on metabolic activity, egg production, and feeding behaviour [51].

Diarrhoea is also an important climate-sensitive disease, because many cases can be attributed to the lack of access to clean drinking water as a result of either drought, flooding, or temperature related bacterial infections in food and water. Diarrhoea is the second leading cause of death in children under five years old; globally, there are in excess of one billion cases of childhood diarrhoeal disease every year resulting in a high death total amongst children under five years old [52].

Because of the gravity of these diseases, and the potential changes in their incidence during ENSO events, we update Kovats et al. [7] with studies published since 2003. In conducting the review of the ENSO malaria/dengue/diarrhoea literature, terms such as ENSO, El Niño, La Niña, and Southern Oscillation Index (SOI), linked with the three diseases, were used to search the Web of Science for the period 2003–2018. Note the search terms were limited to the title. Further, the search term "climate" was not used because this generated a large number of climate change-related studies with little or no reference to ENSO-driven variations in the three diseases. Further, the literature reported on here was restricted, where possible, to that which considered ENSO health links beyond a single El Niño or La Niña event.

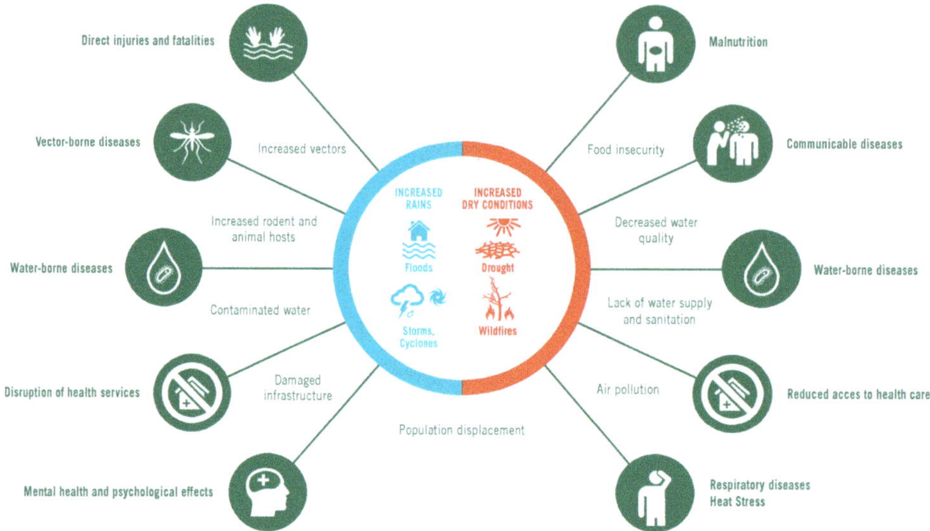

Figure 10. Range of El Niño related health impacts. Sourced and redrawn from the World Health Organisation. http://www.who.int/hac/crises/elnino/who_el_nino_and_health_global_report_21jan2016.pdf.

6.1. Malaria

Because a disproportionately high global malaria burden occurs in the African region (90 percent of global malaria cases and 91 percent of malaria deaths in 2016), and because ENSO influences temperature and precipitation patterns in the continent, there has been interest in the influence of ENSO-related climatic variability on malaria incidence, particularly assessing malaria predictability. For northwest Tanzania, where there are two malaria seasons related to early and late rains, Jones et al. [53] attribute the positive associations identified between rainfall, temperature, and malaria to the influence of El Niño, noting that the 1998 epidemic was associated with El Niño-related heavy rains. For Ethiopia, Bouma et al. [54] demonstrated how El Niño-related above-normal SSTs in the Pacific, via an indirect link to anomalously high SSTs in the western Indian Ocean off the coast of Ethiopia, and thus above-normal winter and spring land surface temperatures in the highlands, are associated with an increased risk of malaria in the subsequent main malaria season. For five countries in Southern Africa, Mabaso et al. [55] used the SOI to assess ENSO malaria associations for the period 1988 to 1999, finding that below (above) normal incidence of malaria corresponded with a negative (positive) SOI; El Niño (La Niña) suppresses (enhances) the chances of malaria incidence via anomalously dry (wet) conditions. Further, there was evidence of possible Indian Ocean-based climate influences on malaria incidence as well as non-climatic factors related to malaria control efforts and response capacity, producing possible non-stationary ENSO malaria associations.

Although Dev [56] reported no association between annualised malaria incidence and annual climate statistics in northeast India, this would be expected given that annualised climate and malaria data will mask seasonally important malaria variations and thus associations with climate variables. Apart from the title, there was no mention of El Niño in the body of the paper, which is symptomatic of the opportunism displayed in some analyses that purport to report on ENSO malaria associations. Zubair et al. [57] reported an association between ENSO phases and malaria for Sri Lanka for the period 1878–2000 that changed over time. Malaria epidemics were associated with El Niño phases over the period of 1880 to 1927. From 1930 to 1980, epidemics had a stronger association with La Niña phases than with El Niño. The authors cite an epochal change in the El Niño-rainfall relationship in Sri Lanka around the 1930's as the likely cause of the shift in the malaria relationship, noting a swing back to the type of association found for the period 1880 to1927, post 1980. This study, like that of Mabaso et al. [55] provides some evidence for the non-stationarity of ENSO phase malaria associations.

For north-eastern Venezuela, Delgado-Petrocelli et al. [58] applied geospatial techniques to the investigation of the influence of ENSO warm, cold, and neutral phases on malaria incidence for the period 1990–2000. They found significant differences in malaria incidence between the three ENSO phases with incidence notably higher during La Niña (cold) phases of moderate intensity. While interesting, this study did not provide an insight into the climate link that ties the ENSO phases to variations in malaria incidence; only a passing reference is made to the possible impacts of El Niño on the ecological system, the state of which is not expanded upon. Using data for French Guiana, Hanf et al. [59] conducted a time series analysis of the association between monthly *Plasmodium* (*P.*) *falciparum* case numbers and ENSO as measured by the Southern Oscillation Index (SOI) for the period 1996 to 2009. While a three-month lagged inverse association was found between the SOI and *P. falciparum* cases (a positive association with El Niño), the SOI only explained four percent of the variation in malaria, with the remaining 96 percent most likely due to non-climatic causes, including population immunity and socio-environmental factors that influence the breeding and ecology of mosquito vectors [59]. As for the climate variables, little insight was provided by the authors, apart from suggesting that ENSO has an impact on the climate that affects the population dynamics of the malaria vectors (mainly *Anopheles darlingi*).

6.2. Dengue

Fuller et al. [60] utilised data on SST anomalies related to ENSO and two vegetation indices to investigate ENSO-related drivers of dengue fever (DF) and dengue haemorrhagic fever (DHF) in Costa Rica from 2003 to 2007. They found that La Niña (ENSO cool phase) conditions were more likely to lead to greater numbers of DF/DHF cases because of La Niña's association with more humid conditions that favour the survival of greater numbers of *Ades. Aegypti*. Using five ENSO indices and two vegetation indices, Fuller et al. [60] were able to explain 64 percent of the variance in DF/DHF cases and reproduce the major epidemic of 2005. They suggest that such associations provide some hope for advanced forecasting of dengue outbreaks.

In a three-country study of the potential relationship between climate and dengue incidence, Johansson et al. [61] reported no systematic association between multi-annual dengue outbreaks and ENSO. In Puerto Rico, on multiyear time scales, temperature, and dengue incidence were only ephemerally associated with ENSO. For Thailand, they found that although ENSO was associated with temperature and precipitation, the association of dengue with precipitation was nonstationary and likely to be spurious. In Mexico, no association between ENSO and dengue was observed. Such findings caused Johannsson et al. [61] to conclude that the evidence for a consistent and reproducible ENSO dengue link was weak. They attribute this to the possible obfuscation of ENSO influences by local small scale climate variations, inadequate data, randomly coincident outbreaks, and other, more substantive non-climatic factors that regulate transmission dynamics.

Using wavelet analysis and the Generalized Additive Model (GAM) approach, Xiao et al. [62] investigated the periodicity of dengue and the dose-response relationship between an ENSO time series, weather variables and dengue incidence in Guangdong Province, China for the period 1988 to 2015. They found an inverted U-shape association for an ocean-based ENSO index-dengue relationship (ENSO index threshold of 0.6 °C), plus evidence for ENSO in the previous 12 months, possibly driving the 1995, 2002, 2006, and 2010 dengue epidemics, and a relatively high dengue incidence during 1997–2001 following the very strong 1997–1998 El Niño event. Although associations between temperature, humidity, and rainfall and dengue were explained in the analysis, an attempt to physically link ENSO-related SST anomalies to local weather variables, and ultimately to dengue incidence was not attempted, making the posited 12-month ENSO dengue lag association difficult to justify on physical climate grounds, notwithstanding the role of possible non-climatic factors.

Similar to Xiao et al. [62], Liyanage et al. [63] also reported ENSO dengue associations. They used the Oceanic Niño Index (ONI) to explore ENSO's impact on dengue incidence over 10 Medical Officer of Health divisions in the Kalutara district of Sri Lanka for the period 2009–2013. The relative risk of dengue increased significantly with rainfall and ONI values in excess of 0.5 °C six months in advance of increases in dengue incidence. This association was likely due to the known lag relationships between ENSO extremes and rainfall, with anomalous high rainfall a characteristic of the inter-monsoon period that follows El Niño-related below normal rainfall. The sensitivity of dengue to ENSO was also apparent in Bangladesh, where Banu et al. [64] suggested the existence of a weak non-linear association between Niño 3.4 temperatures and dengue incidence such that the higher the Niño 3.4 index, the higher dengue incidence at a 4-month lag. The Niño 3.4 to dengue link was explained via the way in which winter El Niño events lead to a general warming of the tropical atmosphere that persists into the next summer. This leads to atmospheric circulation pattern changes over the Indian Ocean region, and greater moisture transport and monsoon rainfall over Bangladesh that extends the breeding season for mosquitoes and their spatial distribution. Banu et al. [64] also noted possible interactive effects between ENSO and the IOD that might influence dengue incidence. In a study on climate and dengue associations in Singapore for the period 2001–2008, Earnest et al. [65] found, using a Poisson model, negative associations between the SOI and dengue, implying that El Niño events engender high dengue incidence. However it is worth noting that weekly SOI values were used in this analysis. From a climatological perspective, this is probably not best practice because SOI values at this time scale are very "noisy" and are more likely to represent weather phenomena other than ENSO.

For Queensland Australia, Hu et al. [66] applied a seasonal auto-regressive integrated moving average model for the period 1993 to 2005 to the analysis of the numbers of notified dengue fever cases and the numbers of postcode areas with dengue fever cases in relation to ENSO as described by the SOI. They found that a decrease in the average SOI (warm phase conditions) during the preceding 3–12 months was significantly associated with an increase in the monthly numbers of postcode areas with dengue fever cases. The SOI dengue links were explained via El Niño's tendency to bring much warmer conditions to Queensland that may enhance dengue fever transmission. This of course assumes that El Niño, which also brings drier, verging on drought, conditions to Queensland, does not affect the number of vectors through the lack of water for suitable breeding sites. That said, the tendency to store water during dry conditions may well provide suitable breeding sites for the dengue vector. In contrast to Johansson et al. [61], Tipayamongkholguln et al. [67], analysing dengue data for Thailand using Poisson regression, found that up to 22% (in eight northern inland mountainous provinces) and 15% (in five southern tropical coastal provinces) of the variation in the monthly incidence of dengue cases were attributable to global ENSO cycles as described by the ENSO multivariate and sea level pressure indices, with the tendency for dengue incidence to increase during El Niño phases. However, the authors noted some geographical heterogeneity in ENSO dengue associations, with not all individual provinces revealing statistically significant associations. In an attempt to explain the ENSO link to dengue epidemics, Tipayamongkholguln et al. [67] pointed to ENSO's warming effect on local temperature such that replication of the dengue virus and the biting behaviour of the mosquito vector *Aedes aegypti* is enhanced. In doing so, and similar to other epidemiological studies of ENSO dengue associations, little attempt is made to discuss the climate linking mechanisms that underpin the statistical relationships described.

Ferreira [68] applied spatial analysis techniques to the exploration of ENSO dengue associations for the countries of the Americas over the period 1995–2004. His results indicated that among the five years with a high number of dengue cases (1997, 1998, 2002, 2001, and 2003), four are associated with El Niño events (see Table 3 above). Furthermore, there appeared to be a spatial trend in the strength of the association between the SOI and dengue occurrence such that warm (cool) or El Niño

(La Niña) phases were associated with high (low) incidence in Mexico, Central America, the northern Caribbean islands, and the extreme north-northwest of South America, while other more poleward regions showed little dengue response to either El Niño or La Niña.

6.3. Diarrhoea

Notwithstanding the complex pathways linking climate anomalies and diarrhoea [69] and the challenges this poses for quantifying the effects of weather and climate on water-associated diseases in general [70–72], diarrhoeal illness is generally sensitive to climate anomalies [73–78] with unusually warm conditions conducive to enhanced pathogen replication and survival rates, while rainfall surpluses may transport faecal matter into water courses with micro-organisms becoming concentrated in water bodies during periods of rainfall deficit. While Demisse and Mengisitie [79] noted that El Niño has an impact on diarrhoea incidence for a number of major geographic regions, many of the cited papers address temperature/rainfall-diarrhoea association as opposed to climate driven variations in diarrhoea moderated by ENSO.

In the Pacific Islands, where diarrhoea is the most significant water-borne disease and ENSO has marked impacts on climate, there is a paucity of evidence for explicit El Niño-diarrhoea associations, although this is implied in a number of studies [80–82]. For West Africa, de Magny et al. [83] suggested associations of diarrhoea with El Niño where ENSO, via the so-called Indian Oscillation and associated variations in large scale rainfall and temperature fields, may well influence cholera dynamics and thus diarrhoea. In a consideration of the spatial dynamics of cholera across the African continent, Moore et al. [84] demonstrated a clear shift in the annual geographic distribution of cholera in El Niño years, with the burden shifting away from Madagascar and parts of southern, Central, and West Africa, to continental East Africa. They found that during El Niño years for East Africa, there were around 50,000 additional cases of cholera in areas with increased rainfall, along with marked increases in some regions with decreased rainfall. Such findings suggest a complex relationship between ENSO, rainfall and cholera, and by implication with diarrhoea incidence. For the Great Lakes Region of Africa, Nkoko et al. [85] applied a multiscale, geographic information system-based approach to assess the association between cholera outbreaks and ENSO. They found that cholera greatly increased during El Niño events, but decreased or remained stable between events because of El Niño-moderated controls on rainfall. For Uganda, Alajo et al. [86] found similar El Niño-moderated impacts on cholera via positive rainfall anomalies.

Building on the earlier work of Pascual et al., [87], who demonstrated associations between cholera and ENSO-related regional temperature anomalies in Bangladesh, Hashizume et al. [88] further investigated climate variability and cholera associations. Based on an analysis of cholera hospitalisations for Dhaka and Matlab in Bangladesh, over the period 1983–2008, they found that the strength of cholera-Indian Ocean Dipole and -ENSO associations changed across time scales, with Dhaka demonstrating little association with ENSO, while in Matlab, the ENSO effect was quite dominant. Based on this finding, Hashizume et al. [88] suggested the existence of non-stationary and possibly non-linear associations between cholera hospitalizations and large-scale modes of climatic variability such as ENSO. This resonates with the conclusions drawn in an earlier study by Rodo et al. [89] for Bangladesh, which found a strong and consistent signature of ENSO in cholera incidence for the period 1980–2001, while for 1893–1920 and 1920–1940, the ENSO-cholera association was weaker and uncorrelated, respectively. They suggested that the switch to more visible ENSO-cholera associations for the period 1980–2001 was related to a change in the background climate state of the Pacific Ocean in the mid-1970s, that resulted in stronger El Niño events and associated health-sensitive climate anomalies. In a purely statistical analysis of the association between ENSO and monthly cholera incidence for an 18-year period, based on power spectral analysis, Ohtomo et al. [90] found that dominant periodic modes of cholera incidence for Dhaka, Bangladesh at $11\cdot0$, $4\cdot8$, $3\cdot5$, $1\cdot6$, and $1\cdot0$ years coincided with similar spectral modes of variability for Pacific Ocean SSTs. Based on this finding, they concluded, without an attempt to put forward a bridging mechanism tying ENSO related climate anomalies to cholera, that cholera incidence in Bangladesh may be influenced by the occurrence of El Niño. Supposedly stimulated by previous work on ENSO-cholera associations for Bangladesh, Martinez et al. [91] developed an El Niño-based forecasting scheme of cholera for Dhaka in an attempt to predict cholera incidence during the 2015–2016 El Niño event.

Peru has received considerable attention in relation to El Niño-diarrhoea associations and cholera, most likely due to the drastic changes in hydroclimate conditions experienced there during El Niño events. For example, Checkley et al. [92] reported that El Niño-related increases in ambient temperature were associated with higher rates of daily admissions for diarrhoeal disease, most likely related to contaminated food and water. Similarly, Bennett et al. [93] found El Niño-diarrhoea associations based on an analysis of daily surveillance data for 367 children in Lima, Peru, for the period 1995 through 1998. Spring diarrhoeal incidence increased by 55% during El Niño compared with before El Niño, pointing to anomalously high temperatures and increased levels of temperature-sensitive pathogens in food and water as the explanation for El Niño-temperature-diarrhoea associations. These findings echo those of Lama et al. [94] who reported associations between El Niño-related elevated air temperatures, cholera, and acute diarrhoea in adults in Lima, Peru for the period 1991–1998. Although focusing strictly on cholera, Ramirez and Grady [95] found increased disease rates in Piura, Peru during El Niño events, but that the association was non-stationary, mediated by local hydrology; the association was evident in the latter part of the 1990s but with little evidence of El Niño-cholera associations in the early 1990s. Lastly, Raszl et al. [96] discussed how *Vibrio parahaemolyticus* outbreaks related to unusually warm coastal waters along the Pacific coast of South America during El Niño events was associated with increases in diarrhea and other similar gastrointestinal-related symptoms as a result of human consumption of infected shellfish.

7. ENSO and Health Forecasting

Due to an improvement in the climate science community's understanding of the large scale mechanisms that influence climate, plus rapid advances in computing technology, seasonal to inter-annual to decadal climate forecasts have become a real prospect [2,97]. This, coupled with an increasing knowledge of the nature of climate-health associations, has spawned a number of attempts to construct disease early warning systems based on seasonal predictions of health-sensitive climate fields, so that potential health threats may be anticipated several months in advance.

A key source of the potential seasonal predictability of health-sensitive climate variables is ENSO. Given this, the hope is that with time, accompanied by an improvement in the understanding of ENSO health associations, effective seasonal forecasting of climate (ENSO)-sensitive health outcomes will become operationally possible [98]. Generally, two broad approaches have been adopted in constructing climate-sensitive disease early warning systems based on known climate and health links, namely numerical and statistical. Numerical schemes take the output from seasonal climate forecast models, usually in the form of a rainfall and/or temperature time series, and ingest this into numerical process-based disease models for diseases such as malaria and dengue (e.g., Liverpool Malaria Model, [99]). Typically the output from disease models includes disease parameters such as disease transmission, size of mosquito population, and disease incidence [100]. Statistical or empirically based forecasting schemes generally draw on a variety of statistical methods and use empirical observations of climate and disease incidence to construct transfer functions that statistically link climate disease associations. Although a simple distinction has been drawn here, between numerical and statistical/empirical models this does not mean to imply that numerical approaches do not draw on statistical methods and vice versa. In most cases, the output from both numerical and statistical models are probabilistic statements about the likelihood of a given climate sensitive disease exceeding a critical threshold and often statistical schemes, when run in forecast mode, will use the numerical output from climate models to force the climate-disease transfer functions so as to gain estimates of disease incidence. Furthermore, many dynamic disease models use statistical functions to model the relationship between disease sensitive climate variables such as temperature, and for example, in the case of mosquito borne diseases, the rate of development of the parasite within the mosquito (the sporogonic cycle) and the mosquito biting/feeding rate (the gonotrophic cycle).

Thomson et al. [101] describe one of the first efforts aimed at seasonal forecasting of malaria in Africa based on ensemble predictions of rainfall and temperature from global coupled ocean-atmosphere climate models, and firmly established statistical climate-malaria links. The models achieved probabilistic predictions of anomalously high and low malaria incidence based on rainfall thresholds for Botswana up to four months in advance. Building on this work, Connor et al. [102] presented a framework for the integration of climate model based seasonal climate forecasts into early warning systems for climate sensitive diseases such as malaria and dengue. This work and that of Thomson [101] has been influential in guiding further endeavours related to the development of operational seasonal forecasts of malaria in southern Africa using outputs from numerical climate models run by forecasting centres such as the European Centre for Medium Range Weather Forecasting (ECMWF) [103,104]. For India, Lauderdale et al. [105] explored the feasibility of malaria forecasting by using a ECMWF seasonal forecast model to drive a numerical process-based dynamic malaria disease model. Using hindcasts from the ECMWF model, simulated forecasts of malaria were produced. These demonstrated probabilistic skill in predicting the spatial distribution of *Plasmodium falciparum* incidence particularly in regions where high seasonal and inter-annual variability of disease incidence is a characteristic. As well as showing some ability to predict the spatial distribution of malaria the seasonal forecast model was able to distinguish between years of "high", "above average" and "low" malaria incidence in the peak malaria transmission seasons with a three month lead time [105].

A number of statistical/empirical seasonal health forecasting models have been developed. For example, Lowe et al. [106] incorporated precipitation, minimum temperature, and Niño 3.4 index forecasts in a Bayesian hierarchical mixed model to make monthly predictions of dengue incidence in Ecuador for 2016. The ENSO element of this forecast system was in the form of Niño 3.4 SSTs derived from a structural time-series SST prediction model. It was found that the dengue forecast model was able to correctly predict an early peak in dengue incidence in March, 2016, with a 90% chance of exceeding the mean dengue incidence for the previous five years. Interestingly, when Lowe et al. [106] controlled for confounding due to chikungunya cases incorrectly recorded as dengue, this improved the prediction of the magnitude of dengue incidence. A similar approach was adopted by Lowe et al. [107] in the development of dengue forecasts for southeast Brazil. Poveda et al. [108] describe how satellite imagery of vegetation activity along with ENSO sensitive climate variables can be used as environmental indicators for malaria occurrence in Columbia. Armed with this knowledge, they demonstrate how statistical models and geographical information systems are applied by the Colombian health authorities to develop early warning systems for malaria. For the Solomon Islands in the western Pacific, where ENSO has clear impacts on rainfall as a disease sensitive climate variable, Smith et al. [109] applied stepwise regression to analyse climate variables and climate-associated malaria transmission at different lag intervals in order to identify rainfall thresholds associated with malaria categorised into three incidence categories. Study results not only revealed clear rainfall thresholds, but significant lag associations between rainfall and increases in malaria incidence such that drier October–December periods are followed by higher malaria transmission periods in January–June. Based on these statistical relationships an experimental early warning system has been proposed for the Guadalcanal region of the Solomon Islands [109]. Chuang et al. [110] used cross-wavelet coherence to evaluate the regional El Nino Southern Oscillation (ENSO) and Indian Ocean Dipole (IOD) effects on dengue incidence and local climate variables for Taiwan. Their work revealed the importance of non-linear and lag effects of minimum temperature and precipitation on dengue. These associations were applied in the successful prediction of dengue transmission between 2013 and 2015 [101].

While the potential for ENSO-based health forecasting is clear, despite improvements in observations and models, ENSO predictability and long-lead seasonal forecast skill, generally taken to mean the extent or lead-time for which boreal winter SST or any other ENSO index can be predicted with measurable skill, remains an issue [111]. A case in point is the 2014 ENSO forecast. Rather than a strong 2014 El Niño event occurring, as forecast, only weak warming in the key El Niño oceanic regions (Figure 5) was observed. This forecast "bust" [112,113] caused the ENSO prediction community to critically examine the efficacy of many of the significant ocean and atmosphere system components drawn on as a source of ENSO predictability [114–116]. A particular challenge for ENSO based health forecasting is the so-called "spring predictability barrier" [117–119]. This is basically a hiatus in ENSO forecasting accuracy for Northern Hemisphere spring when both dynamic and statistical ENSO prediction models display a sharp fall in their ability to predict sea surface temperature fields; following spring, the model ability to predict ENSO improves markedly. It is likely that the spring predictability problem exists because during this season El Niño/La Niña events are often in the stage of decay, following a winter peak, sliding into a neutral phase, which may persist or eventuate in a El Niño/La Niña later in the year. Consequently, the ENSO signal to noise ratio is low. Further during spring, the ocean does not exert a strong influence on the atmosphere because climatological (average) SST gradients in the tropical Pacific Ocean are much reduced and thus strong ocean-atmosphere coupling is compromised [117].

Given the expectations of the broad ENSO forecast user community related to ENSO forecasts as a panacea for climate risk management problems, much effort has been invested in improving predictability [120–122] with seasonal health forecasting scheme developers conscious that validation of predictions is a requisite part of the forecasting development process [104,123]. Further to the issues of predictability, other constraints related to seasonal health forecasting may well bear implications for the operationalisation of ENSO (climate)-sensitive disease early warning systems. Increasingly, seasonal health forecasts are couched in probabilistic terms that have been found to pose communication and uptake problems, making it imperative for forecast developers to think carefully how forecasts are provided to end users [124]. In the context of climate services based on climate forecasts, Ballester et al. [125] provide a sobering review of some of the challenges related to the construction of seasonal health forecasts. These include the capital and human resources and the associated governance arrangements required for development and implementation; the need for forecasting tools to master the complexity of the interactions between climate, disease transmission, socioeconomic disparities, and vulnerability; the imperative for integrated climate and health data sets; and acknowledgement that early warning systems and the climate forecasts on which they are based may only be effective when certain windows of opportunity present themselves, such as during ENSO events when there is a clear climate signature in a range of health responses.

8. Climate Change and ENSO

The recent 2015–2016 El Niño event is a timely reminder of the mammoth impacts that ENSO events can have on ecosystems and society. For instance, extensive forest fires in Indonesia and an associated haze hazard across the wider region, devastating floods in Peru, severe coral bleaching in a number of places across the Pacific, and widespread health issues throughout the Pacific and elsewhere over the course of the 2015–2016 El Niño are similar to the type of impacts that occurred during previous El Niño episodes, such as in 1982–1983 and 1997–1998 [126,127]. Although attention is often directed to El Niño impacts, intense La Niña events can be equally impactful as is evidenced for the 1998–1999 La Niña event that spawned catastrophic flooding in Bangladesh, Venezuela, and China, with a large number of lives lost [128,129]. Understandably consternation associated with such impacts, twinned with the worrisome spectre of anthropogenic climate change has precipitated an immense interest in establishing how ENSO might respond to climate change and the implication this holds for future population health. Two broad approaches have been applied to establish ENSO responses to a warmer world: the analysis of paleoclimate records and the conduct of numerical climate modelling experiments [13,130].

That inter-annual climate variability similar to that associated with ENSO has been a characteristic of the Pacific Basin for millennia is borne out by a number of paleoclimate studies. These revealed not only that strong east to west Pacific contrasts in ocean temperatures, similar to the current climatological difference of 2 °C, existed in the past [131], but that ENSO frequency has not changed significantly since the Pliocene (5.333 to 2.58 million years before present when global temperatures were 2–3 °C higher than present [132]. Similarly there is evidence for ENSO events and associated inter-annual climate variability during the last glacial maximum [133], and the Medieval Climate Anomaly and the Little Ice Age [134]. Paleoclimate studies have also revealed that, compared to previous centuries and millennia, twentieth-century ENSO activity has been considerably stronger [135–137], which has been interpreted as possible evidence for a link between global warming and ENSO response [138]. In brief, the upshot of most paleoclimate studies is that ENSO and marked inter-annual climate variability originating in the Pacific Basin is a characteristic of the global climate system, whether it be in a cooler or warmer state than present.

While it is likely that ENSO will be a feature of a warmer world [13], the question remains as to whether the intensity and frequency of El Niño/La Niña events might change with anthropogenic climate change. About the only way to answer this question is by performing climate model experiments using a range of greenhouse gas concentration scenarios, currently codified as Representative Concentration Pathways (RCP). Cai et al. [130] and Wang et al. [13] provided useful summaries of the current thinking on how ENSO climatology might respond to greenhouse warming based on a review of results from climate models in the Coupled Model Inter-comparison Project phases 3 (CMIP3) and 5 (CMIP5) [139] and the work of others. They concluded that there is some modelling-based evidence for increases in the frequency of ENSO events with global warming. However, in relation to whether future El Niño/La Niña events will become stronger or weaker, Wang et al. [13] were far more cautionary in their conclusions than Cai et al. [130], with the former concluding that evidence for a stronger or weaker El Niño/La Niña under global warming is unclear in contrast to the latter who confidently stated there will be an increased frequency of extreme El Niño and La Niña events. That an unequivocal greenhouse warming response of ENSO in climate models is not apparent stems from a range of factors. These include complex competing ocean-atmosphere feedback processes that have a negating effect on some of the key elements of the ENSO system [13,139], plus general uncertainties related to the ability of climate models to simulate the current ENSO state, the sensitivity of ENSO onset and cessation to global warming, difficulties with parameterizing climate processes that occur at scales less than that resolved in models, and how climate change-related distant influences from the Atlantic and Indian Oceans will affect ENSO [130].

Clearly, the equivocal findings regarding the possible impacts of climate change on ENSO hold important implications for future ENSO-health associations. Given the state of the science, perhaps all that can ventured at this point is that ENSO will be influenced in some way by climate change, with associated implications for health. The direction of such an alteration will depend on a number of climate and non-climate related drivers. The climate drivers include ENSO-related variability in rainfall, temperature, storm activity and ocean currents, layered upon changes to the mean climate state attributable to climate change. Moreover, a factor that makes speculation about the health risks of an altered ENSO phenomenon challenging is the significant inter-event variability of ENSO climate outcomes—is there a canonical El Niño/La Niña—and the decadal scale non-stationary relationship between ENSO and climate and thus health risks. While these generalities might seem inconsequential in terms of furthering our understanding of climate change, ENSO and health relationships, they serve as a reminder that caution is required when telescoping current ENSO health associations into the future in the absence of a firm understanding of how ENSO related climate variability may respond to further greenhouse warming. Lastly, and notwithstanding issues associated with a non-stationary and highly variable ENSO climate system and associated implications for health impacts, if the probability of future ENSO events can be constrained as a result of the convergence of climate modelling results, then estimating future ENSO-related health risks will largely be conditioned on non-climate factors such as the efficacy of early warning systems embedded in wider disaster risk reduction strategies.

9. Conclusions

The El Niño Southern Oscillation (ENSO) is an important of mode of climatic variability that exerts a discernible impact on ecosystems and society. For this reason, ENSO has attracted much interest in the climate and health science community, with many analysts investigating ENSO health links through considering the degree of relationship between an ENSO teleconnection index and a time series of incidence data for a specific climate-sensitive disease. While a plethora of teleconnection indices exist, from single variable atmospheric and oceanic to blended multivariate indices, with many of these applied in ENSO health studies, there remains no common consensus as to which ENSO index best describes ENSO behaviour. Accordingly, we encourage caution in the application of ENSO indices in climate and health studies via a consideration of the appropriateness of a range of teleconnection indices in terms of the geographical location, the climate variable, the disease of interest and, if comparative analyses are of interest, the index-producing agency.

In this review, we emphasised the complexity of ENSO as a physical phenomenon in that it possesses various "flavours", and its long term relationship with climate impacts is non-stationary. Accordingly, perhaps it is no surprise that ENSO health associations are multifarious. The majority of studies considered here did not report an unequivocal association between ENSO and a given health outcome. This review revealed an implicit but unfounded assumption that because a disease is broadly climate sensitive, and ENSO has an impact on climate, then an ENSO disease association should follow. In some ways this constitutes a leap of faith between a large scale mode of climatic variability, and a disease outcome for a specific location or region. That an ENSO signal is not clearly evident in the incidence of some climate-sensitive diseases may be attributed to the varying strength of ENSO-climate links that may be geographically, seasonally as well as climate variable dependent.

A worrying feature of many of the ENSO health studies is that the relationship between ENSO and disease is often viewed through a purely statistical lens. Few plausible explanations are offered as to why ENSO, as represented by a time series of a teleconnection index, might be a driver of disease incidence. This signposts the need to move beyond a purely statistical/mechanical treatment of climatic-health variability associations to one where diagnostic analyses are undertaken to identify the underlying climate mechanisms that form the cascade of processes that link ocean-atmosphere interactions with health. While this might be viewed as unnecessary in some quarters of the climate and health community, in terms of scientific credibility and a holistic understanding of ENSO health links, we suggest that fully integrated all-encompassing analyses are preferable to blunt statistically motivated analyses.

Although not expressed as such, partial and situation dependent evidence of ENSO-health associations has engendered what might be referred to as a post-normal turn in the climate and health science community in that there is a drive to apply the science of ENSO and health linkages to the betterment of society and the achievement of sustainable development goals. This is most evident in the energy applied to the development of climate informed seasonal health forecasts for a range of diseases. Despite the enthusiasm for these, a number of consequential challenges exist in relation to seasonal health forecasts, including the fundament issue of ENSO predictability; only certain windows of opportunity may exist for forecasting; effective ways of communicating ENSO-health warnings to a range of stakeholders remains elusive; and long-term ENSO-health links lack stability. Fully integrated approaches to seasonal forecasting are needed.

The looming spectre of climate change has precipitated much speculation about the associated health risks, with a temptation to project the likely impacts of future ENSO events on health. Conceptually, and notwithstanding the importance of non-climate factors, projecting how future ENSO events could impact health will depend on knowing about future ENSO strength and frequency as well as the future relationship between ENSO and a range of health-sensitive climate variables. While paleoclimatic evidence and climate modelling experiments indicate that ENSO events will remain an important feature of global climate, with ENSO frequency likely to increase, it is difficult to say how population health might respond to a changing ENSO climatology. This is because there are divergent opinions in the climate change modelling literature about possible changes in ENSO strength, which plays a critical role in determining societal impacts, including health, as demonstrated for previous strong ENSO events. Of course, the worst case future ENSO-health scenario is one in which there is an increase in the frequency and strength of both El Niño and La Niña events as a consequence of climate change, with no additional adaptation strategies.

So what of the future for ENSO-health research? An imperative to unravel the complexities of ENSO-health relationships is to build integrated data bases comprising not only climate and health data but "other" environmental data, as well as information on population characteristics including dynamic measures of vulnerability. Achieving this imperative appears a long way off as the research field is still characterised by disparate data sets of variable quality and length, which work against meaningful analyses of climate and health associations. Most epidemiological analyses treated ENSO events as a continuous time series, as represented by a specific teleconnection index. However, there is strong evidence that health responses to El Niño or La Niña events are more often than not restricted to periods within or immediately following such events. Given this, quiescent periods in terms of the climate drivers of disease could be excluded from ENSO health analyses, with analyses based around ENSO phase composites to identify patterns of anomalous disease incidence tied to unusual climate conditions.

Furthermore, most ENSO health analyses treated all El Niño or La Niña events as similar despite strong evidence to the contrary. Accordingly, consideration needs to be given to how health impacts might play out under different ENSO flavours, perhaps starting with exploring the contrasts between eastern Pacific and central Pacific El Niño events. Typically, ENSO health analyses use either the SOI or Niño 3.4 teleconnection indices as indicators of ENSO behaviour. As yet, there has been no attempt to systematically establish which of a range of possible ENSO indices might be best for analysing ENSO health associations for a particular location, region or disease. Research along these lines is needed because some teleconnection indices are likely to be more pertinent for ENSO-health analyses in the Pacific Basin compared to others that might have wider geographical applicability. By default, most ENSO-health analyses focus on the impact of El Niño with the health effects of La Niña, or "exaggerated normal climate conditions" largely ignored. Notwithstanding the asymmetric relationship between El Niño and La Niña, exploring the health impacts of strong La Niña events could shed further light on the nature of the burden of climate-sensitive disease. In the same vein, drawing on classifications of past ENSO events, archival records of disease incidence could be searched for historical evidence of ENSO-related health events.

Effort is also required to move seasonal health forecasting beyond the proof of concept phase through establishing when, where, why, and how ENSO impacts occur in both deterministic and probabilistic frameworks. As the rendering of past ENSO events is improved in climate models, the ENSO and health research community will need to consider how an alteration of ENSO climatology in tandem with changes in non-climate factors might play out in terms of ENSO-related health impacts under climate change. Lastly, future work on ENSO health associations will necessarily involve the deployment of expertise from a range of disciplines, given that forcing of health outcomes via ENSO moderated climate events represents just one dimension of what constitutes a "wicked" research problem.

Funding: This research received no external funding.

Conflicts of Interest: The authors declare no conflicts of interest.

References

1. McMichael, A.J. Globalization, climate change, and human health. *J. Med.* **2013**, *368*, 1335–1343. [CrossRef] [PubMed]
2. McGregor, G.R. Climatology in support of climate risk management. *Prog. Phys. Geogr.* **2015**, *39*, 536–553. [CrossRef]

3. Sheridan, S.; Lee, C. Synoptic climatology and the analysis of atmospheric teleconnections. *Prog. Phys. Geogr.* **2015**, *36*, 548–557. [CrossRef]
4. De Viron, O.; Dickey, J.O.; Ghil, M. Global modes of climate variability. *Geophys. Res. Lett.* **2013**, *40*, 1832–1837. [CrossRef]
5. Kucharski, F.; Kang, I.-S.; Straus, D.; King, M.P. Teleconnections in the atmosphere and oceans. *Bull. Am. Meteorol. Soc.* **2010**, *91*, 381–383. [CrossRef]
6. McGregor, G.R. Hydroclimatology, modes of climatic variability and stream flow, lake and groundwater level variability. *Prog. Phys. Geogr.* **2017**, *41*, 496–512. [CrossRef]
7. Kovats, S.; Bouma, M.J.; Hajat, S.; Worrall, E.; Haines, A. El Niño and Health. *Lancet* **2003**, *362*, 1481–1489. [CrossRef]
8. Glantz, M.H. *Currents of Change: El Nino's Impact on Climate and Society*; Cambridge University Press: Cambridge, UK, 2011.
9. Philander, S.G. *El Nino, La Nina, and the Southern Oscillation*; Academic Press: San Diego, CA, USA, 1990; 293p.
10. Montgomery, R.B. Report on the work of G.T. Walker. *Mon. Weather Rev.* **1940**, *39*, 1–22.
11. Troup, A.J. The southern oscillation. *Q. J. R. Meteorol. Soc.* **1965**, *102*, 490–506. [CrossRef]
12. Bjerknes, J. Atmospheric teleconnections from the equatorial Pacific. *Mon. Weather Rev.* **1969**, *97*, 163–172. [CrossRef]
13. Wang, C.; Deser, C.; Yu, J.-Y.; DiNezio, P.; Clement, A. El Niño-Southern Oscillation (ENSO): A review. In *Coral Reefs of the Eastern Pacific*; Glymn, P., Manzello, D., Enochs, I., Eds.; Springer Science Publisher: New York, NY, USA, 2016; pp. 85–106.
14. Dima, M.; Lohmann, G.; Rimbu, N. Possible North Atlantic origin for changes in ENSO properties during the 1970s. *Clim. Dyn.* **2015**, *44*, 925–935. [CrossRef]
15. Elisio, M.; Vera, C.; Leandro, M. Influences of ENSO and PDO phenomena on the local climate variability can drive extreme temperature and depth conditions in a Pampean shallow lake affecting fish communities. *Environ. Biol. Fishes* **2018**, *101*, 653–666.
16. Girishkumar, M.S.; Prakash, V.P.; Ravichandran, M. Influence of Pacific Decadal Oscillation on the relationship between ENSO and tropical cyclone activity in the Bay of Bengal during October–December. *Clim. Dyn.* **2015**, *44*, 3469–3479. [CrossRef]
17. Krishnamurthy, L.; Krishnamurthy, V. Influence of PDO on South Asian summer monsoon and monsoon-ENSO relation. *Clim. Dyn.* **2013**, *42*, 2397–2410. [CrossRef]
18. Wang, X.; Liu, H. PDO modulation of ENSO effect on tropical cyclone rapid intensification in the western North Pacific. *Clim. Dyn.* **2016**, *46*, 15–28. [CrossRef]
19. Chen, W.; Feng, J.; Wu, R. Roles of ENSO and PDO in the Link of the East Asian Winter Monsoon to the following Summer Monsoon. *J. Clim.* **2013**, *26*, 622–635. [CrossRef]
20. Lestari, R.K.; Koh, T.Y. Statistical evidence for asymmetry in ENSO-IOD interactions. *Atmos. Ocean* **2016**, *54*, 498–504. [CrossRef]
21. Barnston, A. Why Are There So Many ENSO Indexes, Instead of Just One? Available online: https://www.climate.gov/news-features/blogs/enso/why-are-there-so-many-enso-indexes-instead-just-one (accessed on 29 June 2018).
22. Walker, G.T.; Bliss, E.W. World Weather V. *Mem. R. Meteorol. Soc.* **1932**, *4*, 36–53.
23. Rasmusson, E.M.; Carpenter, T.H. Variations in Tropical Sea Surface Temperature and Surface Wind Fields Associated with the Southern Oscillation/El Niño. *Mon. Weather Rev.* **1982**, *110*, 354–384. [CrossRef]
24. Trenberth, K.E.; Stepaniak, D.P. Indices of El Niño Evolution. *J. Clim.* **2001**, *14*, 1697–1701. [CrossRef]
25. Hanley, D.E.; Bourassa, M.A.; O'Brien, J.J.; Smith, S.R.; Spadea, E.R. Quantitative Evaluation of ENSO Indices. *J. Clim.* **2003**, *16*, 1249–1258. [CrossRef]
26. Wright, P.B. Relationships between indices of the Southern Oscillation. *Mon. Weather Rev.* **1984**, *112*, 1913–1919. [CrossRef]
27. Wright, P.B. Homogenized long-period Southern Oscillation indices. *Int. J. Climatol.* **1989**, *9*, 33–54. [CrossRef]
28. Smith, C.A.; Sardeshmukh, P.D. The effect of ENSO on the intraseasonal variance of surface temperatures in winter. *Int. J. Climatol.* **2000**, *20*, 1543–1557. [CrossRef]
29. Allan, R. ENSO and climatic variability in the past 150 years. In *El Niño and the Southern Oscillation: Multiscale Variability and Global and Regional Impacts*; Diaz, H., Markgraf, V., Eds.; Cambridge Univ. Press: Cambridge, UK, 2000; pp. 3–35.

30. Chen, X.Y.; Wallace, J.M. Orthogonal PDO and ENSO Indices. *J. Clim.* **2016**, *29*, 3883–3892. [CrossRef]
31. Henley, B.J.; Gergis, J.; Karoly, D.J.; Power, S.; Kennedy, J.; Folland, C.K. A tripole index for the Interdecadal Pacific Oscillation. *Clim. Dyn.* **2015**, *45*, 3077–3090. [CrossRef]
32. Hu, C.D.; Yang, S.; Wu, Q.G.; Zhang, T.T.; Zhang, C.Y.; Li, Y.N.; Deng, K.Q.; Wang, T.; Chen, J.W. Reinspecting two types of El Nio: A new pair of Nio indices for improving real-time ENSO monitoring. *Clim. Dyn.* **2016**, *47*, 4031–4049. [CrossRef]
33. Jin, F.F.; Kim, S.T.; Bejarano, L. A coupled-stability index for ENSO. *Geophys. Res. Lett.* **2006**, *33*, L23708. [CrossRef]
34. Li, G.; Ren, B.; Yang, C.; Zheng, J. Indices of El Niño and El Niño Modoki: An improved El Niño Modoki index. *Adv. Atmos. Sci.* **2010**, *27*, 1210–1220. [CrossRef]
35. Yu, J.-Y.; Kim, S.T. Identification of central-Pacific and eastern-Pacific types of ENSO in CMIP3 models. *Geophys. Res. Lett.* **2010**, *37*, L15705. [CrossRef]
36. Ren, H.-L.; Jin, F.-F. Niño indices for two types of ENSO. *Geophys. Res. Lett.* **2011**, *38*, L04704. [CrossRef]
37. Takahashi, K.; Montecinos, A.; Goubanova, K.; Dewitte, B. ENSO regimes: Reinterpreting the canonical and Modoki El Niño. *Geophys. Res. Lett.* **2011**, *38*, L10704. [CrossRef]
38. Wolter, K.; Timlin, M.S. Measuring the strength of ENSO events—How does 1997/98 rank? *Weather* **1998**, *53*, 315–324. [CrossRef]
39. Wolter, K.; Timlin, M.S. El Niño/Southern Oscillation behaviour since 1871 as diagnosed in an extended multivariate ENSO index (MEI.ext). *Int. J. Climatol.* **2011**, *31*, 1074–1087. [CrossRef]
40. Ziemke, J.R.; Chandra, S.; Oman, L.D.; Bhartia, P.K. A new ENSO index derived from satellite measurements of column ozone. *Atmos. Chem. Phys.* **2010**, *10*, 3711–3721. [CrossRef]
41. Kulkarni, M.N.; Siingh, D. The atmospheric electrical index for ENSO modoki: Is ENSO modoki one of the factors responsible for the warming trend slowdown? *Sci. Rep.* **2016**, *6*, 24009. [CrossRef] [PubMed]
42. Qu, T.D.; Yu, J.Y. ENSO indices from sea surface salinity observed by Aquarius and Argo. *J. Oceanogr.* **2014**, *70*, 367–375. [CrossRef]
43. Ishii, M.; Shouji, A.; Sugimoto, S.; Matsumoto, T. Objective Analyses of Sea-Surface Temperature and Marine Meteorological Variables for the 20th Century using ICOADS and the Kobe Collection. *Int. J. Climatol.* **2005**, *25*, 865–879. [CrossRef]
44. Japan Meteorological Agency. Characteristics of Global Sea Surface Temperature Analysis Data (COBE-SST) for Climate Use. *Mon. Rep. Clim. Syst.* **2006**, *12*, 116.
45. Huang, B.Y.; Thorne, P.W.; Banzon, V.F.; Boyer, T.; Chepurin, G.; Lawrimore, J.H.; Menne, M.J.; Smith, T.M.; Vose, R.S.; Zhang, H.M. Extended reconstructed sea surface temperature, version 5 (ERSSTv5): Upgrades, validations, and intercomparisons. *J. Clim.* **2017**, *30*, 8179–8205. [CrossRef]
46. Rayner, N.A.; Parker, D.E.; Horton, E.B.; Folland, C.K.; Alexander, L.V.; Rowell, D.P.; Kent, E.C.; Kaplan, A. Global analyses of sea surface temperature, sea ice, and night marine air temperature since the late nineteenth century. *J. Geophys. Res.* **2003**, *108*, 4407. [CrossRef]
47. Santoso, A.; McPhaden, M.J.; Cai, W. The defining characteristics of ENSO extremes and the strong 2015/2016 El Niño. *Rev. Geophys.* **2017**, *55*, 1079–1129. [CrossRef]
48. Johnson, N.A. How Many ENSO Flavors Can We Distinguish? *J. Clim.* **2013**, *26*, 4816–4827. [CrossRef]
49. Paek, H.; Yu, J.Y.; Qian, C.C. Why were the 2015/2016 and 1997/1998 extreme El Niños different? *Geophys. Res. Lett.* **2017**, *44*, 1848–1856. [CrossRef]
50. Campbell-Lendrum, D.; Manga, L.; Bagayoko, M.; Sommerfeld, J. Climate change and vector-borne diseases: What are the implications for public health research and policy? *Philos. Trans. R. Soc. B.* **2015**, *370*, 20130552. [CrossRef] [PubMed]
51. Rohani, P. The Link between Dengue Incidence and El Nino Southern Oscillation. *PLoS Med.* **2009**, *6*, e1000185. [CrossRef] [PubMed]
52. GBD 2016 Risk Factors Collaborators. Global, regional, and national comparative risk assessment of 84 behavioural, environmental and occupational, and metabolic risks or clusters of risks, 1990–2016: A systematic analysis for the Global Burden of Disease Study 2016. *Lancet* **2017**, *390*, 1345–1422.
53. Jones, A.E.; Wort, U.U.; Morse, A.P.; Hastings, I.M.; Gagnon, A.S. Climate prediction of El Nino malaria epidemics in north-west Tanzania. *Malar. J.* **2007**, *6*, 162. [CrossRef] [PubMed]
54. Bouma, M.J.; Siraj, A.S.; Rodo, X.; Pascual, M. El Nino-based malaria epidemic warning for Oromia, Ethiopia, from August 2016 to July 2017. *Trop. Med. Int. Health* **2016**, *21*, 1481–1488. [CrossRef] [PubMed]

55. Mabaso, M.L.H.; Kleinschmidt, I.; Sharp, B.; Smith, T. El Nino Southern Oscillation (ENSO) and annual malaria incidence in Southern Africa. *Trans. R. Soc. Trop. Med. Hyg.* **2007**, *101*, 326–330. [CrossRef] [PubMed]
56. Dev, V. El Niño and malaria transmission in northeast India. *Curr. Sci.* **2010**, *98*, 997–998.
57. Zubair, L.; Galappaththy, G.N.; Yang, H.M.; Chandimala, J.; Yahiya, Z.; Amerasinghe, P.; Ward, N.; Connor, S.J. Epochal changes in the association between malaria epidemics and El Nino in Sri Lanka. *Malar. J.* **2008**, *7*, 140. [CrossRef] [PubMed]
58. Delgado-Petrocelli, L.; Cordova, K.; Camardiel, A.; Aguilar, V.H.; Hernandez, D.; Ramos, S. Analysis of the El Nino/La Nina-Southern Oscillation variability and malaria in the Estado Sucre, Venezuela. *Geospat. Health* **2012**, *6*, S51–S57. [CrossRef] [PubMed]
59. Hanf, M.; Adenis, A.; Nacher, M.; Carme, B. The role of El Nino southern oscillation (ENSO) on variations of monthly Plasmodium falciparum malaria cases at the cayenne general hospital, 1996–2009, French Guiana. *Malar. J.* **2011**, *10*, 100. [CrossRef] [PubMed]
60. Fuller, D.O.; Troyo, A.; Beier, J.C. El Nino Southern Oscillation and vegetation dynamics as predictors of dengue fever cases in Costa Rica. *Environ. Res. Lett.* **2009**, *4*, 014011. [CrossRef] [PubMed]
61. Johansson, M.A.; Cummings, D.A.T.; Glass, G.E. Multiyear Climate Variability and Dengue-El Nino Southern Oscillation, Weather, and Dengue Incidence in Puerto Rico, Mexico, and Thailand: A Longitudinal Data Analysis. *PLoS Med.* **2009**, *6*, e1000168. [CrossRef] [PubMed]
62. Xiao, J.P.; Liu, T.; Lin, H.L.; Zhu, G.H.; Zeng, W.L.; Li, X.; Zhang, B.; Song, T.; Deng, A.P.; Zhang, M. Weather variables and the El Nino Southern Oscillation may drive the epidemics of dengue in Guangdong Province, China. *Sci. Total Environ.* **2018**, *624*, 926–934. [CrossRef] [PubMed]
63. Liyanage, P.; Tissera, H.; Sewe, M.; Quam, M.; Amarasinghe, A.; Palihawadana, P.; Wilder-Smith, A.; Louis, V.R.; Tozan, Y.; Rocklov, J. A Spatial Hierarchical Analysis of the Temporal Influences of the El Nino-Southern Oscillation and Weather on Dengue in Kalutara District, Sri Lanka. *Int. J. Environ. Res. Public Health* **2018**, *13*, 1087. [CrossRef] [PubMed]
64. Banu, S.; Hu, W.; Guo, Y.; Hurst, C.; Tong, S. Projecting the impact of climate change on dengue transmission in Dhaka, Bangladesh. *Environ. Int.* **2014**, *63*, 137–142. [CrossRef] [PubMed]
65. Earnest, A.; Tan, S.B.; Wilder-Smith, A. Meteorological factors and El Nino Southern Oscillation are independently associated with dengue infections. *Epidemiol. Infect.* **2012**, *140*, 1244–1251. [CrossRef] [PubMed]
66. Hu, W.B.; Clements, A.; Williams, G.; Tong, S.L. Dengue fever and El Nino/Southern Oscillation in Queensland, Australia: A time series predictive model. *Occup. Environ. Med.* **2010**, *67*, 307–311. [CrossRef] [PubMed]
67. Tipayamongkholgul, M.; Fang, C.T.; Klinchan, S.; Liu, C.M.; King, C.C. Effects of the El Nino-Southern Oscillation on dengue epidemics in Thailand, 1996–2005. *BMC Public Health* **2009**, *9*, 422. [CrossRef] [PubMed]
68. Ferreira, M.C. Geographical distribution of the association between El Nino South Oscillation and dengue fever in the Americas: A continental analysis using geographical information system-based techniques. *Geospat. Health* **2014**, *9*, 141–151. [CrossRef] [PubMed]
69. Levy, K.; Woster, A.P.; Goldstein, R.S.; Carlton, E.J. Untangling the impacts of climate change on waterborne diseases: A systematic review of relationships between diarrheal diseases and temperature, rainfall, flooding, and drought. *Environ. Sci. Technol.* **2016**, *50*, 4905–4922. [CrossRef] [PubMed]
70. Lo Iacono, G.; Armstrong, B.; Fleming, L.E.; Elson, R.; Kovats, S.; Vardoulakis, S.; Nichols, G.L. Challenges in developing methods for quantifying the effects of weather and climate on water-associated diseases: A systematic review. *PLoS Negl. Trop. Dis.* **2017**, *11*, e0005659. [CrossRef] [PubMed]
71. Mellor, J.E; Levy, K.; Zimmerman, J.; Elliott, M.; Bartram, J.; Carlton, E.; Clasen, T.; Dillingham, R.; Eisenberg, J.; Guerrant, R.; et al. Planning for climate change: The need for mechanistic systems-based approaches to study climate change impacts on diarrheal diseases. *Sci. Total Environ.* **2016**, *548*, 82–90. [CrossRef] [PubMed]
72. Azage, M.; Kumie, A.; Worku, A.; Bagtzoglou, A.C.; Anagnostou, E. Effect of climatic variability on childhood diarrhea and its high risk periods in northwestern parts of Ethiopia. *PLoS ONE* **2017**, *12*, e0186933. [CrossRef] [PubMed]

73. Carlton, E.J.; Eisenberg, J.N.S.; Goldstick, J.; Cevallos, W.; Trostle, J.; Levy, K. Heavy Rainfall Events and Diarrhea Incidence: The Role of Social and Environmental Factors. *Am. J. Epidemiol.* **2014**, *179*, 344–352. [CrossRef] [PubMed]

74. Carlton, E.J.; Woster, A.P.; DeWitt, P.; Goldstein, R.S.; Levy, K. A systematic review and meta-analysis of ambient temperature and diarrhoeal diseases. *Int. J. Epidemiol.* **2016**, *45*, 117–130. [CrossRef] [PubMed]

75. Davies, P.R. The dilemma of rare events: Porcine epidemic diarrhea virus in North America. *Prev. Vet. Med.* **2015**, *22*, 235–241. [CrossRef] [PubMed]

76. Kraay, A.N.M.; Brouwer, A.F.; Lin, N.; Collender, P.A.; Remais, J.V.; Eisenberg, J.N.S. Modeling environmentally mediated rotavirus transmission: The role of temperature and hydrologic factors. *Proc. Natl. Acad. Sci. USA* **2018**, *115*, 2782–2790. [CrossRef] [PubMed]

77. Musengimana, G.; Mukinda, F.K.; Machekano, R.; Mahomed, H. Temperature Variability and Occurrence of Diarrhoea in Children under Five-Years-Old in Cape Town Metropolitan Sub-Districts. *Int. J. Environ. Res. Public Health* **2016**, *13*, 859. [CrossRef] [PubMed]

78. Thiam, S.; Diene, A.N.; Sy, I.; Winkler, M.S.; Schindler, C.; Ndione, J.A.; Faye, O.; Vounatsou, P.; Utzinger, J.; Cisse, G. Association between Childhood Diarrhoeal Incidence and Climatic Factors in Urban and Rural Settings in the Health District of Mbour, Senegal. *Int. J. Environ. Res. Public Health* **2017**, *14*, 1049. [CrossRef] [PubMed]

79. Demissie, S.; Mengisitie, B. The impact of El Niño on diarrheal disease incidence: A systematic review. *Sci. J. Public Health* **2017**, *17*, 446–451. [CrossRef]

80. Singh, R.B.K.; Hales, S.; de Wet, N.; Raj, R.; Hearnden, M.; Weinstein, P. The influence of climate variation and change on diarrheal disease in the Pacific Islands. *Environ. Health Perspect.* **2001**, *109*, 155–159. [CrossRef] [PubMed]

81. McIver, L.; Kim, R.; Woodward, A.; Hales, S.; Spickett, J.; Katscherian, D.; Hashizume, M.; Honda, Y.; Kim, H.; Iddings, S.; et al. Health impacts of climate change in Pacific Island Countries: A Regional Assessment of Vulnerabilities and Adaptation Priorities. *Environ. Health Perspect.* **2016**, *124*, 1707–1714. [CrossRef] [PubMed]

82. Emont, J.P.; Ko, A.I.; Homasi-Paelate, A.; Ituaso-Conway, N.; Nilles, E.J. Epidemiological investigation of a diarrhea outbreak in the South Pacific Island nation of Tuvalu during a Severe La Niña–associated drought emergency in 2011. *Am. J. Trop. Med. Hyg.* **2017**, *96*, 576–582. [CrossRef] [PubMed]

83. De Magny, G.C.; Guegan, J.F.; Petit, M.; Cazelles, B. Regional-scale climate-variability synchrony of cholera epidemics in West Africa. *BMC Infect. Dis.* **2007**, *7*, 20. [CrossRef]

84. Moore, S.M.; Azman, A.S.; Zaitchik, B.F.; Mintz, E.D.; Brunkard, J.; Legros, D.; Hill, A.; Mckay, H.; Luquero, F.J.; Olson, D.; et al. El Nino and the shifting geography of cholera in Africa. *Proc. Natl. Acad. Sci. USA* **2017**, *114*, 4436–4441. [CrossRef] [PubMed]

85. Nkoko, D.B.; Giraudoux, P.; Plisnier, P.D.; Tinda, A.M.; Piarroux, M.; Sudre, B.; Horion, S.; Tamfum, J.J.M.; Ilunga, B.K.; Piarroux, R. Dynamics of Cholera Outbreaks in Great Lakes Region of Africa, 1978–2008. *Emerg. Infect. Dis.* **2011**, *17*, 2026–2034. [CrossRef] [PubMed]

86. Alajo, S.O.; Nakavuma, J.; Erume, J. Cholera in endemic districts in Uganda during El Niño rains: 2002–2003. *Afr. Health Sci.* **2006**, *6*, 93–97. [CrossRef] [PubMed]

87. Pascual, M.; Rodó, X.; Ellner, S.P.; Colwell, R.; Bouma, M.J. Cholera dynamics and El Nino-southern oscillation. *Science* **2000**, *289*, 1766–1769. [CrossRef] [PubMed]

88. Hashizume, M.; Chaves, L.F.; Faruque, A.S.G.; Yunus, M.; Streatfield, K.; Moji, K. A differential Effect of Indian Ocean Dipole and El Niño on cholera dynamics in Bangladesh. *PLoS ONE* **2013**, *8*, e60001. [CrossRef] [PubMed]

89. Rodó, X.; Pascual, M.; Fuchs, G.; Faruque, A.S.G. ENSO and cholera: A nonstationary link related to climate change? *Proc. Natl. Acad. Sci. USA* **2002**, *99*, 12901–12906. [CrossRef] [PubMed]

90. Ohtomo, K.; Kobayashi, N.; Sumi, A.; Ohtomo, N. Relationship of cholera incidence to El Nino and solar activity elucidated by time-series analysis. *Epidemiol. Infect.* **2010**, *138*, 99–107. [CrossRef] [PubMed]

91. Martinez, P.P.; Reiner, R.C., Jr.; Cash, B.A.; Rodo, Â.X.; Shahjahan Mondal, M.; Roy, M.; Yunus, M.; Faruque, A.S.G.; Huq, S.; King, A.A.; et al. Cholera forecast for Dhaka, Bangladesh, with the 2015–2016 El Niño: Lessons learned. *PLoS ONE* **2017**, *12*, e0172355. [CrossRef] [PubMed]

92. Checkley, W.; Epstein, L.D.; Gilman, R.H.; Figueroa, D.; Cama, R.I.; Patz, J.A.; Black, R.E. Effects of El Nino and ambient temperature on hospital admissions for diarrhoeal diseases in Peruvian children. *Lancet* **2000**, *355*, 442–450. [CrossRef]

93. Bennett, A.; Epstein, L.D.; Gilman, R.H.; Cama, V.; Bern, C.; Cabrera, L.; Lescano, A.G.; Patz, J.; Carcamo, C.; Sterling, C.R.; et al. Effects of the 1997–1998 El Niño episode on community rates of diarrhea. *Am. J. Public Health.* **2012**, *102*, e63–e69. [CrossRef] [PubMed]

94. Lama, J.R.; Seas, C.R.; León-Barúa, R.; Gotuzzo, E.; Sack, R.B. Environmental temperature, cholera, and acute diarrhoea in adults in Lima, Peru. *J. Health Popul. Nutr.* **2004**, *22*, 399–403. [PubMed]

95. Ramírez, I.J.; Grady, S.C. El Niño, climate, and cholera associations in Piura, Peru, 1991–2001: A Wavelet Analysis. *Ecohealth* **2016**, *13*, 83–99. [CrossRef] [PubMed]

96. Raszl, S.M.; Froelich, B.; Vieira, C.R.W.; Blackwood, A.D.; Noble, R.T. Vibrio parahaemolyticus and Vibrio vulnificus in South America: Water, seafood and human infections. *J. Appl. Microbiol.* **2016**, *121*, 1201–1222. [CrossRef] [PubMed]

97. Soares, M.B.; Daly, M.; Dessai, S. Assessing the value of seasonal climate forecasts for decision-making. *Wiley Interdiscip. Rev. Clim. Chang.* **2018**, *9*, e523. [CrossRef]

98. Zaitchik, B.F.; Hayden, M.H.; Villela, D.A.M.; Lord, C.C.; Kitron, U.D.; Carvajal, J.J.; Camara, D.C.P. Climate information for arbovirus risk monitoring: Opportunities and challenges. *Bull. Am. Meteorol. Soc.* **2016**, *97*, 1–5. [CrossRef]

99. Ermert, V.; Fink, A.; Jones, A.; Morse, A. Development of a new version of the Liverpool Malaria Model. I. Refining the parameter settings and mathematical formulation of basic processes based on a literature review. *Malar. J.* **2011**, *10*, 35–52. [CrossRef] [PubMed]

100. Metcalf, C.J.E.; Walter, K.S.; Wesolowski, A.; Buckee, C.O.; Shevliakova, E.; Tatem, A.; Boos, W.R.; Weinberger, D.M.; Pitzer, V.E. Identifying climate drivers of infectious disease dynamics: Recent advances and challenges ahead. *Proc. R. Soc. B* **2017**, *284*, 20170901. [CrossRef] [PubMed]

101. Thomson, M.; Doblas-Reyes, F.J.; Mason, S.J.; Hagedorn, R.; Connor, S.J.; Phindela, T.; Morse, A.P.; Palmer, T.N. Malaria early warnings based on seasonal climate forecasts from multi-model ensembles. *Nature* **2006**, *439*, 576–579. [CrossRef] [PubMed]

102. Connor, S.J.; Mantilla, G.C. Integration of seasonal forecasts into early warning systems for climate sensitive diseases such as malaria and dengue. In *Seasonal Forecasts, Climatic Change and Human Health—Health and Climate*; Thomson, M.C., Beniston, M., Garcia-Herrera, R., Eds.; Springer: Dordrecht, The Netherlands, 2008; pp. 71–84.

103. MacLeod, D.A.; Jones, A.; Di Giuseppe, F.; Caminade, C.; Morse, A.P. Demonstration of successful malaria forecasts for Botswana using an operational seasonal climate model. *Environ. Res. Lett.* **2015**, *10*, 044005. [CrossRef]

104. Tompkins, A.M.; Di Giuseppe, F. Potential Predictability of Malaria in Africa Using ECMWF Monthly and Seasonal Climate Forecasts. *J. Appl. Meteorol. Climatol.* **2015**, *54*, 521–540. [CrossRef]

105. Lauderdale, J.M.; Caminade, C.; Heath, A.E.; Jones, A.E.; MacLeod, D.A.; Gouda, K.C.; Murty, U.S.; Goswami, P.; Mutheneni, S.R.; Morse, A.P. Towards seasonal forecasting of malaria in India. *Malar. J.* **2014**, *13*, 310. [CrossRef] [PubMed]

106. Lowe, R.; Stewart-Ibarra, A.M.; Petrova, D.; García-Díez, M.; Borbor-Cordova, M.J.; Mejía, R.; Regato, M.; Rodó, X. Climate services for health: Predicting the evolution of the 2016 dengue season in Machala, Ecuador. *Lancet Planet. Health* **2017**, *1*, e142–e151. [CrossRef]

107. Lowe, R.; Bailey, T.C.; Stephenson, D.B.; Jupp, T.E.; Graham, R.J.; Barcellos, C.; Carvalho, M.S. The development of an early warning system for climate-sensitive disease risk with a focus on dengue epidemics in Southeast Brazil. *Stat. Med.* **2012**, *32*, 864–883. [CrossRef] [PubMed]

108. Poveda, G.; Estrada-Restrepo, O.A.; Morales, J.E.; Hernandez, O.O.; Galeano, A.; Osorio, S. Integrating knowledge and management regarding the climate-malaria linkages in Colombia. *Curr. Opin. Environ. Sustain.* **2011**, *3*, 448–460. [CrossRef]

109. Smith, J.; Tahani, L.; Bobogare, A.; Bugoro, H.; Otto, F.; Fafale, G.; Hiriasa, D.; Kazazic, A.; Beard, G.; Amjadali, A. Malaria early warning tool: Linking inter-annual climate and malaria variability in northern Guadalcanal, Solomon Islands. *Malar. J.* **2017**, *16*, 472. [CrossRef] [PubMed]

110. Chuang, T.W.; Chaves, L.F.; Chen, P.J. Effects of local and regional climatic fluctuations on dengue outbreaks in southern Taiwan. *PLoS ONE* **2017**, *12*, e0178698. [CrossRef] [PubMed]

111. Barnston, A.G.; Tippett, M.K.; L'Heureux, M.L.; Li, S.; DeWitt, D.G. Skill of real-time seasonal ENSO model predictions during 2002–2011: Is our capability increasing? *Bull. Am. Meteorol. Soc.* **2012**, *93*, 631–651. [CrossRef]

112. Magnusson, L. Diagnostic methods for understanding the origin of forecast errors. *Q. J. R. Meteorol. Soc.* **2017**, *143*, 2129–2142. [CrossRef]

113. Lillo, S.P.; Parsons, D.B. Investigating the dynamics of error growth in ECMWF medium-range forecast busts. *Q. J. R. Meteorol. Soc.* **2017**, *143*, 1211–1226. [CrossRef]

114. McPhaden, M.J. Playing hide and seek with El Niño. *Nat. Clim. Chang.* **2015**, *5*, 791–795. [CrossRef]

115. Min, Q.J.; Su, R.; Zhang, R.; Rong, X. What hindered the El Niño pattern in 2014? *Geophys. Res. Lett.* **2015**, *42*, 6762–6770. [CrossRef]

116. Larson, S.M.; Kirtman, B.P. An alternate approach to ensemble ENSO forecast spread: Application to the 2014 forecast. *Geophys. Res. Lett.* **2015**, *42*, 9411–9415. [CrossRef]

117. Duan, W.; Wei, C. The "spring predictability barrier" for ENSO predictions and its possible mechanism: Results from a fully coupled model. *Int. J. Climatol.* **2013**, *33*, 1280–1292. [CrossRef]

118. Larson, S.M.; Kirtman, B.P. Drivers of coupled model ENSO error dynamics and the spring predictability barrier. *Clim. Dyn.* **2017**, *48*, 3631–3644. [CrossRef]

119. Lai, A.W.C.; Herzog, M.; Graf, H.F. ENSO Forecasts near the Spring Predictability Barrier and possible reasons for the recently reduced predictability. *J. Clim.* **2018**, *31*, 815–838. [CrossRef]

120. Gonzalez, P.L.M.; Goddard, L. Long-lead ENSO predictability from CMIP5 decadal hindcasts. *Clim. Dyn.* **2016**, *46*, 3127–3147. [CrossRef]

121. Kumar, A.; Hu, Z.-Z.; Jha, B.; Peng, P.T. Estimating ENSO predictability based on multi-model hindcasts. *Clim. Dyn.* **2017**, *48*, 39–51. [CrossRef]

122. Newman, M.; Sardeshmukh, P.D. Are we near the predictability limit of tropical Indo-Pacific sea surface temperatures? *Geophys. Res. Lett.* **2017**, *44*, 8520–8529. [CrossRef]

123. Jones, A.E.; Morse, A.P. Skill of ENSEMBLES seasonal re-forecasts for malaria prediction in West Africa. *Geophys. Res. Lett.* **2012**, *39*, L23707. [CrossRef]

124. Davis, M.; Lowe, R.; Steffen, S.; Doblas-Reyes, F.; Rodo, X. Barriers to using climate information: Challenges in communicating probabilistic forecasts to decision-makers. In *Communicating Climate-Change and Natural Hazard Risk and Cultivating Resilience: Case Studies for a Multi-Disciplinary Approach*; Drake, J.L., Kontar, Y.Y., Eichelberger, J.C., Rupp, T.S., Taylor, K.M., Eds.; Book Series: Advances in Natural and Technological Hazards Research; Springer: Cham, Switzerland, 2016; Volume 45, pp. 95–113. [CrossRef]

125. Ballester, J.; Lowe, R.; Diggle, P.J.; Rodo, X. Seasonal forecasting and health impact models: Challenges and opportunities. *Ann. N. Y. Acad. Sci.* **2016**, *1382*, 8–20. [CrossRef] [PubMed]

126. Blunden, J.; Arndt, D.S. State of the climate in 2015. *Bull. Am. Meteorol. Soc.* **2016**, *97*, S1–S275. [CrossRef]

127. WHO. El Nino and Health: Global Overview, January 2016. Available online: http://www.who.int/hac/crises/el-nino/who_el_nino_and_health_global_report_21jan2016.pdf (accessed on 18 July 2018).

128. Jonkman, S.N. Global perspectives on loss of human life caused by floods. *Nat. Hazards* **2005**, *34*, 151–175. [CrossRef]

129. Kunii, O.; Nakamura, S.; Abdur, R.; Wakai, S. The impact on health and risk factors of the diarrhoea epidemics in the 1998 Bangladesh floods. *Public Health* **2002**, *116*, 68–74. [CrossRef]

130. Cai, W.J.; Santoso, A.; Wang, G.J.; Yeh, S.W.; An, S.I.; Cobb, K.M.; Collins, M.; Guilyardi, E.; Jin, F.F; Kug, J.S.; et al. ENSO and greenhouse warming. *Nat. Clim. Chang.* **2015**, *5*, 849–859. [CrossRef]

131. Wara, M.W.; Ravelo, A.C.; Delaney, M.L. Permanent El Niño-like conditions during the Pliocene warm period. *Science* **2005**, *309*, 758–761. [CrossRef] [PubMed]

132. Scroxton, N.; Bonham, S.G.; Rickaby, R.E.M.; Lawrence, S.H.F.; Hermoso, M.; Haywood, A.M. Persistent El Niño–Southern Oscillation variation during the Pliocene Epoch. *Paleoceanography* **2011**, *26*, 2215. [CrossRef]

133. Koutavas, A.; Joanidis, S. El Niño during the last glacial maximum. *Geochim. Cosmochim. Acta* **2009**, *73*, A690.

134. Rustic, G.T.; Koutavas, A.; Marchitto, T.M.; Linsley, B.K. Dynamical excitation of the tropical Pacific Ocean and ENSO variability by Little Ice Age cooling. *Science* **2015**, *350*, 1537–1541. [CrossRef] [PubMed]

135. Cobb, K.M.; Westphal, N.; Sayani, H.R.; Watson, J.T.; Di Lorenzo, E.; Cheng, H.; Edwards, R.L.; Charles, C.D. Highly variable El Niño–Southern Oscillation throughout the Holocene. *Science* **2013**, *339*, 67–70. [CrossRef] [PubMed]

136. Li, J.; Xie, S.P.; Cook, E.; Morales, M.S.; Christie, D.A.; Johnson, N.C.; Chen, F.H.; D'Arrigo, R.; Fowler, A.M.; Gou, X.H.; et al. El Niño modulations over the past seven centuries. *Nat. Clim. Chang.* **2013**, *3*, 822–826. [CrossRef]
137. McGregor, S.; Timmermann, A.; England, M.H.; Elison Timm, O.; Wittenberg, A.T. Inferred changes in El Niño–Southern Oscillation variance over the past six centuries. *Clim. Past* **2013**, *9*, 2269–2284. [CrossRef]
138. Bellenger, H.; Guilyardi, E.; Leloup, J.; Lengaigne, M.; Vialard, J. ENSO representation in climate models: From CMIP3 to CMIP5. *Clim. Dyn.* **2014**, *42*, 1999–2018. [CrossRef]
139. Bayr, T.; Latif, M.; Dommenget, D.; Wengel, C.; Harlass, J.; Park, W. Mean-state dependence of ENSO atmospheric feedbacks in climate models. *Clim. Dyn.* **2018**, *50*, 3171–3194. [CrossRef]

Review

Beyond Climate Change and Health: Integrating Broader Environmental Change and Natural Environments for Public Health Protection and Promotion in the UK

Lora E. Fleming [1], Giovanni S. Leonardi [2,3], Mathew P. White [1], Jolyon Medlock [4], Ian Alcock [1,*], Helen L. Macintyre [2,5], Kath Maguire [1], Gordon Nichols [1,2], Benedict W. Wheeler [1], George Morris [1], Tim Taylor [1], Deborah Hemming [6], Gianni Lo Iacono [7], Emma L. Gillingham [2,4], Kayleigh M. Hansford [4], Clare Heaviside [2,3,5], Angie Bone [1,2] and Raquel Duarte-Davidson [2]

[1] European Centre for Environment and Human Health, University of Exeter Medical School, C/O Knowledge Spa RCHT, Truro TR1 3HD, UK; L.E.Fleming@Exeter.ac.uk (L.E.F.); Mathew.White@exeter.ac.uk (M.P.W.); K.Maguire@exeter.ac.uk (K.M.); Gordon.Nichols@phe.gov.uk (G.N.); B.W.Wheeler@exeter.ac.uk (B.W.W.); G.Morris2@exeter.ac.uk (G.M.); timothy.j.taylor@exeter.ac.uk (T.T.); angie.bone19@gmail.com (A.B.)

[2] Centre for Radiation Chemicals and Environmental Hazards, Public Health England, Didcot OX11 0RQ, UK; Giovanni.Leonardi@phe.gov.uk (G.S.L.); Helen.Macintyre@phe.gov.uk (H.L.M.); Emma.Gillingham@phe.gov.uk (E.L.G.); Clare.heaviside@phe.gov.uk (C.H.); Raquel.Duarte-Davidson@phe.gov.uk (R.D.-D.)

[3] Department of Social and Environmental Health Research, London School of Hygiene and Tropical Medicine, 15-17 Tavistock Place, London WC1H 9SH, UK

[4] Medical Entomology & Zoonoses Ecology, Emergency Response Department—Science & Technology, Public Health England (PHE), Porton Down, Salisbury SP4 0JG, UK; Jolyon.Medlock@phe.gov.uk (J.M.); Kayleigh.Hansford@phe.gov.uk (K.M.H.)

[5] School of Geography, Earth and Environmental Sciences, University of Birmingham, Birmingham B15 2TT, UK

[6] Met Office, Hadley Centre, Fitzroy Road, Exeter EX1 3PB, UK; debbie.hemming@metoffice.gov.uk

[7] Department of Veterinary Epidemiology and Public Health, School of Veterinary Medicine, University of Surrey, Vet School Main Building, Daphne Jackson Road, Guildford GU2 7AL, UK; g.loiacono@surrey.ac.uk

[*] Correspondence: I.Alcock@exeter.ac.uk; Tel.: +44-(0)1872-242595

Received: 27 April 2018; Accepted: 20 June 2018; Published: 27 June 2018

Abstract: Increasingly, the potential short and long-term impacts of climate change on human health and wellbeing are being demonstrated. However, other environmental change factors, particularly relating to the natural environment, need to be taken into account to understand the totality of these interactions and impacts. This paper provides an overview of ongoing research in the Health Protection Research Unit (HPRU) on Environmental Change and Health, particularly around the positive and negative effects of the natural environment on human health and well-being and primarily within a UK context. In addition to exploring the potential increasing risks to human health from water-borne and vector-borne diseases and from exposure to aeroallergens such as pollen, this paper also demonstrates the potential opportunities and co-benefits to human physical and mental health from interacting with the natural environment. The involvement of a Health and Environment Public Engagement (HEPE) group as a public forum of "critical friends" has proven useful for prioritising and exploring some of this research; such public involvement is essential to minimise public health risks and maximise the benefits which are identified from this research into environmental change and human health. Research gaps are identified and recommendations made for future research into the risks, benefits and potential opportunities of climate and other environmental change on human and planetary health.

Keywords: demographic change; infectious diseases; vector-borne diseases; aerosolized exposures; pollen; well-being; public health; land management; patient and public involvement (PPI); land-use

---

1. Introduction

There is a growing awareness of how biotic (i.e., all animal and plant life) and abiotic (e.g., geological, weather and climate) natural systems interact to affect human socio-cultural-economic activities, and ultimately human and planetary ecosystem survival [1,2]. Although there has been a significant focus on how human activities both affect the climate and are affected by it, climate change is only one example of how broader patterns of environmental change are both caused by and influence human behavior and the health and well-being of global populations [3–7]. For instance, changes in land-use (e.g., increased urbanization), farming practices, industrial activities, and transportation networks, all interact with changes in climate to produce complex threats to health from both natural sources (e.g., changes in the distribution and prevalence of allergenic pollens, vector-borne diseases, and harmful algal blooms) and from anthropogenic sources (e.g., persistent organic pollutants, heavy metals, and an over-abundance of nutrients in surface waters) [3–5,8].

At the same time, there is an increasing evidence base and appreciation of the potential benefits of natural environments for human health and well-being [9,10]. Humans have actively destroyed, degraded, and impacted natural environments for millennia, yet increasingly the potential and realized value of these environments as 'natural capital' (especially natural environments that are of high quality and/or well managed), are being noted economically and culturally [11]. Thus, environmental change does not necessarily need to be bad. Good management, underpinned by state-of-the-art science, might actually be able to promote and support health and well-being, with evidence suggesting that the benefits may be strongest for some of the most vulnerable in society who are often the most exposed to environmental threats [12,13]. Ultimately, these insights can inform significant international efforts at producing sustainable, as opposed to unsustainable, growth (e.g., UN Sustainable Development Goals (SDGs), https://sustainabledevelopment.un.org) [9,14–16].

The aim of the current paper is twofold. First, it summarises and uses as exemplars selected research from a six-year cross-sector multi-centre UK funded initiative, the National Institute of Health Research (NIHR) Health Protection Research Unit (HPRU) in Environmental Change and Health (http://www.hpru-ech.nihr.ac.uk). The initiative was explicitly developed to improve our understanding of these complex interactions between different types of environmental change and human health. The HPRU project as a whole examines health in the UK within, and across, three core themes: Climate Resilience (Theme 1), Healthy and Sustainable Cities (Theme 2), and the Natural Environment (Theme 3); the focus of the current paper is Theme 3. "Natural environment" in this UK context includes all nature which has been impacted on by anthropogenic influences, across the urban and rural landscapes. Topics examined in this paper include a range of risks and threats to health, such as changes in the distributions of allergenic pollens and vector-borne diseases under a changing climate and other environmental change, as well as opportunities for health promotion through changes in the salutogenic use of green and blue spaces (e.g., for physical activity and/or well-being enhancement).

The second aim of the current paper is to draw these various strands together. In particular, the paper presents how an integrated understanding of the complex and multi-faceted interconnections between humans and their environment, including an improved understanding of how to balance the risks and benefits, is needed to identify and support opportunities to develop practical solutions able to protect and promote public health in a changing environment. And as a corollary (as we discuss below), essential to this area of research going forward is the integration of community involvement throughout the research process in identifying and understanding the impacts of environmental change on human health and well-being. This section builds on: (a) the growing awareness of the interconnections between the health of both the environment and humans in the medical, public health, environmental, and economic sciences, the arts and humanities, and among diverse communities including government, business, non-governmental organizations (NGOs), and communities; and (b) new perspectives and conceptual frameworks attempting to understand and articulate these ideas such as: One Health, Planetary Health and Planetary Boundaries, Environmental Global Health, Evolutionary Health, the Overview, EcoHealth, and Ecologic Public Health, as well as more general calls for 'systems thinking' [17,18].

The research we summarize and illustrate from the HPRU in Environmental Change and Health extends from earlier work by using a more systematic approach to integrate climate and other environmental change within a single interdisciplinary research programme. Additionally, the paper attempts to: indicate the global relevance and interconnections beyond the UK; identify areas of knowledge and research gaps; and stress the need for continuous assessment and monitoring of the risks and benefits of ongoing environmental change on the interactions between humans and the natural environment for their mutual health and future existence.

2. Background

Climate change is a major future (and increasingly current) threat to the health of humans and the planet. There are many ways in which climate change can impact human health and well-being through the natural environment, including more frequent and intense extreme weather events (e.g., hurricanes/cyclones), temperature changes, sea level rise, etc. (Table 1). Although accepted for several decades in the wider environmental science community, research into these effects of climate change on human health are relatively recent [3–5], but growing [18–21]. There is also increasing recognition that climate change effects that may appear to be distal in time and/or space can still have major effects on ecosystems and human health in areas such as the UK in the present or near future [22].

These climate change factors and effects cannot be viewed in isolation, nor as the sole drivers of effects on the natural environment or on current and future impacts on ecosystem and human health. In particular, other types of environmental change, ranging from natural and manmade contamination/pollution to changes in land-use need to be factored into any consideration of the effects of climate change on human and environment health (Table 1 and Figure 1). These other examples of environmental changes illustrate how humans (and their attempts to adapt to and mitigate these changes) currently affect the "health" of the natural environment, and how this can have ongoing (often unintended) consequences by impacting on current and future health and well-being.

An example of this is the contamination of the natural environment with manmade pharmaceuticals. For instance, antibiotics have been over-used in both human and veterinary healthcare, leading to the phenomenon of antimicrobial resistance (AMR), considered to be as big a threat as climate change to the future of humankind [23,24]. When AMR was initially described, it was identified as originating in hospitals, and then spreading into the community. However, due to ongoing antibiotic contamination of the natural environment, together with normal evolutionary processes, AMR is now developing widely in the natural environment and the community, and then entering back into hospital environments with potentially extremely serious direct consequences for human and animal health [24].

An especially challenging concept for the public health and research community is that of social complexity in the determinants of health and well-being. In the "socio-ecological model of health", health and disease are viewed as products of a complex interaction between societal-level factors (e.g., the physical environment) and characteristics specific to the individual (e.g., individual behavior). Although it is a recasting of much older ideas, this socio-ecological model transformed what was often a very siloed public health world at the end of the 20th century. Furthermore, acceptance of this model is especially challenging because it demands recognition of a much more complex real-world and policy context for research and action in environment and human health. Many of the highest profile public health challenges (e.g., mental health issues, well-being, and increasing inequalities) are recognized as being driven by multiple interacting determinants, including the natural environment [25].

Table 1. Climate and other environmental changes with potential for large scale population health impacts, highlighting contextual, adapting, and mitigating factors.

Climate Change Factors	Other Environmental Change Factors	Other Context Factors	Possible Mitigating/ Adapting Factors
• Variable weather with temperature extremes • Extreme weather events • Sea level rise • Ocean acidification	• Decreasing planetary resources (e.g., phosphorus, "rare earths" used in electronics, fossil fuels) • Land-use changes • Decreasing biodiversity • Loss of species and habitats • Anthropogenic Pollution (POPs, nutrients, pharmaceuticals, plastics, nano particles) • Decreasing availability of potable water • Continuing emissions of air pollutants from combustion • Invasive species (including vectors for disease) • Continuing emissions of nutrients and increasing harmful algal bloom risk • Changes in planting patterns (and associated pollen risks)	• Human demographic change (older, living longer, growing population) • Increasing chronic diseases (physical and mental) in human populations • Increasing inequalities within and between countries • Increasing migration • Increasing international trade in food • Increasing Tourism • Increasing international transport • Chronic conflicts	• Increasing use of renewable energy • Increasing Internet/ Globalized communication • Increasing access to Big Data • Increasing education of women/girls • Increasing community co-creation/participation • Pro-environmental behavior (e.g., recycling, active travel) • Technology • Increasingly taking climate and other environmental change into account in international finance, policy and governmental planning (e.g., Paris Accords)

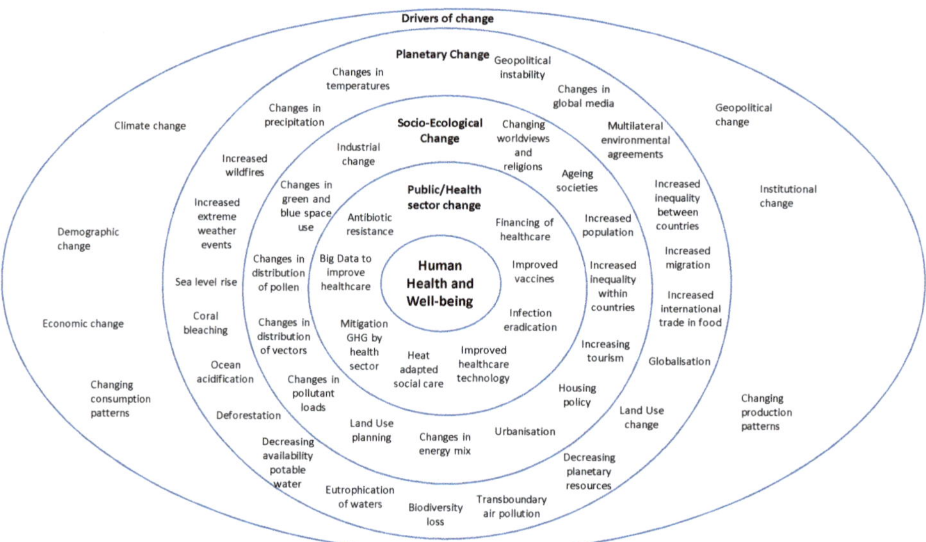

Figure 1. Environmental Change and Health: Global, Socio-ecological and Public/Health Sector changes that impact on human health and well-being.

In particular, theory and research on environment and human health from the public health community have been limited by thinking and perspectives which have focused on the health of only human populations, to the exclusion of the health of the natural environment and other organisms, including a lack of appreciation of the impacts of ecosystem "health" on human health and well-being and resilience [15,16]. More recently, the evolution of thinking in this area has been dominated by a wider understanding of the health, equity and existential relevance of the exceedance of "planetary boundaries" for both ecosystems and humans [26,27]. This development requires both the environment and health research communities to think and act together on vastly extended temporal and spatial scales, and to embrace further layers of complexity in analyses and investigation [25].

Ultimately, the result of the complex interactions (both known and unknown) of climate and other environmental change with all the other factors mentioned above impacting on ecosystem and human health, will determine all of our "Planetary Health" [27]. The consequences of some of these drivers on environmental change are discussed briefly below, particularly in the context of other factors which may lead to both risks and benefits for humans and the natural environment. As stated above, ongoing research from the Natural Environment Theme of the HPRU in Environmental Change and Health is used as exemplars below.

3. Climate and Other Environmental Changes, and Infectious Disease Risk

There are many historic examples of changing weather patterns and their impacts on the health of the humans and ecosystems of historic civilisations [28]. All sorts of drivers can affect infectious diseases [29], and most known and emerging infectious disease outbreaks are not directly attributable to changes in weather/climate. From the perspective of responding to emergencies, there are recent international examples of unexpected events that have initiated outbreaks including post-disaster outbreaks of cholera, and emerging infectious diseases such as Ebola and Zika. In these examples, a change in climate was not an important factor directly contributing to the increased cases, but they do illustrate the difficulties of predicting and managing such events. However, because some historic events have been associated with sudden changes in the burden of infectious diseases, it seems reasonable to be prepared for what might change in the coming years to better understand the weather and other drivers, and how these can influence the behaviour of individual pathogens. Globally, the historic overview suggests those most likely to be affected will be the poor rather than the rich, and those in low-income, rather than high-income countries.

3.1. Water-Borne Infectious Disease Risk

A variety of techniques have been used to examine the association between weather and various infectious diseases [30,31]. Water-borne infectious disease outbreaks have been associated with climate/weather, such as sporadic cryptosporidiosis [32,33]. A systematic review examined 24 papers on outbreaks of water-borne infectious disease and found associated factors included low rainfall, increasing temperature and heavy rainfall before the outbreak [34]. The impact of weather on drinking water quality is reduced with the provision of modern drinking water supplies, but is more marked in private water supplies [35]. Although modern treatment should still be able to cope with most extreme weather events, drinking water quality and supply may come under increased pressure with changes in climate. Flooding can impact on the infrastructure of water treatment, sewage disposal, and electricity supply. In the Baltic and beyond, seawater temperature may also be playing a role in the increase in diseases such as cholera related to the bacteria, *Vibrio* spp. [36].

The supply of food may be adversely influenced by drought or floods affecting agriculture [37,38]. As part of the HPRU in Environmental Change and Health Theme 1 research, we have explored the health and well-being consequences of flood exposures in the UK [39] and found that the seasonality of a number of foodborne pathogens can be influenced by weather, particularly temperature [40]. We have also reviewed the seasonality of over 2000 distinct infectious diseases reported in the UK [41] and the co-occurrence of weather conditions with human infection cases of individual pathogens provides some indication that these might be related. However, because such links can be associated with a range of other variables, we have developed techniques that link local weather to individual patients so that the weather components can be partly decoupled from seasonality; methods for extracting and linking these data locally have also been tested, based on linking the geographical location of where diagnosis data were collected and analyzed, with weather data for that location [42].

One of the problems with linking human infections with the weather across large areas is the differences in monitoring (including methods, reporting and completeness), particularly over a sufficiently long timescale [43]. To facilitate a wider analysis of pathogens, we have also reviewed methods used in linking water-associated diseases with climate [30,31], developed methods based on the infectious disease Campylobacter as a case-study (see Box 1), and produced a website for visualising data and generating hypotheses about the relationships between weather parameters and disease (https://www.data-mashup.org.uk/research-projects/climate-weather-infectious-diseases/).

Box 1. Campylobacter Case Study.

Campylobacter represents a good case example of an organism to study the effects of climate and weather on human infections in the UK and beyond. This is because the organism has a seasonality which is different from all other pathogens [44], numerous previous studies have been unable to identify the driving factors for this seasonal distribution [45–48], and since chickens appear to be an important source of infection, the association with weather is likely to be multi-factorial and indirect.

Within the Health Protection Research Unit (HPRU) in Environmental Change and Health research, the comparison of data based on the weekly incidence of campylobacter infections in the UK at different temperature and rainfall values has been useful in partially separating out the weather component from seasonality [49,50]. One traditional way of examining a time-series analysis is to use an autoregressive integrated moving average (ARIMA) model, and this has been adapted to include seasonality (SARIMA). A range of methods have been used to examine these, including: a novel Comparative Conditional Incidence (CCI) wavelet analysis [51]; hierarchical clustering [52]; generalized additive models for location, scale and shape (GAMLSS); and generalized structural time series (GEST) models [49,50]. This area of work is going to be further developed by future HPRU research so that it can be applied to a wider range of pathogens, weather/climate change and other complex factors.

3.2. Vector-Borne Disease Risk

The status of vector-borne diseases (i.e., infections transmitted by arthropod vectors) has changed significantly in the UK and Europe over the early 21st century [53]. During the early 20th century outbreaks of dengue and malaria were common in the Mediterranean region. Both malaria and the dengue mosquito vector, *Aedes aegypti*, were largely eradicated later that century so interest in vector-borne disease in Europe waned. However, the discovery in the early 1980s that the bacteria, *Borrelia burgdorferi*, in *Ixodid* ticks, which causes Lyme borreliosis (i.e., 'Lyme disease'), increasingly raised concerns over the role of ticks as disease vectors in Europe. Ticks are also known to be efficient vectors diseases, including the tick-borne encephalitis virus, the Crimean-Congo Haemorrhagic fever virus (rickettsial bacterial infection,) and other pathogens such as *Babesia*, *Anaplasma*, *Neoehrlichia*, and Louping Ill (infectious encephalomyelitis of sheep).

In 2012, dengue fever returned to Europe with >2000 cases in Madeira [54], and each year there are local cases of either dengue and/or chikungunya viruses across the Mediterranean Basin, including >400 cases of chikungunya in Italy in 2017 [55]. This transmission is associated with the importation, establishment, and spread of non-native invasive mosquitoes, *Aedes aegypti* and *Aedes albopictus* [56]. Since the start of the 21st century, West Nile virus has also consistently been reported in Eastern and Southern Europe, and more recently there have been local outbreaks of malaria in Greece and Italy [57]. The large outbreak of Zika virus in the Americas in 2015–2016 has further highlighted the threat posed by imported non-native mosquitoes and the potential for continental and global spread of mosquito-borne arboviruses, thus ensuring that vector-borne diseases will be an ever present public and veterinary health issue in coming years.

The changing status of vector-borne disease risk in the UK and Europe is partly attributable to globalisation and climate change; however, there are also various other environmental change factors that could equally be playing a role, and over a much shorter timescale. The invasive *Aedes* mosquito vectors of dengue, chikungunya and Zika are peri-domestic species, exploiting containers provided by humans, and dispersing in vehicles along highway systems. Thus, they are responsive to water storage and drought, and extreme precipitation, as well as changes in climate that can impact their development and capacity as vectors. In contrast, the habitat suitability of native mosquitoes and of native and non-native ticks are also driven by habitat availability and connectivity, and animal movements, which can be impacted by environmental change [53].

In the UK context in particular, the HPRU in Environmental Change and Health has focused on the sheep/deer tick, *Ixodes ricinus*, as the primary vector of Lyme borreliosis. It is able to feed on a variety of animal hosts during its three active life stages: larva, nymph and adult. This non-specialist feeding behaviour means that they can feed equally on wildlife, companion animals, livestock and humans, depending upon availability. These ticks thrive in the moist mild UK climate; however, their ability to survive off host is primarily determined by habitat structure, as well as animal diversity. Whilst historically it is a tick of upland sheep pasture and lowland woodland, these ticks are now found in lowland grazed grassland and urban green-space. Their dispersal throughout the countryside is contingent on their movement by animals.

It follows therefore that any environmental management of habitats that increase habitat connectivity or coverage, leads to greater dispersal and greater abundance of ticks. Recent published surveillance data by Public Health England (PHE) reported a significant change in the distribution of *Ixodes ricinus* [58], with particular expansion in the southern UK counties. Habitat management that favours ride (path) management in woodland, the creation of field margins, the impact of increased ecotonal habitat (i.e., habitat between two distinct habitats), changes in management of grassland habitat, or urban greenspace, have all been shown to influence the survival and abundance of ticks. In turn, the prevalence of the *Borrelia* bacteria within ticks is also influenced by both habitat and animal diversity [59–61].

This complex interplay between habitat, climate, animal population dynamics, vector density and pathogen prevalence demonstrates remarkable spatial and temporal heterogeneity, even at fine spatial scales. Although any scheme that modifies and enhances habitat structure may impact tick abundance in the UK, there are opportunities for managing these habitats to both maximise biodiversity and minimise human exposure and public health impacts [60], which can be particularly challenging in an urban setting (see Box 2). Further studies to develop empirical data both to inform such interventions and to understand the ecological aspects of disease transmission cycle are now a priority.

Box 2. Urban greenspace and ticks in the city of Bath (UK).

Urban greenspace has been linked to improved human health and well-being, and such spaces are a focal point of the UK government adaptation plans to mitigate the effects of our changing climate. The benefits of urban greenspace and the arguments for increasing and improving access are clear (as discussed later in this paper), but it is important to also investigate the potential risks posed by tick-borne disease. To date, few studies on ticks in urban greenspace have been conducted in the UK. This research aimed to fill these gaps, and helps inform guidance development. Methodology involves identifying a range of urban greenspace habitats such as woodland, woodland edge, meadows, parkland and short grassland. In each habitat and at each site, standard tick collection is conducted using flagging a cloth over vegetation. Ticks are counted and identified, with abundance of ticks calculated for unit area. All ticks are tested for *Borrelia* bacteria using PCR and sequenced to genospecies to inform Lyme risk. Following initial surveys in Salisbury (UK) on the suitability of urban greenspace for ticks and the *Borrelia* bacteria (causing Lyme disease), comparable surveys were commenced in the city of Bath, working closely with Bath and North East Somerset (BANES) local authority. BANES have a network of natural and managed urban greenspace, and this involves the creation of urban meadows. BANES have also led the way in working with Public Health England (PHE) with regard to assessing and mitigating the risks posed by urban ticks. Integrating the findings of academic and public health research on tick and Lyme risk enables better informed assessment of risk and appropriate public health awareness strategies, and this is proving a model that needs to be replicated across local authorities in the UK where urban ticks are being reported.

As part of the HPRU, field sampling of urban greenspace in Bath has taken place over the last two years, and found high suitability for ticks in some areas, mainly in woodland and woodland edge habitat, and the presence of *Borrelia*-infected ticks. So far, based upon preliminary results, prevalence rates appear lower (at 4.5%; unpublished data) than Salisbury (18.1%) [61], and appear restricted to only parts of the city. This work has been integral to enabling BANES to provide accurate tick awareness material, target their interventions, and assess the risks posed to humans by ticks and the impacts of their management. However further evidence is required, and as part of the HPRU in Environment and Human Health research, further sampling is taking place in Bath as well as in other cities such as Bristol and Southampton. Based upon this collective field research, PHE will be in a better position to advise local authorities, and to begin developing guidelines that can be used for management and in targeting tick awareness.

In contrast to tick-borne disease, which currently causes 2000–3000 cases of human Lyme borreliosis each year, the disease threat posed by native UK mosquitoes is currently insignificant; however, this should not lead to complacency. The incidence of several arbovirus diseases in the rest of Europe, together with the facts that malaria was once endemic in the UK and that mosquitoes are still a serious pest in some localities, suggest there is a need to understand how climate and other environmental change might affect future disease risk.

For example, in the UK, there is a large-scale programme for the development of new wetlands. The aims of this programme include: the managed re-alignment of the coasts to create new saltmarsh and mudflat habitat to mitigate coastal flooding and storm surges, as well as creating new protected habitats; the reversion of arable land to flooded grassland, as part of large-scale wetland expansion projects; and the creation of urban wetlands to provide sustainable urban drainage and to create mitigation habitat under the European Habitats Directive [62]. Field studies in the UK in each of these habitats has shown that mosquitoes do exploit newly created habitats, but their ability to colonise is largely dependent upon the design of new wetlands, the management of tidal waters, the flooding regime of wet grassland and the design of ditches, scrapes and sewage treatment reed beds [63–65]. The UK has 36 recorded species of mosquito, three of which have only been detected in the last 5–10 years [66] and one is known to be non-native and very invasive [67].

Although there is no current health risk, a warming climate and changes in animal movement and human interactions with the natural environment, coupled with an increase in the numbers of infected travelling pets, mean that any habitat management strategy that favours mosquito habitat could be significant. As with ticks, research that focusses on the how these vectors may be managed within the environment, particularly in protected habitats, is a key requirement for future contingency planning. Ensuring that we can progress environmental change and increase biodiversity, without the unexpected negative impact on vector-borne disease risk, is a priority area for PHE and the HPRU in Environmental Change and Health.

4. Aeroallergens/Aerosols Risk: The Case of Pollen

Aeroallergens contribute to the increasing burdens of asthma and allergies on society. The financial costs associated with asthma (excluding wider societal costs) in the UK have been estimated at £1.1 billion per year [68]. Seasonal allergies including allergic rhinitis ('hay fever') have been increasing in the UK, with some reports suggesting that as many as 40% of UK children suffer from hay fever [69]. Recent results from the "Britain Breathing" project, show that self-reported seasonal allergy symptoms across the UK are strongly correlated with reduced well-being, and these trends also correlate with the number of antihistamines prescribed by general practitioners [70].

Pollen from certain plant taxa or species, particularly grasses and some tree species (e.g., Betulaceae (birch)), can exacerbate allergenic conditions, including hay fever [71,72] and asthma [73,74]. Although the scale of individual sensitivity is highly variable, it is associated with exposure to pollen grains and the allergenicity of the pollen. Therefore, to help understand and manage individual exposure to allergenic pollen, it is important to: (a) quantify the health impacts of short-term increases in pollen exposure; (b) know where the major concentrations of allergenic plant species are located; and (c) forecast with reasonable accuracy, the timing, amount and dispersion characteristics of pollen emissions. Climate change is likely to affect pollen exposure through changes in plant productivity, the geographical range of allergenic species, the amount and allergenicity of pollen produced by each plant, and the timing and duration of the pollen season [75]. At the same time, there are ongoing efforts to increase the amount of and access to green space in urban areas to promote physical and mental well-being, which may result in increasing human exposure to mixtures of pollen with air pollutants (as discussed below).

There is the potential for research to inform planting, land management and development practices in order to reduce allergy risk [76]. For example, there is currently a widespread preference for planting only male trees along roadsides to avoid street litter from seeds and fruit produced by female trees. However, this can increase allergenic pollen exposure due to the pollen produced by the male trees. Tree planting might also be promoted to mitigate allergic exacerbation in urban areas where co-exposure to air pollutants is high. Grass cutting regimes can be modified to cut grass before it flowers and produces pollen.

The exacerbation of allergic respiratory conditions from pollen exposure may be intensified by air pollutants. Laboratory and field experiments suggest that air pollution and allergenic pollen exposures may interact [77]. For example, studies show that ozone [78,79] and nitrogen dioxide [80,81] can affect pollen morphology and change the pollen protein content or protein release processes, thus increasing the risk of allergic reaction following inhalation, with effects being species and concentration dependent [82]. Furthermore, grass pollen can attach to particulate air pollution (e.g., diesel exhaust particles), allowing allergenic particles of combined pollen and air pollutants to become concentrated in polluted air [83].

To address this research gap, studies conducted by the HPRU in Environmental Change and Health have focused on mapping the location of allergenic plant species across the UK [84] and relating pollen and land cover with health outcomes, including asthma exacerbation-related hospital admissions [85]. Results of this latter study showed that daily concentrations of grass pollen were significantly associated with adult hospital admissions for asthma in London, with a 4–5 day lag from increased pollen levels to hospital admissions. Increased hospital admissions were also associated with grass pollen concentrations categorized using the Met Office's 'pollen alert' levels, which range from 'very high' to 'low' days, with a lag in this case of three days. Additional research has examined the complex relationship between air pollution, land-use/cover and human health (see Box 3).

Box 3. Aero-allergens, land-cover and asthma.

Relationships between vegetative land cover and respiratory conditions are potentially influenced by multiple causal pathways. For example, whilst the presence of trees may increase local risk through their emission of allergenic pollen, those same trees may reduce local air pollutant concentrations [86,87] which also exacerbate conditions such as asthma [88,89]. In a similar way, areas of greenspace may result in localised seasonal exacerbation due to grass pollen emissions, but may also contribute to the reduction of asthma risk factors (e.g., obesity and stress) by providing a setting which promotes exercise and psychological restoration.

In recognition of both the effects of multiple and apparently opposing pathways, and also the potential interaction between pollutant and pollen exposures, another strand to the HPRU in Environmental Change and Health research has aimed to clarify the net effects of residential area exposure to trees and greenspace at different levels of background air pollutant exposure [90]. A comparison of UK asthma hospitalisation rates across 26,000 small urban areas showed that trees were associated with greater reductions in asthma hospitalisations when air pollutant levels were higher, but had no effect when air pollutant levels were very low; whereas greenspace was associated with greater reductions in hospitalisation when air pollutant levels were lower, but had no effect when air pollutant levels were very high.

Further work is underway to understand and model the environmental determinants of the key allergenic species or taxa across the UK. This will enable prediction of their spatial ranges, timing of pollen emissions, and concentration of pollen grains of the different species/taxa in the atmosphere at any time and location across the UK (both in the current climate, and under future climate and other environmental change scenarios).

5. Benefits of Natural Environments

As noted above, there is a growing recognition that the sustainable management of the world's natural resources is vital for human health, well-being and resilience, often referred to as 'Natural Capital' or 'ecosystem services'. Even at a very basic level, the air we breathe, the water we drink, the food we eat, and the sources of energy we rely on all depend on natural processes that support and regulate the environmental systems that support all life on the planet [2,91,92]. Given the extensive literature available on the direct and indirect benefits of natural ecosystems for human health and well-being (e.g., food, water, fuel etc.), the HPRU in Environmental Change and Health has focused on researching the potential benefits from natural environments in relation to two of the most pressing current public health issues poor mental health [93] and the lack of physical activity [94]. In particular, the aim has been to explore whether the availability and utilization of 'natural' spaces for recreation could: (a) help mitigate the increasing pressures of modern living on people's cognitive and emotional health; and (b) encourage individuals to engage in levels of physical activity conducive to good physical and mental health in an increasingly sedentary society.

With respect to mental health, the aim was to go beyond typical analyses to explore how different kinds of natural setting might influence different kinds of mental health outcome. Traditionally, research in this area has demonstrated that living within a neighbourhood with more generic 'greenspace' (e.g., as measured using satellite images) is associated with a lower risk of common mental health problems such as anxiety and depression [13,95–99]. The focus has been in exploring the following areas: (a) different types of greenspace in urban areas (e.g., street trees [10,100]; (b) greenspace land cover in rural areas (e.g., managed grasslands, deciduous woodlands, moorland, [101]); and (c) well-being outcomes in relation to different exposure types (e.g., evaluative, eudaimonic (i.e., feelings of meaning and achievement), and experiential well-being [102]). In Theme 2 of the HPRU, we have also quantified the effects of urban infrastructure on local urban temperatures compared the potential co-benefits with temperatures in green environments and quantified the related health impacts [103], as well as the impact of factors such as the deprivation status of people in relation to exposure to the highest urban temperatures [104].

With respect to urban street trees, an ecological analysis controlling for indices of area level greenspace, income, deprivation, and smoking rates, showed that London boroughs with higher densities of street trees had lower anti-depressant prescription rates ([100]; see [105] for similar findings in the US). A review of the urban tree literature reported a range of other potential benefits to health, such as reduced particulate air pollution, plus some challenges such as increased exposure to pollen [10]. Alcock et al. [101] used 18 years of longitudinal data to model symptoms of anxiety and depression using the General Health Questionnaire (GHQ) (a screening device for identifying minor psychiatric disorders in the general population) among individuals who moved between rural areas with different land cover mixes. Results suggested that the mental health of rural dwellers was better during the years when they lived in areas with a higher proportion of managed grasslands, uplands, and coastal habitats, relative to land cover associated with buildings and infrastructure. For other land cover types that might have been expected to have similar benefits (such as deciduous woodland), no such benefits were found, although some benefits have been reported in the US [106].

This research has gone beyond simply identifying (reduced) symptoms of common mental health disorders, to better understand the potential benefits to mental health and well-being from interacting with natural environments. International coordination efforts [107] have identified four key types of well-being: (a) positive and (b) negative, experiential well-being (i.e., emotions and moods); (c) evaluative well-being (i.e., global life satisfaction); and (d) eudaimonic well-being (i.e., feelings of meaning and achievement). Controlling for a range of potential confounders, the study found that the difference in levels of eudaimonic well-being between individuals who visited natural environments for recreation at least once a week, compared to those who rarely visited, was similar to the difference between individuals who were in a long-term relationship vs. those who were single/divorced or widowed. In other words, the findings raised the possibility that spending time in nature can promote the kinds of well-being normally associated with close personal relationships. Results also suggest that daily emotional states can be greatly improved by even relatively short visits to different kinds of natural settings.

In terms of physical activity, the focus of the research has moved beyond the traditional approach of demonstrating that people who lived in 'greener areas' tended to engage in higher levels of physical activity [108–111]. First, given the previous interest in the relationships between coastal and other 'blue space' environments and health [12,111–113], the research began by investigating whether individuals who lived nearer the coast in England, were also more likely to meet recommended levels of physical activity. Although there was strong support for this hypothesis in the West of the country, including both South and North West regions, there was no support in Eastern coastal regions [114]. Possible factors may have included differences in coastal types, access, populations and weather. In order to try and control for these factors, a repeat cross-over experimental lab-based study was conducted where individuals used a stationary exercise bike while viewing a large video projection of either coastal, rural, urban, or neutral gym settings [115]. Findings suggested individuals enjoyed their ride in the coastal setting more than the other settings, were more willing to do the coastal ride again, and importantly, had different perceptions of the passing of time in the coastal setting, suggesting that they might be willing to exercise for longer periods by the sea, with associated benefits to health (Box 4).

Box 4. Natural environments: the Monitor of Engagement with the Natural Environment (MENE) survey.

> The HPRU in Environmental Change and Health research examining the potential health and well-being benefits of interacting with the natural environment has used a range of methods and datasets to explore the factors which may be related to the use of natural environments for physical activity. Several studies have used the Monitor of Engagement with the Natural Environment (MENE) survey of how people use the natural environment in England which has collected data on over 40,000 individuals per year since 2009. For instance, using data from over 280,000 individuals we estimated that the total amount of physical activity conducted in all natural environments in England had a social value of over £2 billion per year, in terms of Quality Adjusted Life Years (QALYs) [116]. A second study suggested that while local greenspace is positively associated with increased levels of physical activity among dog owners, there is no relationship with non-dog owners [117].
>
> Additional research using the MENE dataset has identified the key reasons people report for rarely or never visiting natural environments for recreation, and the factors predicting those reasons [118]. For instance, nearly 20% of infrequent visitors stated that they were 'not interested' or had 'no particular reason' for not visiting these sites more often, and an analysis of the socio-demographic profiles of these individuals allowed the identification of a predominantly younger urban cohort from lower socio-economic status backgrounds who were more likely to report these reasons. Further experimental work within the HPRU suggests that this subpopulation might be an audience for interventions to raise interest or awareness in the possibility of using natural environments for health promotion in the future [119].

6. Public Involvement

An important lesson learned in this Environmental Change and Health HPRU, and of relevance to all other public health research, is the value of involving communities and other stakeholders in the exploration and co-creation of the research. Minimally, this engagement will broaden the scope of the research inquiry to include perspectives of those who are directly affected by climate and other environmental change impacts. In addition to asking different questions and instilling greater creativity, such co-creation and involvement can make the research more directly impactful and applicable to the affected communities. Furthermore, involving communities would be highly beneficial to some areas of science traditionally perceived as "less suitable" for the involvement of communities, such as mathematical modelling. For instance, insights from the communities can be crucial to the model assumptions, formulation and critical scrutiny of model findings [120,121]. There are also direct benefits in terms of training and research support by these communities (Box 5).

Public participation frequently features in strategies for addressing inequalities and enabling communities to improve health outcomes [123]. It has also been argued that addressing complex interactions between natural environments and human behaviour requires a social learning approach which goes beyond the provision of information, public consultation and community participation, to the development of 'situated and collective engagement' [8,124]. This approach speaks to what Jasanoff [125] described as the development of 'civic epistemology', the collective evaluation of knowledge to inform public policy and social action.

Yet there are a number of barriers to effective public involvement in public health research such as the HPRU which can be difficult for individual researchers or small teams to address [126]. The competences needed for effective involvement by academics and other researchers are still rarely taught as part of the research curriculum [127]. Structural support for public involvement from universities, funders and public bodies (e.g., www.ecehh.org/about-us/engagement/; www.invo.org. uk/; the PHE People's Panel) can help address barriers, but if time and resources for involvement are not integrated into research programmes from the outset, it can be difficult for researchers to access this support. Better integration of public involvement in all our research into environmental change and human health remains both a challenge and an opportunity.

Box 5. Health and Environment Public Engagement.

The Health and Environment Public Engagement (HEPE) group is a public forum of "critical friends" who have worked with the European Centre for the Environment and Human Health since 2013 (www.ecehh. org/about-us/engagement/; [122]). At quarterly meetings, HEPE discuss projects taking place in the Centre, including the HPRU in Environmental Change and Health research and support annual public workshops with researchers. These meetings use a range of participatory techniques to support inclusive and focused discussions.

The first workshop focused on prioritising issues the researchers focusing on possible salutogenic interactions with natural environments could address by linking large data sets. Researchers 'pitched' their ideas, and there was an initial individual vote by all workshop participants. Discussion of these choices was followed by a second vote which prioritised Mental Health and Social Relations. Notes from discussions showed some topics gained few votes because they were seen as integral to all the research rather than as individual projects (e.g., economic costs and benefits); this is a good example of why it is important to have multiple ways of recording workshop activities.

The second workshop used scenarios about individuals accessing natural environments. The group discussed whether this access made an important difference to the individuals and would any difference change their responses to self-reported health questionnaires widely used in the large secondary "big data" analysed by the HPRU team. The researchers were impressed by the magnified importance very small differences may have for people with multiple physical or mental health and social problems when answering these types of questionnaires.

The third workshop again used a scenario, this time alongside survey questions on partner/spouse relationships. Discussions explored which were most relevant as measures of social relations, and whether these could plausibly be used to help explore whether natural environments improve health directly, or whether this occurs through the promotion of better inter-personal relationships. The group members were concerned that some of the standard questions used in research in this area might introduce bias because of different cultural expectations and different interpretations of the survey questions. This insight helped the researchers to identify which data to use in the secondary data analyses for the HPRU research.

Involving the public has substantially helped to shape the European Centre's work within the HPRU. Having HEPE as a standing group has greatly facilitated that public involvement. Public interest in this HPRU research has also attracted five new members to HEPE, with several members participated in the most recent Annual Meeting of the HPRU in Environmental Change and Health, creating a cycle of mutual interest and support.

There is a growing variety of methods and approaches for working with communities in research collaboration. For example, one method which can directly involve communities and other stakeholders is the *ecosystems-enriched Drivers, Pressures, State, Exposure, Effects, Actions* or *'eDPSEEA'* model [128]. The *eDPSEEA* model can also help to populate and expand our understanding of a theoretical model of the research question which takes into account the health of both the environment and humans by integrating ecosystem services into the theoretical model, whereas the more traditional DPSEEA model has focused more on a unidirectional link between environment and human health. There are many other approaches which can engage communities and lead to improved co-creation of research and training, many of which emphasise the importance of trying to appreciate and understand the complex systems underlying interactions between human health, climate and other environmental change [17]. These are essential activities as we face a world in which human-driven actions, both intentional and unintentional, can have planetary impacts.

7. Discussion

In this paper, we have provided a brief overview of the growing subject area of human health and well-being with reference not just to climate change, but to other broader patterns of environmental change. There is growing evidence that environmental change is both caused by and influences human behavior and the health and well-being of global populations. In particular, we identified the emerging knowledge gaps in this area which are the focus of research by the HPRU in Environmental Change and Health and others [6,7]. However, there was no intention to imply that this selection of topics encompasses the entire research area of environmental change and health; rather the ongoing HPRU research provides a range of examples to demonstrate the complexity and importance of this growing research area. In this paper, we have highlighted the research on various aspects of human health and well-being, environmental change and the 'natural environment', including both threats and opportunities that rural and managed urban natural spaces have for a range of health and well-being outcomes. A key focus of the research has been on how a true appreciation of the complex and multi-dimensional nature of environmental change and health requires a deep integration of research approaches and findings from often disparate perspectives (including active community involvement), in order to better inform the trade-offs needed at a public health policy level.

The experience of the UK with regards to the impacts on the health of both humans and the natural environment from climate and other environmental change is not unique. As noted above, it is important to broaden our view of the potential effects of climate and other environmental change to include phenomena taking place distant in time and/or space from the UK (i.e., a more planetary perspective). Furthermore, the research presented in this paper demonstrates that there are lessons to share and to learn from others, particularly with the most vulnerable populations likely to be affected by climate and other environmental change, such as the tropical developing world, island states, and the polar regions, as well as our more local communities.

If we compare the achievements reviewed in this paper with research agendas such as those promoted by the World Health Organization (WHO) and other international agencies, the task has only barely been outlined by work conducted so far [5,6,129–131]. The assessment of health risks attributable to climate and other environmental change has only just started, including some initial evaluation of indirect effects such as infectious diseases. However, more remains to be done in terms of the assessment of effectiveness and cost-effectiveness of health protection and promotion strategies and measures, the assessment of the health impacts of potential adaptation and mitigation measures, and the assessment of the likely financial costs necessary to protect public health from climate and other environmental changes.

The main research gaps can be categorized as outlined below. There is a need to undertake research to:

- identify the most effective interventions, by systematic reviews of the evidence base for interventions, and/or by using methodological research to improve analytical tools for analysis of cost-effectiveness;
- identify the health co-benefits of mitigation and adaptation actions to address climate and other environmental change in non-health sectors. This includes improved methods for the assessment of the health implications of decisions in other sectors (such as in the energy and transport sectors and in the water, food and agriculture sectors), and the improved integration of climate change mitigation, adaptation and health through "settings-based" research;
- improve decision-support, for example, research to improve vulnerability and adaptation assessments, operational predictions, and the understanding of decision-making processes;
- improve the public's understanding of these issues, and into what interventions and mitigation strategies will be deemed acceptable to which constituencies;
- estimate the costs of protecting health from the effects of climate change, including the characterisation of harmonized methods to estimate costs and benefits, the assessment of the health costs of inaction and the costs of adaptation, and improved economic assessment of the health co-benefits of climate and other environmental change mitigation [25,130–133].

Recommendations for research highlighted by the WHO are relevant to public health services development: (1) evidence of interactions of climate and other environmental change with other health determinants and trends; (2) knowledge of the direct and indirect effects of climate and other environmental change; (3) evidence on the effectiveness of short-term interventions; (4) evidence of health impacts of policies in non-health sectors; and (5) general public health skills deployed to strengthening public health systems to address the health effects of climate and other environmental change [132]. The inclusion of environmental public health in mainstream public health services, with attendant capacity building of the service workforce, is likely required for development of a valid research evidence base in this area at the rapid rate at which it is needed [133].

In addition, in order to understand and follow the consequences of the complex interactions between human health and well-being and climate and other environmental change, it is important to have appropriate monitoring and data collecting systems in place. Environmental public health tracking at all levels (locally, nationally, regionally and globally) joined up through linkage of databases and standardized collection systems is an essential ingredient to pushing forward the science in this area [134]. A greater appreciation is required of the importance of big data involved in environment and human health collected by these environmental public health tracking systems [135]. The affected communities across the world require access to these data, while respecting issues of confidentiality, privacy and data governance [136]. Additionally, an important implication of both big data and the surveillance/monitoring systems is the need for new ways of thinking about and analyzing these data, learning from the experience of other scientific disciplines (such as oceanography or climate change modelling) in terms of improving modelling expertise over much broader time and spatial scales [135–137].

To gain an appropriate understanding of the impacts of wider environmental change on health, there is a need for a systematic attempt to assess the risks and benefits using a common framework. Addressing environmental change requires an inter-sectoral response—bridging health, environmental and industrial sectors. The need for risks from climate change to be assessed is mandated under the UK Climate Change Act (2008). A wider assessment of risks and benefits of broader environmental change and other factors is needed so that resources are appropriately allocated to adapting to our changing planetary environment [25].

Author Contributions: Conceptualization, Writing—review & editing, L.E.F., G.S.L., M.P.W., J.M., I.A., H.L.M., K.M., G.N., B.W.W., G.M., T.T., D.H., G.L.I., E.L.G., K.M.H., C.H., A.B. and R.D.-D.; Formal analysis, M.P.W., J.M., I.A., G.N. and G.L.I.; Investigation, L.E.F., M.P.W., J.M., I.A., K.M., G.N., B.W.W., G.L.I. and K.M.H.; Supervision, L.E.F., G.S.L., E.L.G., A.B. and R.D.-D.; Writing—original draft, L.E.F., G.S.L., M.P.W., J.M., I.A., H.L.M., K.M., G.N., G.M., T.T., D.H., G.L.I. and K.M.H.

Funding: The research was funded in part by the National Institute for Health Research Health Protection Research Unit (NIHR HPRU) in Environmental Change and Health at the London School of Hygiene and Tropical Medicine in partnership with Public Health England (PHE), and in collaboration with the University of Exeter, University College London, and the Met Office (HPRU-2012-10016); the UK Medical Research Council (MRC) and UK Natural Environment Research Council (NERC) for the MEDMI Project (MR/K019341/1, https://www.data-mashup.org.uk); the Economic and Social Research Council (ESRC) Project (ES/P011489/1); and the NIHR Knowledge Mobilisation Research Fellowship for Maguire.

Conflicts of Interest: The authors declare no conflict of interest. The founding sponsors had no role in the design of the study; in the collection, analyses, or interpretation of data; in the writing of the manuscript, and in the decision to publish the results. The views expressed are those of the authors not necessarily those of the NHS, the NIHR, the Department of Health and Social Care or Public Health England.

References

1. Johnson, D.L.; Ambrose, S.H.; Bassett, T.J.; Bowen, M.L.; Crummey, D.E.; Isaacson, J.S.; Johnson, D.N.; Lamb, P.; Saul, M.; Winter-Nelson, A.E. Meanings of Environmental Terms. *J. Environ. Qual.* **1997**, *26*, 581–589. [CrossRef]

2. Millennium Ecosystem Assessment (MEA). *Ecosystems and Human Well-Being: Synthesis*; Island Press: Washington, DC, USA, 2005.

3. Haines, A.; Fuchs, C. Potential impacts on health of atmospheric change. *J. Public Health Med.* **1991**, *13*, 69–80. [PubMed]

4. Epstein, P. Climate Change and Human Health. *N. Engl. J. Med.* **2005**, *353*, 1433–1436. [CrossRef] [PubMed]

5. McMichaels, A.J. Globalization, Climate Change, and Human Health. *N. Engl. J. Med.* **2013**, *368*, 1335–1343. [CrossRef] [PubMed]

6. Ebi, K.L.; Hess, J.J. The past and future in understanding the health risks of and responses to climate variability and change. *Int. J. Biometeorol.* **2017**, *61* (Suppl. 1), 71–80. [CrossRef] [PubMed]

7. Ebi, K.L.; Ogden, N.H.; Semenza, J.C.; Woodward, A. Detecting and Attributing Health Burdens to Climate Change. *Environ. Health Perspect.* **2017**, *125*, 085004. [CrossRef] [PubMed]

8. Bardosh, K.L.; Ryan, S.J.; Ebi, K.; Welburn, S.; Singer, B. Addressing vulnerability, building resilience: Community-based adaptation to vector-borne diseases in the context of global change. *Infect. Dis. Poverty* **2017**, *6*, 166. [CrossRef] [PubMed]

9. Hartig, T.; Mitchell, R.; de Vries, S.; Frumkin, H. Nature and health. *Ann. Rev. Public Health* **2014**, *35*, 207–228. [CrossRef] [PubMed]

10. Salmond, J.A.; Tadaki, M.; Vardoulakis, S.; Arbuthnott, K.; Coutts, A.; Demuzere, M.; Dirks, K.N.; Heaviside, C.; Lim, S.; Macintyre, H.; et al. Health and climate related ecosystem services provided by street trees in the urban environment. *Environ. Health* **2016**, *15*, 95–111. [CrossRef] [PubMed]

11. Natural Capital Committee. *The State of Natural Capital: Protecting and Improving Natural Capital for Prosperity and Well-Being*; 3rd Report of the NCC; Defra: London, UK, 2015.

12. Wheeler, B.W.; White, M.; Stahl-Timmins, W.; Depledge, M.H. Does living by the coast improve health and wellbeing? *Health Place* **2012**, *18*, 1198–1201. [CrossRef] [PubMed]

13. Mitchell, R.J.; Richardson, E.A.; Shortt, N.K.; Pearce, J.R. Neighborhood Environments and Socioeconomic Inequalities in Mental Well-Being. *Am. J. Prev. Med.* **2015**, *49*, 80–84. [CrossRef] [PubMed]

14. UNEP. Embedding the Environment in Sustainable Development Goals. UNEP Post-2015 Discussion Paper 1. United Na ons Environment Programme (UNEP), Nairobi. 2013. Available online: www.unep.org/pdf/embedding-environments-in-SDGs-v2.pdf (accessed on 15 June 2018).

15. Pruss Ustun, A.; Wolf, J.; Corvalan, C.; Neville, T.; Niera, M. Diseases due to unhealthy environments: An updated estimate of the global burden of disease attributable to environmental determinants of health. *J. Public Health* **2016**, *39*, 464–475.

16. Pruss Ustun, A.; Wolf, J.; Corvalan, C.; Niera, M. *Preventing Disease through Healthy Environments: A Global Assessment of the Environmental Burden of Disease from Environmental Risks*; World Health Organization: Geneva, Switzerland, 2016; Available online: http://apps.who.int/iris/bitstream/10665/204585/1/9789241565196_eng.pdf?ua=1 (accessed on 15 June 2018).

17. Rutter, H.; Savona, N.; Glonti, K.; Bibby, J.; Cummins, S.; Finegood, D.T.; Greaves, F.; Harper, L.; Hawe, P.; Moore, L.; et al. The need for a complex systems model of evidence for public health. *Lancet* **2017**, *390*, 2602–2604. [CrossRef]

18. Buse, C.G.; Oestreicher, J.S.; Ellis, N.R.; Patrick, R.; Brisbois, B.; Jenkins, A.P.; McKellar, K.; Kingsley, J.; Gislason, M.; Galway, L.; et al. Public health guide to field developments linking ecosystems, environments and health in the Anthropocene. *J. Epidemiol. Community Health* **2018**. [CrossRef] [PubMed]

19. Vardoulakis, S.; Heaviside, C. (Eds.) *Health Effects of Climate Change in the UK 2012–Current Evidence, Public Health Recommendations and Research Gaps*; Health Protection Agency, Centre for Radiation, Chemical and Environmental Hazards: London, UK, 2012; ISBN 978-0-85951-723-2.

20. Hajat, S.; Vardoulakis, S.; Heaviside, C.; Eggen, B. Climate change effects on human health: Projections of temperature-related mortality for the UK during the 2020s, 2050s and 2080s. *J. Epidemiol. Community Health* **2014**, *68*, 641–648. [CrossRef] [PubMed]

21. Woodward, A.; Smith, K.R.; Campbell-Lendrum, D.; Chadee, D.D.; Honda, Y.; Liu, Q.; Olwoch, J.; Revich, B.; Sauerborn, R.; Chafe, Z.; et al. Climate change and health: On the latest IPCC report. *Lancet* **2014**, *383*, 1185. [CrossRef]

22. Morris, G.P.; Reis, S.; Beck, S.; Fleming, L.E.; Adger, W.N.; Benton, T.G.; Depledge, M.H. Scoping the Proximal and Distal Dimensions of Climate Change on Health and Well-being. *Environ. Int.* **2017**, *16* (Suppl. 1).

23. UK Chief Medical Officer (CMO). CMO Annual Report Volume Two, Infections and the Rise of Antimicrobial Resistance. 2011. Available online: https://www.gov.uk/government/uploads/system/uploads/attachment_data/file/138331/CMO_Annual_Report_Volume_2_2011.pdf (accessed on 15 June 2018).

24. Roca, I.; Akova, M.; Baquero, F.; Carlet, J.; Cavaleri, M.; Coenen, S.; Cohen, J.; Findlay, D.; Gyssens, I.; Heure, O.E.; et al. The global threat of antimicrobial resistance: Science for intervention. *New Microbes New Infect.* **2015**, *6*, 22–29. [CrossRef] [PubMed]

25. Morris, G.P.; Saunders, P. The Environment in Health and Wellbeing. Oxford University Encyclopaedia of Environmental Science. 2017. Available online: http://environmentalscience.oxfordre.com/view/10.1093/acrefore/9780199389414.001.0001/acrefore-9780199389414-e-101?rskey=tGYyxX&result=8 (accessed on 5 April 2018).

26. Rockström, J.W.; Steffen, K.; Noone, Å.; Persson, F.S.; Chapin, E., III; Lambin, T.M.; Lenton, M.; Scheffer, C.; Folke, H.; Schellnhuber, B.; et al. Planetary boundaries: Exploring the safe operating space for humanity. *Ecol. Soc.* **2009**, *14*, 32. [CrossRef]

27. Steffen, W.; Richardson, K.; Rockström, J.; Cornell, S.E.; Fetzer, I.; Bennett, E.M.; Biggs, R.; Carpenter, S.R.; de Vries, W.; de Wit, C.A.; et al. Planetary boundaries: Guiding human development on a changing planet. *Science* **2015**, *347*, 1259855. [CrossRef] [PubMed]

28. Whitmee, S.; Haines, A.; Beyrer, C.; Boltz, F.; Capon, A.G.; de Souza Dias, B.F.; Ezeh, A.; Frumkin, H.; Gong, P.; Head, P.; et al. Safeguarding human health in the Anthropocene epoch: Report of The Rockefeller Foundation-Lancet Commission on planetary health. *Lancet* **2015**, *386*, 1973–2028. [CrossRef]

29. McMichael, A.J.; Woodward, A.; Muir, C. *Climate Change and the Health of Nations: Famines, Fevers, and the Fate of Populations*; Oxford University Press: New York, NY, USA, 2017; p. 370.

30. Ayres, J.G.H.; Maynard, R.L.; McClellan, R.O.; Nichols, G.L. Environmental medicine in context. In *Environmental Medicine*; Ayres, J.G.H., Nichols, G.L., Maynard, R.L., Eds.; Hodder Arnold: London, UK, 2010; pp. 3–21.

31. Lo Iacono, G.; Armstrong, B.; Fleming, L.E.; Elson, R.; Kovats, S.; Vardoulakis, S.; Nichols, G.L. Challenges in developing methods for quantifying the effects of weather and climate on water-associated diseases: A systematic review. *PLoS Negl. Trop. Dis.* **2017**, *11*, e0005659. [CrossRef] [PubMed]

32. Lo Iacono, G.; Nichols, G.L. Modeling the Impact of Environment on Infectious Diseases. *Environ. Sci.* **2017**. [CrossRef]

33. Nichols, G.; Lane, C.; Asgari, N.; Verlander, N.Q.; Charlett, A. Rainfall and outbreaks of drinking water related disease and in England and Wales. *J. Water Health* **2009**, *7*, 1–8. [CrossRef] [PubMed]

34. Lake, I.R.; Bentham, G.; Kovats, R.S.; Nichols, G.L. Effects of weather and river flow on cryptosporidiosis. *J. Water Health* **2005**, *3*, 469–474. [CrossRef] [PubMed]

35. Guzman Herrador, B.R.; de Blasio, B.F.; MacDonald, E.; Nichols, G.; Sudre, B.; Vold, L.; Semenza, J.C.; Nygård, K. Analytical studies assessing the association between extreme precipitation or temperature and drinking water-related waterborne infections: A review. *Environ. Health* **2015**, *14*, 29. [CrossRef] [PubMed]

36. Said, B.; Wright, F.; Nichols, G.L.; Reacher, M.; Rutter, M. Outbreaks of infectious disease associated with private drinking water supplies in England and Wales 1970–2000. *Epidemiol. Infect.* **2003**, *130*, 469–479. [PubMed]

37. Semenza, J.C.; Trinanes, J.; Lohr, W.; Sudre, B.; Lofdahl, M.; Martinez-Urtaza, J.; Nichols, G.L.; Rocklöv, J. Environmental Suitability of Vibrio Infections in a Warming Climate: An Early Warning System. *Environ. Health Perspect.* **2017**, *125*, 107004. [CrossRef] [PubMed]

38. Boxall, A.B.; Hardy, A.; Beulke, S.; Boucard, T.; Burgin, L.; Falloon, P.D.; Haygarth, P.M.; Hutchinson, T.; Kovats, R.S.; Leonardi, G.; et al. Impacts of climate change on indirect human exposure to pathogens and chemicals from agriculture. *Environ. Health Perspect.* **2009**, *117*, 508–514. [CrossRef] [PubMed]

39. Betts, R.A.; Alfieri, L.; Bradshaw, C.; Caesar, J.; Feyen, L.; Friedlingstein, P.; Gohar, L.; Koutroulis, A.; Lewis, K.; Morfopoulos, C.; et al. Changes in climate extremes, fresh water availability and vulnerability to food insecurity projected at 1.5 °C and 2 °C global warming with a higher-resolution global climate model. *Philos. Trans. A Math. Phys. Eng. Sci.* **2018**, *376*, 20160452. [CrossRef] [PubMed]

40. Jermacane, D.; Waite, T.D.; Beck, C.; Bone, A.; Amlot, R.; Reacher, M.; Kovats, S.; Reacher, M.; Armstrong, B.; Leonardi, G.; Rubin, G.J.; et al. The English National Cohort Study of Flooding and Health: The change in the prevalence of psychological morbidity at year two. *BMC Public Health* **2017**, *17*, 129. [CrossRef] [PubMed]

41. Lake, I.R.; Gillespie, I.A.; Bentham, G.; Nichols, G.L.; Lane, C.; Adak, G.K.; Threlfall, E.J. A re-evaluation of the impact of temperature and climate change on foodborne illness. *Epidemiol. Infect.* **2009**, *137*, 1538–1547. [CrossRef] [PubMed]

42. Cherrie, M.P.C.N.; Lo Iacono, G.; Sarran, C.; Hajat, S.; Fleming, L. Pathogen seasonality and links with weather in England and Wales: A big data time series analysis. *BMC Public Health* **2018**, in press.

43. Djennad, A.L.I.; Sarran, C.; Fleming, L.E.; Kessel, A.; Haines, A.; Nichols, G.L. A comparison of weather variables linked to infectious disease patterns using laboratory addresses and patient residence addresses. *BMC Infect. Dis.* **2018**, in press. [CrossRef] [PubMed]

44. Nichols, G.L.; Andersson, Y.; Lindgren, E.; Devaux, I.; Semenza, J.C. European monitoring systems and data for assessing environmental and climate impacts on human infectious diseases. *Int. J. Environ. Res. Public Health* **2014**, *11*, 3894–3936. [CrossRef] [PubMed]

45. Nichols, G.L.; Richardson, J.F.; Sheppard, S.K.; Lane, C.; Sarran, C. Campylobacter epidemiology: A descriptive study reviewing 1 million cases in England and Wales between 1989 and 2011. *BMJ Open* **2012**, *2*, e001179. [CrossRef] [PubMed]

46. Nichols, G.L. Fly transmission of Campylobacter. *Emerg. Infect. Dis.* **2005**, *11*, 361–364. [CrossRef] [PubMed]

47. Nylen, G.; Dunstan, F.; Palmer, S.R.; Andersson, Y.; Bager, F.; Cowden, J.; Feierl, G.; Galloway, Y.; Kapperud, G.; Megraud, F.; et al. The seasonal distribution of campylobacter infection in nine European countries and New Zealand. *Epidemiol. Infect.* **2002**, *128*, 383–390. [CrossRef] [PubMed]

48. Kovats, R.S.; Edwards, S.J.; Charron, D.; Cowden, J.; D'Souza, R.M.; Ebi, K.L.; Gauci, C.; Gerner-Smidt, P.; Hajat, S.; Hales, S.; et al. Climate variability and campylobacter infection: An international study. *Int. J. Biometeorol.* **2005**, *49*, 207–214. [CrossRef] [PubMed]

49. Spencer, S.E.; Marshall, J.; Pirie, R.; Campbell, D.; Baker, M.G.; French, N.P. The spatial and temporal determinants of Campylobacteriosis notifications in New Zealand, 2001–2007. *Epidemiol. Infect.* **2012**, *140*, 1663–1677. [CrossRef] [PubMed]

50. Djennad, A.; Rigby, R.; Stasinopoulos, D.; Voudouris, V. Beyond location and dispersion models: The Generalized Structural Time Series Model with Applications. *Munich Pers. REPIC Repos.* **2015**, 1–33. [CrossRef]

51. Djennad, A.L.I.; Sarran, C.; Lane, C.; Elson, R.; Höser, C.; Lake, I.; Colón-González, F.J.; Kovats, S.; Semenza, J.C.; Bailey, T.C.; et al. Methods to examine the seasonality and effects of weather on Campylobacter infections. *BMC Infect. Dis* **2018**, in press.

52. Cazelles, B.; Cazelles, K.; Chavez, M. Wavelet analysis in ecology and epidemiology: Impact of statistical tests. *J. R. Soc. Interface* **2014**, *11*, 20130585. [CrossRef] [PubMed]

53. Everitt, B.S.; Landau, S.; Leese, M.; Stahl, D. *Cluster Analysis*, 5th ed.; John Wiley & Sons: Hoboken, NJ, USA, 2011.

54. Medlock, J.M.; Leach, S.A. Effect of climate change on vector-borne disease risk in the UK. *Lancet Infect. Dis.* **2015**, *15*, 721–730. [CrossRef]

55. Schaffner, F.; Medlock, J.M.; Van Bortel, W. Public health significance of invasive mosquitoes in Europe. *Clin. Microbiol. Infect.* **2013**, *19*, 685–692. [CrossRef] [PubMed]

56. European Centre for Disease Prevention and Control (ECDC). *Clusters of Autochthonous Chikungunya Cases in Italy*; First Update–9 October 2017; ECDC: Stockholm, Sweden, 2017.

57. Medlock, J.M.; Hansford, K.M.; Schaffner, F.; Versteirt, V.; Hendrickx, G.; Zeller, H.; Van Bortel, W. A review of the invasive mosquitoes in Europe: Ecology, public health risks, and control options. *Vector Borne Zoonotic Dis.* **2012**, *12*, 435–447. [CrossRef] [PubMed]

58. Piperaki, E.T.; Daikos, G.L. Malaria in Europe: Emerging threat or minor nuisance? *Clin. Microbiol. Infect.* **2016**, *22*, 487–493. [CrossRef] [PubMed]

59. Cull, B.; Pietzsch, M.E.; Hansford, K.M.; Gillingham, E.L.; Medlock, J.M. Surveillance of British ticks: An overview of species records, host associations, and new records of *Ixodes ricinus* distribution. *Ticks Tick-Borne Dis.* **2018**, *9*, 605–614. [CrossRef] [PubMed]

60. Medlock, J.M.; Pietzsch, M.E.; Rice, N.V.P.; Jones, L.; Kerrod, E.; Avenell, D.; Los, S.; Ratcliffe, N.; Leach, S.; Butt, T. Investigation of ecological and environmental determinants for the presence of questing *Ixodes ricinus* (Acari: Ixodidae) on Gower, South Wales. *J. Med. Entomol.* **2008**, *45*, 314–325. [CrossRef] [PubMed]

61. Medlock, J.M.; Shuttleworth, H.; Copley, V.; Hansford, K.M.; Leach, S. Woodland biodiversity management as a tool for reducing human exposure to *Ixodes ricinus* ticks: A preliminary study in an English woodland. *J. Vector Ecol.* **2012**, *37*, 307–315. [CrossRef] [PubMed]

62. Hansford, K.M.; Fonville, M.; Gillingham, E.L.; Coipan, E.C.; Pietzsch, M.E.; Krawczyk, A.I.; Vaux, A.G.C.; Cull, B.; Sprong, H.; Medlock, J.M. Ticks and *Borrelia* in urban and peri-urban green space habitats in a city in Southern England. *Ticks Borne Dis.* **2017**, *8*, 353–361. [CrossRef] [PubMed]

63. Medlock, J.M.; Vaux, A.G. Impacts of the creation, expansion and management of English wetlands on mosquito presence and abundance-developing strategies for future disease mitigation. *Parasit. Vectors* **2015**, *8*, 142. [CrossRef] [PubMed]

64. Medlock, J.M.; Vaux, A.G. Colonization of UK coastal realignment sites by mosquitoes: Implications for design, management, and public health. *J. Vector Ecol.* **2013**, *38*, 53–62. [CrossRef] [PubMed]

65. Medlock, J.M.; Vaux, A.G. Colonization of a newly constructed urban wetland by mosquitoes in England: Implications for nuisance and vector species. *J. Vector Ecol.* **2014**, *39*, 249–260. [CrossRef] [PubMed]

66. Medlock, J.M.; Vaux, A.G. Seasonal dynamics and habitat specificity of mosquitoes in an English wetland: Implications for UK wetland management and restoration. *J. Vector Ecol.* **2015**, *40*, 90–106. [CrossRef] [PubMed]

67. Harbach, R.E.; Dallimore, T.; Briscoe, A.G.; Culverwell, C.L.; Vaux, A.G.C.; Medlock, J.M. *Aedes nigrinus* (Eckstein, 1918) (Diptera, Culicidae), a new country record for England, contrasted with *Aedes sticticus* (Meigen, 1838). *Zookeys* **2017**, *671*, 119–130. [CrossRef]

68. Medlock, J.M.; Vaux, A.G.; Cull, B.; Schaffner, F.; Gillingham, E.; Pfluger, V.; Leach, S. Detection of the invasive mosquito species *Aedes albopictus* in southern England. *Lancet Infect. Dis.* **2017**, *17*, 140. [CrossRef]

69. Mukherjee, M.; Stoddart, A.; Gupta, R.; Nwaru, B.; Farr, A.; Heaven, M.; Fitzsimmons, D.; Bandyopadhyay, A.; Aftab, C.; Simpson, C.R.; et al. The epidemiology, healthcare and societal burden and costs of asthma in the UK and its member nations: Analyses of standalone and linked national databases. *BMC Med.* **2016**, *14*, 113. [CrossRef] [PubMed]

70. Pawankar, R. Allergic diseases and asthma: A global public health concern and a call to action. *World Allergy Organ. J.* **2014**, *7*. [CrossRef] [PubMed]

71. Vigo, M.; Hassan, L.; Vance, W.; Jay, C.; Brass, A.; Cruickshank, S. Britain Breathing: Using the experience sampling method to collect the seasonal allergy symptoms of a country. *J. Am. Med. Inform. Assoc.* **2018**, *25*, 88–92. [CrossRef] [PubMed]

72. Burbach, G.J.; Heinzerling, L.M.; Edenharter, G.; Bachert, C.; Bindslev-Jensen, C.; Bonini, S.; Bousquet, J.; Bousquet-Rouanet, L.; Bousquet, P.J.; Bresciani, M.; et al. GA2LEN skin test study II: Clinical relevance of inhalant allergen sensitizations in Europe. *Allergy* **2009**, *64*, 1507–1515. [CrossRef] [PubMed]

73. Greiner, A.N.; Hellings, P.W.; Rotiroti, G.; Scadding, G.K. Allergic rhinitis. *Lancet* **2012**, *378*, 2112–2122. [CrossRef]

74. DellaValle, C.T.; Triche, E.W.; Leaderer, B.P.; Bell, M.L. Effects of ambient pollen concentrations on frequency and severity of asthma symptoms among asthmatic children. *Epidemiology* **2012**, *23*, 55–63. [CrossRef] [PubMed]

75. Ito, K.; Weinberger, K.R.; Robinson, G.S.; Sheffield, P.E.; Lall, R.; Mathes, R.; Ross, Z.; Kinney, P.L.; Matte, T.D. The associations between daily spring pollen counts, over-the-counter allergy medication sales, and asthma syndrome emergency department visits in New York City, 2002–2012. *Environ. Health* **2015**, *14*, 71. [CrossRef] [PubMed]

76. Lake, I.R.; Jones, N.R.; Agnew, M.; Goodess, C.M.; Giorgi, F.; Hamaoui-Laguel, L.; Semenov, M.A.; Solomon, F.; Storkey, J.; Vautard, R.; et al. Climate Change and Future Pollen Allergy in Europe. *Environ. Health Perspect.* **2017**, *125*, 385–391. [CrossRef] [PubMed]

77. Cariñanos, P.; Casares-Porcel, M. Urban green zones and related pollen allergy: A review. Some guidelines for designing spaces with low allergy impact. *Landsc. Urban Plan.* **2011**, *101*, 205–214. [CrossRef]

78. Motta, A.C.; Marliere, M.; Peltre, G.; Sterenberg, P.; LaCroix, G. Traffic-related air pollutants induce the release of allergen-containing cytoplasmic granules from grass pollen. *Int. Arch. Allergy Immunol.* **2006**, *139*, 294–298. [CrossRef] [PubMed]

79. Suárez-Cervera, M.; Castells, T.; Vega-Maray, A.; Civantos, E.; del Pozo, V.; Fernández-González, D.; Moreno-Grau, S.; Moral, A.; López-Iglesias, C.; Lahoz, C.; et al. Effects of air pollution on Cup a 3 allergen in *Cupressus arizonica* pollen grains. *Ann. Allergy Asthma Immunol.* **2008**, *101*, 57–66. [CrossRef]

80. García-Gallardo, M.V.; Algorta, J.; Longo, N.; Espinel, S.; Aragones, A.; Lombardero, M.; Bernaola, G.; Jauregui, I.; Aranzabal, A.; Albizu, M.V.; et al. Evaluation of the effect of pollution and fungal disease on *Pinus radiata* pollen allergenicity. *Int. Arch. Allergy Immunol.* **2013**, *160*, 241–250. [CrossRef] [PubMed]

81. Chassard, G.; Choël, M.; Gosselin, S.; Vorng, H.; Petitprez, D.; Shahali, Y.; Tsicopoulos, A.; Visez, N. Kinetic of NO_2 uptake by *Phleum pratense* pollen: Chemical and allergenic implications. *Environ. Pollut.* **2015**, *196*, 107–113. [CrossRef] [PubMed]

82. Zhao, F.; Elkelish, A.; Durner, J.; Lindermayr, C.; Winkler, J.B.; Ruëff, F.; Behrendt, H.; Traidl-Hoffmann, C.; Holzinger, A.; Kofler, W.; et al. Common ragweed (*Ambrosia artemisiifolia* L.): Allergenicity and molecular characterisation of pollen after plant exposure to elevated NO_2. *Plant Cell Environ.* **2015**, *39*, 147–164. [CrossRef] [PubMed]

83. Frank, U.; Ernst, D. Effects of NO_2 and Ozone on Pollen Allergenicity. *Front. Plant Sci.* **2016**, *7*, 91. [CrossRef] [PubMed]

84. Knox, R.B.; Suphioglu, C.; Taylor, P.; Desai, R.; Watson, H.C.; Peng, J.L.; Bursill, L.A. Major grass pollen allergen. Lol p 1 binds to diesel exhaust particles: Implications for asthma and air pollution. *Clin. Exp. Allergy* **1997**, *27*, 246–251. [CrossRef] [PubMed]

85. McInnes, R.N.; Hemming, D.; Burgess, B.; Lyndsay, D.; Osborne, N.J.; Skjøth, C.A.; Thomas, S.; Vardoulakis, S. Mapping allergenic pollen vegetation in UK to study environmental exposure and human health. *Sci. Total Environ.* **2017**, *599–600*, 483–499. [CrossRef] [PubMed]

86. Osborne, N.; Alcock, I.; Wheeler, B.; Hajat, S.; Sarran, C.; Clewlow, Y.; McInnes, R.; Hemming, D.; White, M.; Vardoulakis, S.; et al. Pollen exposure and hospitalization due to asthma exacerbations: Daily time series in a European city. *Int. J. Biometeorol.* **2017**, *61*, 1837–1848. [CrossRef] [PubMed]

87. Wuyts, K.; De Schrijver, A.; Staelens, J.; Gielis, L.; Vandenbruwane, J.; Verheyen, K. Comparison of forest edge effects on throughfall deposition in different forest types. *Environ. Pollut.* **2008**, *156*, 854–861. [CrossRef] [PubMed]

88. Nowak, D.J.; Hirabayashi, S.; Bodine, A.; Hoehn, R. Modeled PM2.5 removal by trees in ten U.S. cities and associated health effects. *Environ. Pollut.* **2013**, *178*, 395–402. [CrossRef] [PubMed]

89. Wilhelm, M.; Meng, Y.; Rull, R.P.; English, P.; Balmes, J.; Ritz, B. Environmental public health tracking of childhood asthma using California health interview survey, traffic, and outdoor air pollution data. *Environ. Health Perspect.* **2008**, *116*, 1254–1260. [CrossRef] [PubMed]

90. Roberts, S.E.; Button, L.A.; Hopkin, J.M.; Goldacre, M.J.; Lyons, R.A.; Rodgers, S.E.; Akbari, A.; Lewis, K.E. Influence of social deprivation and air pollutants on serious asthma. *Eur. Respir. J.* **2012**, *40*, 785–788. [CrossRef] [PubMed]

91. Alcock, I.; White, M.; Cherrie, M.; Wheeler, B.; Taylor, J.; McInnes, R.; Otte, I.; Eveline, K.; Vardoulakis, S.; Sarran, C.; et al. Land cover and air pollution are associated with asthma hospitalisations: A cross-sectional study. *Environ. Int.* **2017**, *109*, 29–41. [CrossRef] [PubMed]

92. Costanza, R.; d'Arge, R.; De Groot, R.; Farber, S.; Grasso, M.; Hannon, B.; Limburg, K.; Naeem, S.; O'neill, R.V.; Paruelo, J.; et al. The value of the world's ecosystem services and natural capital. *Nature* **1997**, *387*, 253. [CrossRef]

93. The UK National Ecosystem Assessment (UK NEA). *The UK National Ecosystem Assessment Technical Report*; UNEP-WCMC: Cambridge, UK, 2011.

94. UK Chief Medical Officer (CMO). *Annual Report of the Chief Medical Officer 2013 Public Mental Health Priorities: Investing in the Evidence*; CMO: London, UK, 2013.

95. World Health Organisation (WHO). Physical Activity: Factsheet (Updated February 2018). Available online: http://www.who.int/mediacentre/factsheets/fs385/en/ (accessed on 15 June 2018).

96. Alcock, I.; White, M.P.; Wheeler, B.W.; Fleming, L.E.; Depledge, M.H. Longitudinal effects on mental health of moving to greener and less green urban areas. *Environ. Sci. Technol.* **2014**, *48*, 1247–1255. [CrossRef] [PubMed]

97. De Vries, S.; Verheij, R.A.; Groenewegen, P.P.; Spreeuwenberg, P. Natural environments—Healthy environments? An exploratory analysis of the relationship between greenspace and health. *Environ. Plan. A* **2003**, *35*, 1717–1731. [CrossRef]

98. Gascon, M.; Triguero-Mas, M.; Martínez, D.; Dadvand, P.; Forns, J.; Plasència, A.; Nieuwenhuijsen, M.J. Mental health benefits of long-term exposure to residential green and blue spaces: A systematic review. *Int. J. Environ. Res. Public Health* **2015**, *12*, 4354–4379. [CrossRef] [PubMed]

99. Maas, J.; Van Dillen, S.M.; Verheij, R.A.; Groenewegen, P.P. Social contacts as a possible mechanism behind the relation between green space and health. *Health Place* **2009**, *15*, 586–595. [CrossRef] [PubMed]

100. White, M.P.; Alcock, I.; Wheeler, B.W.; Depledge, M.H. Coastal proximity, health and well-being: Results from a longitudinal panel survey. *Health Place* **2013**, *23*, 97–103. [CrossRef] [PubMed]

101. Taylor, M.S.; Wheeler, B.W.; White, M.P.; Economou, T.; Osborne, N.J. Research note: Urban street tree density and antidepressant prescription rates—A cross-sectional study in London, UK. *Landsc. Urban Plan.* **2015**, *136*, 174–179. [CrossRef]

102. Alcock, I.; White, M.P.; Lovell, R.; Higgins, S.L.; Osborne, N.J.; Husk, K.; Wheeler, B.W. What accounts for 'England's green and pleasant land'? A panel data analysis of mental health and land cover types in rural England. *Landsc. Urban Plan.* **2015**, *142*, 38–46. [CrossRef]

103. White, M.P.; Pahl, S.; Wheeler, B.W.; Depledge, M.H.; Fleming, L.E. Natural environments and subjective wellbeing: Different types of exposure are associated with different aspects of wellbeing. *Health Place* **2017**, *45*, 77–84. [CrossRef] [PubMed]

104. Heaviside, C.; Vardoulakis, S.; Cai, X.-M. Attribution of mortality to the Urban Heat Island during heatwaves in the West Midlands, UK. *Environ. Health* **2016**, *15* (Suppl. 1), 27. [CrossRef] [PubMed]

105. Macintyre, H.; Heaviside, C.; Taylor, J.; Picetti, R.; Symonds, P.; Cai, X.; Vardoulakis, S. Assessing urban population vulnerability and environmental risks across an urban area during heatwaves-implications for health protection. *Sci. Total Environ.* **2018**, *610–611*, 678–690. [CrossRef] [PubMed]

106. Beyer, K.M.; Kaltenbach, A.; Szabo, A.; Bogar, S.; Nieto, F.J.; Malecki, K.M. Exposure to neighborhood green space and mental health: Evidence from the survey of the health of Wisconsin. *Int. J. Environ. Res. Public Health* **2014**, *11*, 3453–3472. [CrossRef] [PubMed]

107. Akpinar, A. How is quality of urban green spaces associated with physical activity and health? *Urban For. Urban Green.* **2016**, *16*, 76–83. [CrossRef]

108. Organisation for Economic Development and Cooperation (OECD). OECD Guidelines on Measuring Subjective Well-Being. Retrieved 10 June 2015. OECD Publishing. 2013. Available online: https://read.oecd-ilibrary.org/economics/oecd-guidelines-on-measuring-subjective-well-being_9789264191655-en#page1 (accessed on 15 June 2018).

109. Astell-Burt, T.; Feng, X.; Kolt, G.S. Green space is associated with walking and moderate-to-vigorous physical activity (MVPA) in middle-to-older-aged adults: Findings from 203,883 Australians in the 45 and Up Study. *Br. J. Sports Med.* **2014**, *48*, 404–406. [CrossRef] [PubMed]

110. Coombes, E.; Jones, A.P.; Hillsdon, M. The relationship of physical activity and overweight to objectively measured green space accessibility and use. *Soc. Sci. Med.* **2010**, *70*, 816–822. [CrossRef] [PubMed]

111. Giles-Corti, B.; Broomhall, M.H.; Knuiman, M.; Collins, C.; Douglas, K.; Ng, K.; Lange, A.; Donovan, R.J. Increasing walking: How important is distance to, attractiveness, and size of public open space? *Am. J. Prev. Med.* **2005**, *28*, 169–176. [CrossRef] [PubMed]

112. Wendel-Vos, G.C.W.; Schuit, A.J.; Feskens, E.J.M.; Boshuizen, H.C.; Verschuren, W.M.M.; Saris, W.H.M.; Kromhout, D. Physical activity and stroke. A meta-analysis of observational data. *Int. J. Epidemiol.* **2004**, *33*, 787–798. [CrossRef] [PubMed]

113. White, M.; Smith, A.; Humphryes, K.; Pahl, S.; Snelling, D.; Depledge, M. Blue space: The importance of water for preference, affect, and restorativeness ratings of natural and built scenes. *J. Environ. Psychol.* **2010**, *30*, 482–493. [CrossRef]

114. White, M.P.; Alcock, I.; Wheeler, B.W.; Depledge, M.H. Would you be happier living in a greener urban area? A fixed-effects analysis of panel data. *Parapsychol. Sci.* **2013**, *24*, 920–928. [CrossRef] [PubMed]

115. White, M.P.; Wheeler, B.W.; Herbert, S.; Alcock, I.; Depledge, M.H. Coastal proximity and physical activity: Is the coast an under-appreciated public health resource? *Prev. Med.* **2014**, *69*, 135–140. [CrossRef] [PubMed]

116. White, M.P.; Pahl, S.; Ashbullby, K.J.; Burton, F.; Depledge, M.H. The effects of exercising in different natural environments on psycho-physiological outcomes in post-menopausal women: A simulation study. *Int. J. Environ. Res. Public Health* **2015**, *12*, 11929–11953. [CrossRef] [PubMed]

117. White, M.P.; Elliott, L.R.; Taylor, T.J.; Wheeler, B.W.; Spencer, A.; Bone, A.; Depledge, M.H.; Fleming, L. Recreational physical activity in natural environments and implications for health: A population based cross-sectional study in England. *Prev. Med.* **2016**, *91*, 383–388. [CrossRef] [PubMed]

118. White, M.P.; Elliott, L.R.; Wheeler, B.W.; Fleming, L.E. Neighbourhood greenspace is related to physical activity, but only among dog owners. *Landsc. Urban Plan.* **2018**, *174*, 18–23. [CrossRef]

119. Boyd, F.; White, M.P.; Bell, S.; Burt, J. Who doesn't visit natural environments for recreation and why: A population representative analysis of spatial, individual and temporal factors among adults in England. *Landsc. Urban Plan.* **2018**, *175*, 102–113. [CrossRef]

120. Elliott, L.R.; White, M.P.; Taylor, A.H.; Abraham, C.; Fleming, L.E. Promoting intentions to walk in natural environments through the application of behaviour change theories to recreational walking brochures. *Environ. Behav.* **2018**, in press.

121. Scoones, I.; Jones, K.; Lo Iacono, G.; Redding, D.W.; Wilkinson, A.; Wood, J.L.N. Integrative modelling for One Health: Pattern, process and participation. *Philos. Trans. R. Soc. B Biol. Sci.* **2017**, *372*, 20160164. [CrossRef] [PubMed]

122. Grant, C.; Lo Iacono, G.; Dzingirai, V.; Bett, B.; Winnebah, T.R.A.; Atkinson, P.M. Moving interdisciplinary science forward: Integrating participatory modelling with mathematical modelling of zoonotic disease in Africa. *Infect. Dis. Poverty* **2016**, *5*, 17. [CrossRef] [PubMed]

123. Maguire, K.; Garside, R.; Poland, J.; Fleming, L.E.; Alcock, I.; Taylor, T.; Macintyre, H.; Lo Iacono, G.; Green, A.; Wheeler, B. White Public involvement in research about environmental change and health: A case study. *Health* **2018**. under review.

124. Morgan, L. Community participation in health: Perpetual allures, persistent challenge. *Health Policy Plan.* **2001**, *16*, 221–230. [CrossRef] [PubMed]

125. Jassanoff, S. *Designs on Nature: Science and Democracy in Europe and the United States*; Princeton University Press: Princeton, NJ, USA, 2005; pp. 247–271.

126. Collins, K.; Ison, R. Jumping off Arnstein's Ladder: Social Learning as a New Policy Paradigm for Climate Change Adaptation. *Environ. Policy Gov.* **2009**, *19*, 358–373. [CrossRef]

127. Brunton, G.; Thomas, J.; O'Mara-Eves, A.; Jamal, F.; Oliver, S.; Kavanagh, J. Narratives of community engagement: A systematic review-derived conceptual framework for public health interventions. *BMC Public Health* **2017**, *17*, 944. [CrossRef] [PubMed]

128. De Wit, M.; Beurskens, A.; Piškur, B.; Stoffers, E.; Moser, A. Preparing researchers for patient and public involvement in scientific research: Development of a hands-on learning approach through action research. *Health Expect.* **2018**. [CrossRef] [PubMed]

129. Reis, S.; Morris, G.; Fleming, L.E.; Beck, S.; Depledge, M.H.; Steinle, S.; Sabel, C.E.; Cowie, H.; Hurley, F.; Dick, J.; et al. Integrating Health and Environmental Impact Analysis. *J. Public Health* **2015**, *129*, 1383–1389. [CrossRef] [PubMed]

130. World Health Organization (WHO). Protecting Health from Climate Change: Connecting Science, Policy and People. 2009. Available online: http://apps.who.int/iris/bitstream/handle/10665/44246/9789241598880_eng.pdf?sequence=1 (accessed on 10 April 2018).

131. World Health Organization (WHO). Protecting Health from Climate Change: Global Research Priorities. 2009. Available online: http://apps.who.int/iris/bitstream/10665/44133/1/9789241598187_eng.pdf (accessed on 10 April 2018).

132. Bouzid, M.; Hooper, L.; Hunter, P.R. The effectiveness of public health interventions to reduce the health impact of climate change: A systematic review of systematic reviews. *PLoS ONE* **2013**, *8*, e62041. [CrossRef] [PubMed]

133. Leonardi, G. Adaptation to climate change. In *Environmental Medicine*; Ayres, J.G., Harrison, R.M., Nichols, G.L., Maynard, R.L., Eds.; Hodder Education: London, UK, 2010; Available online: https://s3.amazonaws.com/academia.edu.documents/5253859/Environmental_CH49.pdf?AWSAccessKeyId=AKIAIWOWYYGZ2Y53UL3A&Expires=1523380339&Signature=qy946LpDPk1xcHwCH0laqHtacog%3D&response-content-disposition=inline%3B%20filename%3DAdaptation_to_climate_change._Chapter_49.pdf (accessed on 10 April 2018).

134. Leonardi, G.; Rechel, B. Environmental health. In *Facets of Public Health in Europe*; Rechel, B., McKee, M., Eds.; Open University Press: Maidenhead, UK, 2014; pp. 93–112.

135. Behbod, B.; Lauriola, P.; Leonardi, G.; Crabbe, H.; Close, R.; Staatsen, B.; Knudsen, L.E.; de Hoogh, K.; Medina, S.; Semenza, J.C.; et al. Environmental and public health tracking to advance knowledge for planetary health. *Eur. J. Public Health* **2016**, *26*, 900. [CrossRef] [PubMed]

136. Fleming, L.E.; Tempini, N.; Gordon-Brown, H.; Nichols, G.; Sarran, C.; Vineis, P.; Leonardi, G.; Golding, B.; Haines, A.; Kessel, A.; et al. Big Data in Environment and Human Health: Challenges and Opportunities. *Oxf. Res. Encycl. Environ. Sci.* **2017**. [CrossRef]

137. Independent Expert Advisory Group on a Data Revolution for Sustainable Development (IEAG). A World that Counts: Mobilising the Data Revolution for Sustainable Development. Report Prepared at the Request of the United Nations Secretary-General. 2014. Available online: http://www.undatarevolution.org/wp-content/uploads/2014/12/A-World-That-Counts2.pdf (accessed on 15 June 2018).

MDPI

St. Alban-Anlage 66

4052 Basel

Switzerland

Tel. +41 61 683 77 34

Fax +41 61 302 89 18

www.mdpi.com

Atmosphere Editorial Office

E-mail: atmosphere@mdpi.com

www.mdpi.com/journal/atmosphere

9 783039 367405